Henry Hampton

WOLVES OF MINONG

DURWARD L. ALLEN

Wolves of Minong

Isle Royale's Wild Community

Ann Arbor

THE UNIVERSITY OF MICHIGAN PRESS

First edition as an Ann Arbor Paperback 1993

Published in the United States of America by
The University of Michigan Press
Manufactured in the United States of America

1996 1995 1994 4 3 2

A CIP catalogue record for this book is available from the British Library.

Allen, Durward Leon, 1910–
Wolves of Minong : Isle Royale's wild community / Durward L. Allen, — 1st ed. as an Ann Arbor paperback.
p. cm. — (Ann Arbor paperbacks)
Originally published: Boston : Houghton Mifflin, 1979.
Includes bibliographical references (p.) and index.
ISBN 0-472-08237-X (alk. paper)
1. Wolves—Michigan—Isle Royale National Park—Behavior. 2. Predation (Biology) 3. Natural history—Michigan—Isle Royale National Park. 4. Isle Royale National Park (Mich.) I. Title.
QL737.C22A44 1993
599.74'442—dc20 93-8192
CIP

". . . this shard of a continent becalmed in the green fresh-water sea is indeed royal, sovereign, isolate, supreme."

T. MORRIS LONGSTRETH

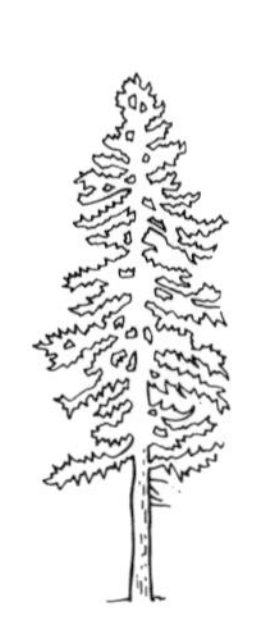

Foreword

THIS ACCOUNT OF WOLVES on Isle Royale, a national park, describes their lives and doings through 18 years of nearly unique freedom — freedom from traps, snares, poison, and airborne gunnery. Free they were, but bound in an ancient accord between the wild hunters and the hunted, those who live by killing and those who die in service by being eaten. The plan was established long before there were men to meddle and confuse.

Some gentle citizens of this man-centered world could regard events I recount as brutish skullduggery, a kind of villainy we should not tolerate in a peaceful wilderness set aside, as affirmed by law, to be a pleasuring ground for the people. A few have urged that carnivore atrocities be brought under some kind of civilized control. In fact, they would require creatures of nature to conform to moral standards that refined human beings have sought vainly to enforce upon their own kind.

This view could keep a threadbare controversy alive. It propagates the righteous ignorance that made the wolf a villain in fable and legend. Human retribution, by methods that outbrutalize the innocent carnage of the wild, has brought many of our flesh-feeding mammals and birds to the edge of oblivion. In an age when science has become respectable, one must ask: Can we now apply some intellectual maturity? Is it not time to outgrow the idyll of the lion and the lamb and learn about the wolf and the moose? They need no one to lead them, but only a place to be left alone.

The natural design that requires both predator and prey has had almost infinite testing. It is basically involved with the health of ecosystems we must preserve and prudently use. The native tenants of our out-of-doors make it the complete habitat for human retreat and relaxation. It could always be so, for under enlightened surveillance wildlife values are largely self-renewing.

The inside story of the meat-eaters and their victims must be known in truth if enough people are to sponsor a husbandry of nature that will work. Further, as a personal gain for each of us, it is through discovery and understanding that we enjoy in greatest measure the things about us. The mysteries of elemental life furnish endless challenge; rewards abound for those who learn to see the inner side of that green landscape. There, inevitably, subsisting on the annual yield of verdure, is a realm of creatures fated to be eaten. They provision the higher levels of an earth-based pyramid of life.

In few spots on earth can we still seek out the highest level of all, the top of the carnivore line. But here I tell of such a place. The throng of us cannot go there to observe, nor would we wish to. For that matter, events of the past 18 years could never be repeated. No doubt important things were missed; it is always so. But for you, the reader, and for myself I skimmed the long record, and I found new adventure.

Enjoy it with me.

D. L. A.

Contents

Illustrations

All photographs not otherwise credited were taken by the author.

MAPS, CHARTS, AND GRAPHS

TABLES

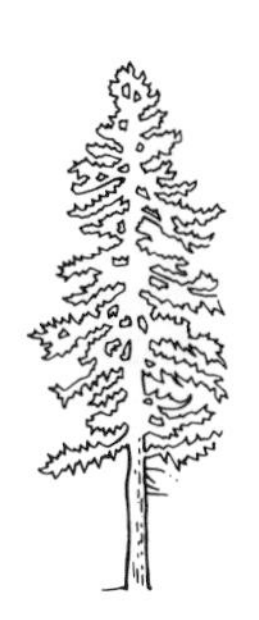

Prelude: How It All Got Started

MY FASCINATION with the north country had its roots far back in childhood, before I ever knew of a great rocky island in Lake Superior. Some intuitive person read to me portions of the epic of Hiawatha. Thereafter I was fatally ensnared by Longfellow's mystical tale of the Indian boy's wilderness home on a shore where

> Dark behind it rose the forest,
> Rose the black and gloomy pine-trees,
> Rose the firs with cones upon them;
> Bright before it beat the water,
> Beat the clear and sunny water,
> Beat the shining Big-Sea-Water.

This was Lake Superior country, as I realized later. As the poet evoked the image, it would have been somewhere on the mainland, because it involved animal life that was not present on Isle Royale. But the lilting roundelay comes back betimes as I sit, hard at work, on a rotting log and watch the sparkling waves in their patient nibbling at the rocks.

My story is of an island national park and the creatures that inhabit it. It is a blend of natural and human history. The chronicle involves somewhat the real Indians who mined copper there four thousand years ago, and more recently the voyageurs, whose canoes, for two centuries, coasted the big lake on the fur trade route from Sault Sainte Marie to Grand Portage and on to the northwest.

Ojibwas called the island "Minong," meaning place of blueberries. Today it bears the century-old signs of prospectors who burned off its ridges to inspect the rocks for copper and, perchance, for better things. The excavations and dumps of serious mining efforts are there, under slow reclamation by the forest. There were burnings in the 30s and a brief time of forest cutting, but much of the best pine had already been destroyed. Resources of the pure northern lake held up better, and families of hardy commercial fishermen made Isle Royale their home. The last of them provide a cultural feature of the modern park.

Isle Royale became known to me about 1930, when I was an undergraduate in zoology at the University of Michigan. Its abounding moose herd had achieved worldwide fame. Walter Hastings, photographer of the Michigan Department of Conservation, was showing spectacular films of the great beasts that could be seen on every bayshore of the big island that was Michigan's northernmost county. While no wolves were on Isle Royale, a few persisted in the state's upper peninsula. I was enchanted to examine two large males, taken under the bounty system, that arrived at the museum in Ann Arbor. Each was said to weigh more than a hundred pounds.

At that time Adolph Murie and Paul Hickie had just finished

Bright before it beat the water.

their field work on Isle Royale moose. Their findings were published by the museum in Murie's bulletin of 1934. Hickie joined the staff of the Michigan Game Division and was sent back to Isle Royale for three winters. Helped by the Civilian Conservation Corps (CCC), from 1934 to 1937 he trapped 71 moose, which were used in an unsuccessful attempt to revive the almost-gone moose population in the state's upper peninsula.

As shall be related in more detail, during the early 30s Isle Royale moose met disaster from the accumulating ills of overpopulation and food depletion. The island became a national park in 1940. Moose were already protected by the state, and now any shooting to control the island herd was out of the question. A national park was a place for nature to "take its course." Well, how much of this was natural?

In 1951 I found myself working in the Interior Building in Washington, D.C., just down the hall from the office of the chief biologist of the National Park Service, Victor H. Cahalane. The two of us conferred frequently on the spectacle of another moose build-up on Isle Royale and another devastation of the food supply. In primitive times the natural check probably would have been the wolf, and we were bemused by reports

In the 20s, Isle Royale supported the most famous moose herd in the world.

Prospectors of a century ago burned off the ridges, and in many openings the bedrocks are still unreclaimed by vegetation.

that wolf tracks had been found on the island. But coyotes certainly were there, and records of wolves had to be treated with reservation.

One morning Cahalane came into my office in the Division of Wildlife Research of the Fish and Wildlife Service. He laid on my desk the cast of a large canid footprint.[1]

"Vic," I said, "there are wolves on Isle Royale for sure!"

Both of us were delighted, since this event introduced the greatest of all experiments in predator-prey relations. Potentially at least, the wolves could build up, stabilize the moose herd, and bring some protection to the vegetation. It was the pattern of primitive times, now to be replayed in a world where such patterns are confused and obscured by the almost universal hunting of moose and the wiping out or heavy control of wolves.

I was especially interested in islands as research areas, where the animals you are counting and studying do not wander away. Isle Royale seemed to offer all the advantages. It was big enough (210 square miles) to support a major moose herd, and, quite likely, a viable population of wolves. It was unthinkable to waste this opportunity to find out what was happening, to verify or disprove all the truths and fictions that have been handed down in the natural and unnatural history of man's relationship with the wolf. Truth or fiction: Which was which?

My division of the Fish and Wildlife Service was the legal and logical agency to do this kind of research on federal lands. As assistant chief under Logan J. Bennett, I prepared the budgets, and we tried for three years to get funding for a major predator-prey study on Isle Royale. Our requests never got into the agency's budget, and after the election of 1952 there was no chance — we could not even hold what funding we had. There were no public pressures for wildlife research.

In 1953 Bennett went back home to become executive director of the Pennsylvania Game Commission. I carried on alone in the office for a year; positions were not being filled on a permanent basis. Then I too left, to join the staff of Forestry and Conservation at Purdue University in my home state of Indiana. I had a growing idea there was another route to follow in getting something done on Isle Royale.

As it worked out there was a way, and this book recounts our principal results in 18 years. "Our" is proper terminology, and from here on the first-person often will be we and us. In substantial degree this report skims the findings of my students and colleagues who carried the burden of the research. This does not mean they agree with everything I say, for there are differences in both appraisal and interpretation. Necessarily much of this version is my own, with occasional recognition that there could be another viewpoint.

The participants in this study need an early introduction, since I will be referring to them from the start. To round out the list I include myself and our pilot:

Old timers

Durward L. Allen: 1957–76. Director, fund raiser, field assistant, and chief cook in the winter camp.

Donald E. Murray: 1959–76. Winter field pilot, wolf observer, and willing helper in all we had to do.

Principal investigators

L. David Mech (pronounced Meech): 1958–62. A Cornell

graduate in wildlife science. Completed work for the Ph.D. on "Ecology of the timber wolf in Isle Royale National Park."

Philip C. Shelton: 1960–64. A graduate in wildlife technology at Montana State University. Doctoral thesis at Purdue, "Ecological studies of beavers, wolves, and moose in Isle Royale National Park."

Peter A. Jordan: 1963–66. Came to this work as post-doctoral research associate from studies of the mule deer in California (Ph.D., University of California, Berkeley, 1967).

Wendel J. Johnson: 1966–69. Received master's degree in herpetology from Michigan State University. In the Purdue program completed doctoral thesis on "Food habits of the Isle Royale red fox and population aspects of three of its principal prey species."

Michael L. Wolfe: 1967–70. A Cornell wildlife graduate, received doctorate on history of game management in Germany, University of Göttingen, 1966. Joined Purdue research on Isle Royale as post-doctoral associate.

Rolf O. Peterson: 1970–76. Graduate in zoology at the University of Minnesota at Duluth. Completed requirements for Ph.D. on "Wolf ecology and prey relationships on Isle Royale," 1974. Post-doctoral associate, 1975. Moved to Michigan Technological University (Houghton) and became director of the program, 1975.

This is the official roster, but it is not complete. With the exception of Shelton, each of these men had a wife of distinguished capabilities who was doing heroic deeds to carry his work forward. Betty Ann Mech, Martha Jordan, Marilyn Johnson, Marieluise Wolfe, and Carolyn Peterson lived close to the land (more properly, rocks), dealt with baffling family logistics, spent long days at the typewriter, fed and cared for summer assistants, and in three cases produced children during the 3 to 4 years of their hitch on the project. After some orientation in 1971, Carolyn Peterson also put a pack on her back and hiked many hundreds of miles over the ridges and through the swamps of Isle Royale.

Most of our research results have been recorded in scientific literature, where the percentage of error is stated (when calculable) or allowed for. I drew freely on these sources, as well as from unpublished reports and a long shelf of everyone's field notes, which were copied for the permanent file through the years. However, this book has no final word on many questions, and sometimes it is appropriate to interpret tentatively on the basis of what we now have. My hypotheses (far-fetched guesses) and theories (plausible interpretations that could be wrong) will be labeled for what they are, but I present as fact the things I judge to be right beyond doubt or with only minor possibilities for error. If some of these statements are found to be awry, it will not alter the general understanding we have developed.

This summary of findings aims at a broad range of usefulness. As a natural history, it is an interpretive record of many kinds of observations. I hope it will be a convenient source of information in a field where humankind clings to faulty traditions and has been slow to learn: the management of carnivores. It should add to the useful framework of knowledge being constructed by the Department of the Interior on wildlife problems of parks and other public lands. Most important of all, these chapters should be readable for pleasure by that increasing public who are charmed by the world of living things — people who own and insist on preserving our remaining samples of primitive wilderness and the life communities they support.

In line with this objective, some of the more routine supporting information is relegated to the appendices. Documentation is "informal." Where I mention an author's name, this can be looked up in the alphabetically arranged listing under "Literature Cited," pp. 471-483. If the writer has more than one entry, a clue will be given to the proper one. References are numbered, and sometimes the number(s) of works will simply be inserted parenthetically at the end of a sentence or paragraph to indicate the source of information used. Although I frequently quote field notes, letters, and unpublished reports, these are not included in the references section, nor are newspaper articles, the date and source of which are mentioned. Quotations from field notes may be edited slightly to avoid symbols and shorthand, and they are used without the authors' permission. Pertinent discussions not strictly necessary to the text are given on some subjects in the section on "Notes," pp. 443-469.

In identifying living things, I use the name that seems most common and correct in our area. The technical identification will be found in a separate index of plant and animal names (Appendix V, p. 439).

Our operations on Isle Royale were greatly dependent on the good will and cooperation of administrative personnel. Appendix I (p. 419) names, in chronological order, the park superintendents, chief rangers, and chief naturalists with whom we worked over the years. We gratefully acknowledge the many times they made special provisions on our behalf or bailed us out of difficulty.

Appendix I also gives a schedule of our periods afield, with a record of the people primarily involved. This includes the temporary student assistants from Purdue or elsewhere who helped during most summers and several winters. A main purpose of the schedule is to identify systematically the staff members of Isle Royale and other parks who were with us each winter in approximately seven weeks of field work. The crew of (usually) four constituted the entire human population of the island at that time of year. Ordinarily, park men joined us one at a time and changed off at intervals of a week or 10 days.

Our flight contractor from 1959 to 1963 was Northeast Airways, of Eveleth, Minnesota. Arthur C. Tomes, president, took a personal interest in our work and piloted on many of the exchange flights to the island. Beginning in 1964, these transportation flights, carrying supplies and personnel from the Eveleth-Virginia Airport to our camp on Washington Harbor, were made for nine years by the late William J. Martila, whose memory we cherish. Bill also piloted for us in fall "antler and calf" counts and beaver colony surveys, as did Jack Burgess, Robert R. Mohr, Dale Chilson, William Coza, and John Brandrup. In 1973 we shifted our winter flight base to Grand Marais, Minnesota, which is on the north shore, only 40 miles (as we flew it from Devil's Track Lake) from Isle Royale. There we enjoyed many favors from Superintendent Sherman W. Perry and the staff of Grand Portage National Monument. From this location our flights to the island were contracted with Wilderness Wings Airways of Ely, Minnesota, Pat Magie, managing owner.

This entire book is a testimonial to the critical part played in these studies by Donald E. Murray of Mountain Iron, Minnesota. As our winter field pilot for 18 years, he has been an integral member of the team and will emerge as a principal character in the chapters to follow.

For all of us who have known them, there is particular pleasure in acknowledging our debt to two individuals who held the position of park naturalist before that position was lost in 1966. Robert M. Linn was our firm supporter from the time I first visited the island in 1957. He left the park in 1961 to become chief scientist of the National Park Service, from which position he had surveillance of all studies such as ours. Following Linn at Isle Royale was William W. Dunmire, who came to our winter camp, literally, with his skis on and a gleam in his eye. Bill left us to become chief naturalist at Yellowstone in 1966. Naturalist duties on the island were then taken over, successively, on a part-time basis by Warner M. Forsell, Alan D. Eliason, and Ivan R. Tolley, each of whom was outstandingly helpful. In the chapters that follow the participation of other park staff members will be mentioned frequently.

In 1973 Bob Linn was succeeded in the Office of the Chief Scientist, National Park Service, by Theodore W. Sudia. The understanding and many good turns of these men, particularly with reference to publications and grants, were of great importance to us, both technically and logistically.

Our studies were fortunate in having the interest and blessing

View across Duncan Bay westward toward Five Finger Bay and Amygdaloid Island. Beyond waters of the north channel lie Thunder Bay and the Canadian Lakehead.

of two secretaries of the Interior, Stewart L. Udall and Rogers C. B. Morton, and three successive directors of the National Park Service, George B. Hartzog, Jr., Ronald H. Walker, and Gary E. Everhardt. It is pleasant to recall the stimulation and encouragement we received in recent years from the personal interest in this work of Mrs. Rogers Morton and Nathaniel P. Reed, Assistant Secretary of the Interior for Fish and Wildlife, and Parks.

With the passage of time, I grow increasingly appreciative of the administrative indulgence that permitted me — three years after joining the Purdue staff — to venture north from Indiana's cornfields and establish a research program near the Canadian border. The university has contributed my time as I chose, and I could have asked for nothing better than the friendly understanding and backing I have had, in turn, from two directors of the Agricultural Experiment Station, Norman J. Volk and Herbert H. Kramer, and two heads of the Department of Forestry and Natural Resources, William C. Bramble and Mason C. Carter.

We have received aid on many problems from members of the Purdue staff. John W. Moser, of our department, participated regularly as statistical consultant, and Erich Klinghammer, of the Department of Psychological Sciences, has been our close collaborator in behavior problems. We have had frequent need for counsel from the staff of the School of Veterinary Medicine, and Sayed M. Gaafar, Edward O. Haelterman, Raymond L. Morter, and George M. Neher have been particularly helpful.

For the first 10 years of this program principal support came from four National Science Foundation grants. Since the mid-60s we have received funds from contracts with the National Park Service, our chief support in recent years. The course of the study has been marked by fairly frequent fiscal emergencies, and we have been helped over periods of stress by timely contributions — mentioned in chronological order — from the National Wildlife Federation, Purdue Research Foundation, National Geographic Society, Wildlife Management Institute, Carnegie Museum, Defenders of Wildlife, National Audubon Society, Boone and Crockett Club, and National Rifle Association.

Individuals who have contributed equipment or funds to this

project are: Robert M. Linn, Houghton, Michigan; Bayard W. Read, Rye, New York; Lee J. Smits and Andrew W. Barr, Detroit, Michigan; George W. McCullough, Minneapolis, Minnesota; Hugh McMillan, Jr., Sharon, Connecticut; Anna M. de la Cueva, Baldwin Park, California.

The slowly disintegrating cabin of the late Jack Bangsund stands among the spruces and firs (with cones upon them) on the south side of Rock Harbor directly across from Daisy Farm campground. This has been our summer base since Dave and Betty Ann Mech first occupied it in 1960.

On protected water, with ready access to trails and the Mott Island headquarters — distance 2.5 miles — the location and accommodations have been ideal. In keeping with the mission and natural character of the park, many old structures have been eliminated; isolated tumbledowns have a way of catching fire in midwinter. After the admittedly hazardous dock and fish house were torn down, the Bangsund Cabin was somewhat marginal as a historic monument. Thus it has been under profane scrutiny by successive generations of superintendents and chief rangers — the latter are a particularly destructive lot.

In defense of squatters' rights, each of our devoted investigators has had to protest the critical importance to science and the surpassing conveniences offered by the cabin: How well the hill-and-dale floor is held together by layered patches of variegated linoleum; how nicely the roof admits the starlight and yet keeps out much of the rain; how smoothly the door opens when you lift firmly on the knob; how ideally the two one-room accessory structures serve as "slave" quarters for summer assistants. And what would befall the local ward of dependent woodmice if the cabin commissary were to fail?

These supplications have managed to stall the march of progress, and we continue to occupy the finest summer research facility on the largest island in the greatest freshwater lake in the world. We salute the memory of fisherman Jack Bangsund, whom we never knew. He did his work well.

WOLVES OF MINONG

CHAPTER ONE

The Founders

ON A NIGHT IN FEBRUARY 1949 ice of the north channel glowed shadow-white under a partial moon. Drifting floes had locked-in solidly during a week of calm and cold while a midwinter "high" lay over the upper lakes. Around the end of Ontario's Sibley Peninsula the floating crust was heaved and shattered where it ground against the rocky shore.

Eastward lights winked dimly from a fisherman's home at Silver Islet, where ice-making was in progress. Eighteen miles south in the darkness lay Isle Royale. To the west beyond Thunder Bay a tiara of luminous sky hovered above Port Arthur and Fort William. Now and again the snow-muffled quiet gave way to a jarring thud, as ice adjusted to cyclopean pressures. Or a high-key metallic strike resounded through brittle floes from some hidden fault line.

Movement in the gloom of the woods edge — a line of doglike figures came lunging through drifts among tumbled ice chunks and trotted out onto the surface of the shallow bay. On better footing the rearmost animals romped ahead, tongues dangling and breaths pluming in the frosty air.

There were seven, perhaps an average wolf pack. For half a mile they straggled out from shore, then stood around for a time or flopped briefly on the firm snow.

A robust, square-jawed male stood at the shoulder of his mate. The thick brush of his tail waved slowly at an angle above horizontal. In the semi-dark his light throat and ear patches contrasted with the apparent blackness of his back and sides. A slender puppish animal sidled toward him in a half crouch, poking playfully with a paw, then fell on its side as the male turned and, with open jaws, pinned its head momentarily to the snow. As he stood above the subordinate, the other wolves closed around him, nose to nose, tails wagging. It was the greeting ceremony of the pack.

The male broke away and ran a few steps, catching up with the she-wolf, who had started off through the darkness. They settled into a trot, as the rest of the pack followed, a loose and changing line that sometimes strung out several hundred yards to the rear. The moon rose to zenith and declined toward a gathering cloud bank in the west. The wolves moved on toward a shore that had been darkly visible at sundown.

The dominant pair were not always ahead. On occasion they sauntered aside to socialize. It was the season when sexual interest is high. Then another wolf led for a time, trotting to the southeast across the frozen channel. An ancient wanderlust led the pack away in feckless abandonment of familiar scenes. Over the ice they walked, gamboled, idled a time, and bounded on in gladsome pursuit of a rendezvous with destiny.

In darkness before sunrise they moved steadily ahead. Then graying light from the east revealed a string of snow-mounded rocks protruding above the ice. The alpha male turned aside to inspect one of them. He nosed it carefully, raised his leg, and

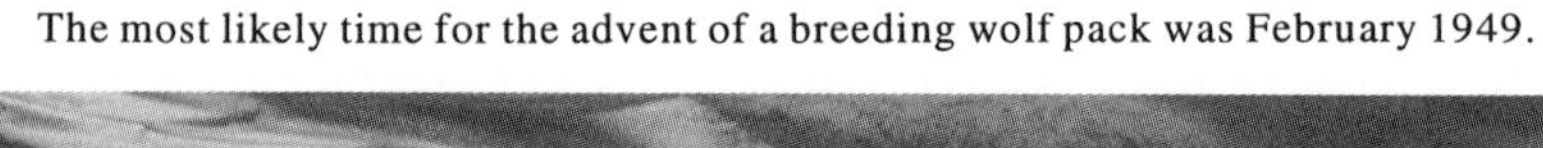

The most likely time for the advent of a breeding wolf pack was February 1949.

sprinkled it with urine. Scratching with all four feet, he threw a shower of snow to the rear. Other wolves, as they came up to it, carefully inspected the scent station and followed on.

Athwart the line of travel loomed the black silhouette of a shore set with spires of a conifer forest. Stragglers were catching up as the pack crossed the mouth of a channel stretching southwest between a ridged island and one of Isle Royale's eastern peninsulas. They slowed to a walk, passed a low headland on either side, and entered a protected cove. Here the snow was softer, level, and unmarked. Overshadowed by low ridges, the cove and its surrounding shores were slowly revealed as the sky brightened.

The wolves stopped on the ice and milled about uncertainly, testing the wind, ears pricked toward the tree line 200 yards ahead. The environs were strange; there was no frolicking now. The alpha pair settled themselves on the snow, head up and passively alert. Half an hour later the big male curled down to sleep, and in time the others followed suit. A freshening breeze stirred their fur and swept a dusting of snow over the dark forms.

It was midmorning when a wolf extended his legs in a rigid stretch and raised his head. He yawned, sat up, and looked about. Before a steady wind low clouds were quartering across the island from the west. Ridge-top trees bowed before it; a muttering of sound came from the ice field to the north.

The wolf walked around, rousing several of his companions. Soon the pack were moving about, and one of them started for the shore. The dominant pair followed; in deep snow at the forest edge the bitch moved up to break trail. The line of wolves threaded their way among down timber into a thick grove of white cedars, where the wind was stilled and hare tracks padded the loose snow. They crossed the white expanse of a lake and a rocky hogback. Then they were climbing through drifts to an old glacial beachline that contoured the higher ridge. There a beaten trail led westward and they followed in the easier footing. A moose had passed this way, and later a couple of coyotes and a fox. At the base of a misshapen birch the pack gathered, nosing and scenting some yellow marks of urine. They added their own and continued ahead, winding through thickets and leaping over log barriers where the moose had dragged its belly and left tufts of coarse hair clinging to stubs and sprinkling the snow. The oval brown pellets of the moose were scattered for a dozen feet in the trail.

The sky darkened under thickening clouds; a powerful windflow whined through the forest, obscuring the tracks ahead. The file of wolves turned into a ravinelike cleft in the abrupt north slope and wallowed on toward the crest of the ridge. Breasting the bottomless drifts was enervating, and soon the female lay panting on the snow. The male plodded into the lead and they labored up the acclivity, nose to tail, leaving a deep furrow in the powdery surface.

As they emerged onto the summit the frigid blast of the storm combed their dense coats of agouti gray and cinnamon brown. Oblivious to cold and the pelting snow, they ambled across the wind-lashed height of land, stopping occasionally to regard a scene no wolf could comprehend.

A continuous growling roar came from the north channel, as the ice mass disintegrated and drove eastward into the open lake. Black leads of open water showed where the plain of white had bound Isle Royale to the mainland hours before. The land mass of Ontario was hidden behind miles of cloud and mist. South of the narrow island the surging combers of Lake Superior frothed brilliantly over shallow reefs and spent themselves against the granitic bastions of coastline that stretched out of sight into the weather westward.

Not before another winter would there be an ice bridge from the island to Canada. The immigrants were now at home. Isle Royale had a socially organized wolf pack — the potential founders of a resident population that must crowd in readjustment into the community of moose, beaver, coyote, fox, and hare.

The south slope was a gradual descent, and the pack worked their way downward into a clump of spruce near the end of a long, fjordlike bay. Out of the gale that bounded off the heights a thickening fall of snow swirled down onto the lee of the ridge. Under its insulating blanket the wolves bedded down again.

It was another morning — one of empty stillness, glinting ice prisms, and brilliant sunrays slanting across the narrow avenue of far-reaching Rock Harbor. In the woods a few chickadees called. Higher on the slope a pileated woodpecker hammered a short cadence against the shell of a long-dead stub.

The pack trotted westward, the bonded pair ahead, tails wav-

As the island appeared from Landsat Satellite, 570 miles up in space (NASA photo).

ing, the male running close to the shoulder of the bitch. As they stood for a moment he laid a paw over her back and made mounting gestures. She growled and snapped at him, then walked on along the cedar-lined shore. The five followers scattered over the snow-covered ice, the skein of their tracks showing faintly on the hard bas-relief of storm-cut patterns.

Farther up the harbor ravens were hooting from a spruce near the shoreline. As the wolves approached, several of the alert birds flew out to meet them, circling and swooping. The wolves sniffed at a cluster of coyote tracks and a scattering of moose hair, then followed a fresh trail inland. Beside the bare trunk of a fir a coyote leaped up from its bone-chewing and bounded off through the yielding drifts. Three wolves lunged after it. Within a dozen yards it was stretched between savage jaws that tore at its vitals and crushed out its life. In sportive mood, the three romped back to the woods edge, as a raven flew down to a branch above the limp bundle of fur.

Under a fluffy overlay from the recent storm, the dismembered skeleton of a moose calf was embedded in a floor of trampled snow. Probing noses quickly located the bones, which pack members pulled forth and carried about. Little had been left by the coyotes and foxes. However, the dominant male made a find. Probing under a snow-weighted limb of spruce, he retrieved the well-picked head of the calf. Anchoring his ivory canines in the base of the skull, he split the bones with a powerful bite — something the lesser carnivores could not do. Growling and snarling as the other animals approached, he licked out the tidbit of the brain.

As the pack moved up the harbor, a single wolf stayed by the calf remains, searching for something edible. From the snow it lifted a leg bone with hoof attached and carried it out onto the ice. Shearing with cleaverlike carnassials, it began to pare away the bone and tendon of the hock. Then it raised its head and gazed after the pack, now more than half a mile away. It stood up and walked a few steps, its tucked tail, flattened ears, drawn lips, and trembling hind quarters betraying apprehension.

There would be a time when this, the least of lesser wolves, would stay behind and accept a precarious living on the leftovers at kills. Social tensions in the pack and the dominance of other males were becoming a torment of stress. But the time to become a true "loner" was not now. These ridges, bays, and woodlands were not the Sibley range of familiar roads, deer trails, and landmarks. This was alien country, where the presence of the pack gave reassurance to a subordinate who knew and accepted his socially inferior role. At a rapid trot the wolf followed the trail to the west.

On a late afternoon two moose, a cow and her calf, arose from their beds in the forest edge and moved slowly along the shore of Intermediate Lake, near the eastern edge of the great burn. Vegetation of the lake margin had been heavily browsed, with fragments of foliage sprinkling the packed glaze under overhanging cedars. The cow reached high to glean yet a few more of the pungent green sprays and to nip sparingly of the plentiful alder. She passed the spruces with no show of interest. Only substandard foods remained, and the trailing calf found nothing within reach. Its rough coat and sunken flanks betrayed a desperate state of malnourishment. This was a winter of reckoning, when Isle Royale's increasing moose had taken the bulk of available browse in the region of the burn.

Had there been some timely urge in early winter to lead many moose away to the east, over frozen basins and snow-clogged woodlands, they could have found lightly used thickets of young balsams to sustain them through the cold season. But this was not part of the design. Why those acres of seedling and sapling balsams to the east are not used more by wintering moose is an open question. Perhaps something in the topography and lay of the land could explain it. But in the year of hardship there was no succor for the cow and calf. Where they were, the survival problem bore heavily.

From soft drifts of the shoreline the cow turned out onto the wind-whipped lake. There she stopped, ears pricked toward the backtrail. A quarter-mile away, a cluster of unfamiliar figures approached rapidly over the white expanse. As they neared, a primordial unrest alerted the cow, and she moved to the rear of her calf.

It was now four days since the wolves had fed. As they advanced on their prey, the strong scent came to them, excitement spread, and all senses were sharpened. Scattering widely, they broke into a run.

The cow received them with lowered head and flattened ears, the ruff of her neck and shoulders bristling defiance. She lunged to meet the nearest wolf, and it veered quickly away from her lashing hoofs. As several of the pack circled toward the calf, it headed inland through the deep snow. A poor tactic; it had no stamina for breaking trail. The cow tried to follow, but a wolf leaped at her rump, driving deeply with fangs that opened a gash in the tough hide. She whirled on her assailants, scattering them again, as she backed into the thick branches of a spruce.

The exhausted calf was floundering. A wolf seized it by the shoulder and bore it down, as another dived at its throat. The pack swarmed over it, tearing and chewing. Again the cow drove them off, but the calf lay bleeding and helpless in the trodden drift. For half an hour she stood over the remains, countering every threat or advance of the wolves, who sauntered about or lay patiently on the ice. The sun had dipped below Greenstone Ridge. Shadows were deepening as the cow gave up her vigil and continued west toward the end of the lake.

The time of feeding had come, and the wolves closed in.

In some such manner a breeding wolf population became established on Isle Royale. The most probable time is early in 1949, but it is an open question why this did not happen sooner. The reason undoubtedly was related to events on the mainland north of Lake Superior, which shall be treated in more detail in the next chapter. In 1950 de Vos obtained significant information on wolves of the Sibley Peninsula from a local resident.

Mr. J. Cross, who was born and raised at the tip of the Peninsula, took accurate notes on the early abundance of different mammals seen and trapped in the area. He reports that there were no wolves at all on Sibley Peninsula during his childhood and that they appeared after the turn of the century, shortly after moose and deer arrived. About thirty years ago Mr. Cross trapped and poisoned from thirty to forty wolves, in one winter, within the area and on the bays around it. The largest wolf pack he ever encountered on Sibley was one of thirteen to fourteen individuals, on the basis of a track count.

Although this informant thought the coyote appeared in his area later than the wolf, there is little question that the smaller animal arrived on Isle Royale at about the same time as did the moose. Krefting summarized coyote records indicating that it reached the Sibley area about 1900 and was being trapped on Isle Royale a few years later.

It is probable that wolves — especially wandering loners — reached Isle Royale at times in the past when winter ice was favorable.[1] We may assume that a breeding pack did not get there, or the population would have built up long before it did. This could imply also that loners of opposite sex were not present at the same time.

At midcentury there still were wolves on all sides of Lake Superior, although they were everywhere subject to bounty hunting, trapping, snaring, and poisoning. Shooting from aircraft was becoming important, especially in the lake country of Minnesota and Ontario. The species was long gone from Michigan's lower peninsula; according to Lee Smits the last specimen of record was in Saginaw County in 1909. He summarized the situation in the upper peninsula, as of 1955:

The remaining Michigan wolves are in the wilder parts of the upper peninsula. The state of Michigan pays a bounty of 20 dollars for female wolves, 15 dollars for males. The number of wolves annually presented for bounty has been a little under thirty for some years. In addition, five or six wolves are killed each year in the fall by deer hunters. The average take of about thirty-five . . . is believed to represent the progeny of no more than six pairs . . . It is probable that Michigan's total wolf population at this moment is between twenty-five and fifty, with twenty-five the more likely estimate.

The state's bounty law, of 122 years duration, was finally rescinded in 1960, and the wolf got full protection in 1965, at which time only a few nonbreeding individuals were left. The same can be said for the contiguous range to the west in northern Wisconsin, although the state anticipated Michigan by eight years in giving protection to a few straggling survivors.

The eastern timber wolf has continued to maintain its numbers well in the border country of northern Minnesota, showing its greatest decline in areas where the maturation of forests has reduced the deer herd on which it largely subsists. These wolves — possibly replenished by movements across the international boundary — constitute the one major viable population of wolves in the conterminous 48 states. The only other known breeding population is on that small, northern satellite of Michigan, Isle Royale.

When a few Ontario wolves made their epochal crossing to a new island range, they were in a park that was unique in states south of Canada — one where the long-persecuted carnivore could find undisturbed sanctuary. By this time the species had been practically eliminated from the West, although the public was not widely aware of this. As a principal reason, the taking of many "wolves" was still being reported in annual budget requests to Congress by the Division of Predator and Rodent Control of the Fish and Wildlife Service. The rationale behind such statistics was that a few of the dwindling red wolves were still being trapped and poisoned in Texas and possibly other adjacent states, and in some areas the coyote is called the "brush wolf." Obviously, with the long history of the Biological Survey behind it, one would expect the Fish and Wildlife Service to be definitive and accurate in its reports to Congress (8).

The history of the war against the gray wolf, *Canis lupus*, in the West has been described in detail by Stanley P. Young, who had extensive personal experience in the campaign to wipe out the stockman's chief competitor. By 1950 it was uncertain whether *any* indigenous wolves survived from the seven subspecies that probably had inhabited large or small ranges in areas from the high plains west during primitive times. It appeared that occasional stragglers crossed the border from western Canada and from Mexico, and more recently there are indications that a few of the (now nearly vanished) subspecies *irremotus* hung on in the Yellowstone area.

Yellowstone, Glacier, and other national parks had been drained of their wolves by heavy control operations in adjacent public and private lands as well as by direct control conceived to be a legitimate part of park wildlife management. When, in 1895, General Hiram M. Chittenden published his famous book on Yellowstone National Park, he cited this and other major parks for their outstanding value as "game preserves." From earliest times big game had been considered scarce in the Yellowstone region, but it was building up late in the century. There were heavy inroads from hunting and poaching even after establishment of the park by act of Congress in 1872. Killing wildlife was first prohibited in 1883 and confirmed by the Protective Act of 1894.[2] About that time a few wolves were appearing in the park.

In a book on the wildlife of Yellowstone, the chief field naturalist of the Biological Survey, Vernon Bailey, said, "The large gray wolves at times have become abundant in the park and wrought great havoc among game animals, but at present [1930] they are being hunted so persistently over much of the outside range as well as in the park that they have become extremely scarce." He noted that in 1914 and 1915 they had been "especially destructive" among the elk, and Mr. Frazier of the buffalo ranch had killed two pups and captured two more. In 1916 in the Lamar and Yellowstone valleys four old wolves and a family of seven pups were killed.

On 20 March 1923 Henry W. Shoemaker, publisher of the Altoona (Pennsylvania) *Tribune,* sent the following note attached to a letter to E. H. McCleery, of Kane, Pennsylvania:

Mr. M. P. Skinner, Park Naturalist, Yellowstone Park, Wyoming, says, "I had hoped that the present management of the Park would write you about the predatory animals and give you the latest data thereon. But since they are so slow, I shall be glad to do the best I can and give you my general impression. There are presumably five hundred coyotes here now ranging in all parts of the park. In spite of our Rangers killing from one to two hundred each year, I am afraid they are increasing in destructiveness. We have a hundred or more of the big timber wolves, and, curious to say, a very large proportion are black wolves and the rest are gray wolves. They are very destructive and we hope they are not increasing in number. These wolves cruise all over the Park in summer but are all on the elk ranges in winter. Not more than fifteen or twenty are killed each year."[3]

Skinner left Yellowstone, which he had first visited in 1895, and published a bulletin on the "Predatory and fur-bearing animals of Yellowstone Park." As of that time (1927) he indicated that there were about 20 wolves in the park and that they were decreasing.

In his National Parks Fauna on the ecology of the coyote in Yellowstone, Adolph Murie cited reports of the early superintendents frequently requesting support for predator control operations, particularly on the cougar and coyote. He said the last cougar was killed in 1925, and all predator control came to an end in the winter of 1934–35.

Previous to the establishment of the National Park Service in 1916 there was little controversy about the control of predators because nearly everyone agreed that game had to be "protected." This view came not only from the backwoods but from the scientific community. The great naturalist C. Hart Merriam, first director of the Biological Survey (1905), had stated in *Vertebrates of the Adirondacks*:

The wolf is one of the most cowardly and wary of our mammals, always taking good care to keep out of sight; and he is so crafty and sagacious that it is almost impossible to allure him into any kind of trap.

When opportunity affords he is one of the most destructive and wasteful of brutes, always killing as much game as possible, regardless of the condition of his appetite, and he used to be the greatest enemy that our deer had to contend with. During the deep snows a small pack of wolves would sometimes kill hundreds of deer, taking here and there a bite, but leaving the greater number untouched.

In reviewing the development of predator control in the national parks, Cahalane said that poison was being used in Yellowstone in 1898, and probably before. There were records of its use in Glacier, Yosemite, Rocky Mountain, and other parks until 1922. Steel traps were employed in control work until the early 30s, and shooting was the most widespread method of all. After 1918 government hunters of the Biological Survey carried on control operations in many of the parks.

In terms of today's ideas, it all sounds quite primitive. However, early administrators of the National Park Service had major problems directly concerned with the existence and buildup of the young agency to which they were clearly devoted. It is not surprising if they did not immediately promote a new philosophical attitude toward predatory wildlife. Unquestionably, they thought they were carrying out the intent of Congress as expressed in the most quoted and variously interpreted passage in Public Law 235 of 1916, by which the National Park Service was created:

> The Service . . . shall promote and regulate the use of . . . national parks [and] monuments . . . by such means and measures as conform to the fundamental purpose of the said parks [and] monuments . . . which purpose is to conserve the scenery and the natural and historic objects and wild life therein and to provide for the enjoyment of the same in such manner and by such means as will leave them unimpaired for the enjoyment of future generations.

In the first decades of this century the impetus of predator control in public agencies reflected simplistic attitudes toward the natural world common since pioneer times (21). But during the 20s viewpoints were changing both in the parks and among the public. Cahalane said that from 1928 to 1930 the Service received many complaints from scientists and scientific organizations who condemned predator control not preceded by adequate investigation. In 1931 director Horace M. Albright went on record in the *Journal of Mammalogy* defining, in effect, a new policy under which predators were recognized as having a real place in nature and thus in the parks. Predators were to be protected, and no widespread campaigns of destruction would be countenanced.[4] This led logically to the forthright policy recommendations in the first of the National Parks Faunas by Wright, Dixon, and Thompson in 1933:

> That no management measure or other interference with biotic relationships shall be undertaken prior to a properly conducted investigation.
>
> That every species shall be left to carry on its struggle for existence

unaided, as being to its greatest ultimate good, unless there is real cause to believe that it will perish if unassisted . . .

That the rare predators shall be considered special charges of the national parks in proportion that they are persecuted everywhere else.

That no native predator shall be destroyed on account of its normal utilization of any other park animal, excepting if that animal is in immediate danger of extermination, and then only if the predator is not itself a vanishing form.

By the year of the founders, 1949, the wolves that came to Isle Royale were in no danger of persecution by responsible authorities. As a matter of fact, difficulties might have been avoided if they had revealed their presence at once by some convincing howling near the Mott Island headquarters after the staff returned to the park in early May.

Michigan's long-term friend of the wolf was Lee J. Smits of Detroit, widely known newspaperman, outdoor writer, and for 4 years a member of the State Conservation Commission. He fought the wolf bounty and was an eager promoter of the protective legislation that came too late. Smits knew the north woods well. As a young man, he worked in lumber camps of upper Michigan before the turn of the century. His experience with captive wolves spanned a period of nearly 40 years.

The idea that Isle Royale could be a sanctuary for the eastern timber wolf undoubtedly occurred to various people both inside and outside the National Park Service. In Michigan, Smits attributed the suggestion originally to A. M. Stebler, who made the first studies of Michigan wolves at the Cusino Wildlife Experiment Station in the upper peninsula. By 1951 a plan was afoot to pay trappers 50 dollars for every pup they could obtain from dens in the wild so these could be sent to Isle Royale in a stocking experiment. Smits summed up the thinking of supporters of the plan (256).

It is hard to see how any objections can be raised to the proposal that Isle Royale, a true wilderness park, 45 miles long and 9 miles wide, become a permanent sanctuary for timber wolves. It is not to be expected that tourists visiting the island would ever catch a glimpse of wolves, since they are totally unlike the bears that come to feed at garbage dumps and become fearless — and, in some degree, dangerous to humans. But it might be that hikers in the interior of the island would hear a wolf, and certainly there would be naturalists grateful for the opportunity to study this highly-intelligent and elusive creature in the wild.

The park staff and summer visitors were getting increasing indications that wild wolves were on the island. In a memorandum to the regional director dated 5 April 1951, superintendent Charles E. Shevlin said he thought they had wolves (soon to be verified) and that a solid ice bridge to Canada had formed in each of the two preceding winters.

Shevlin was skeptical about the plan to bring in more wolves, especially on the basis of public relations. Although a primary objective was to have a refuge where this species could be protected, the thought that wolves might prevent another moose irruption was also in the picture. The superintendent pointed out that

> The reaction of the general public, particularly in the Lake States, would be distinctly adverse to the introduction of wolves at Isle Royale. At the present time, there is a good sized bounty on wolves in Michigan and Minnesota and possibly Wisconsin and the attitude of most people is that they should be exterminated. They are considered as vicious beasts of prey on all types of game animals and kill for the sheer joy of killing. If the public learned that they were being introduced at Isle Royale to control moose, the Service would have to answer many questions. It would be extremely difficult to explain why this method of control was used in place of regulated hunting. We run the chance too, of a severe [moose] die-off from disease brought on by malnutrition but the wolf would get the credit.

Notwithstanding this counsel, the plan to stock wolves on Isle Royale gained momentum, abetted by an article by Edwin D. Neff in the November 1951 issue of *Natural History*. Lee Smits took the initiative. In January 1952 he queried the National Park Service and by early April had received permission to go ahead with the project. Director Conrad L. Wirth received word from Michigan conservation director Gerald E. Eddy that the plan had the approval of the State Conservation Commission. All concerned were encouraged by favorable resolutions and statements of citizen groups in Michigan and national conservation organizations.

Despite the proffered reward, trappers had failed to find any breeding dens, so wild pups were not available. As a next-best the decision was made to use wolves that Smits had been in-

The evidence had become convincing.

strumental in bringing together at the Detroit Zoo. There was a recognized hazard in using animals that might become a nuisance. Island summer residents and others would find cause to demand that *all* wolves be removed from the park. Smits agreed to pay for any damage his wolves did, and he contributed other expenses of the operation.

On 9 August 1952 head keeper Bromley of the Detroit Zoo arrived on the island with two pairs of young adult wolves, all reared in captivity and appearing to the park staff to be "comparatively tame." The animals were placed in a hogwire pen at Peter Edisen's commercial fishing base on Rock Harbor, where fish offal would be available as a food source. The two small females were named Lady and Queenie, and a black male was called Adolph. The fourth and largest wolf was a gray male weighing 90 pounds. "Big Jim" merits special comment.

Lee and Peggy Smits received this wolf from the zoo on 16 May 1951, when it was a day old. They reared it on a bottle. Mrs. Smits bore the burden of wolf care night and day. According to Lee, when the pup was six weeks old "she kept a pair of twelve-quart galvanized buckets handy to stand in while Jimmy was with her to protect her ankles from his playful attentions"(257).

The Smits had a water spaniel, Junie, to whom the young wolf was greatly attached. In imitation of the spaniel, Jimmy learned to retrieve ducks and thereby became a television personality.

> That attracted the attention of a New York magazine publisher. He sent a photographer to Detroit with a station wagon full of equipment to photograph the retrieving wolf. He worked all day with Jim. He retrieved everything, including the photographer's tripod — everything but a duck. He refused to touch a duck on that particular day.

Jim was of mixed parentage. He was sired by an eastern timber wolf, *Canis lupus lycaon,* from Michigan. His mother, of the black color phase, was from Saskatchewan — presumably of the subspecies *Canis l. griseoalbus*. Northern wolves are large, and this one was true to type. At eight months of age he weighed 85 pounds. He outgrew the Smits household, was returned to the zoo, and became known as "Big Jim." Shy of strangers, he would go into transports of joy when visited by the Smits; he was especially fond of Mrs. Smits. Lee thought this wolf would avoid people if he had the opportunity and that, above all, "His one ambition is to run far and wide." Thus Big Jim got his great chance in life when he became a part of the Isle Royale expedition.

Subsequent events developed rapidly and were detailed in a report of 26 August by chief ranger Floyd A. Henderson.

Peter Edisen had been engaged to care for the wolves. Within 24 hours the small female, Lady, got through the fence and escaped. In similar efforts the other animals were getting tangled in the wire, and so on 11 August they too were released.

The wolves seemed congenial, Jim having definitely established his dominance. They played around like a group of friendly, but very rough, dogs. For the Edisens it became a full-time job, one they had not bargained for. The romping pack tore clothes off the line, chewed up hooked rugs, and destroyed a 40-dollar nylon fishnet.

It was evident that the wolves must be removed from the vicinity of people, and traps were built. The animals were becoming wary, but one of the females was caught and taken to Locke Point, an isolated peninsula at the east end of the island. The two males eventually entered the pen, where they had to be "snared and dragged to the cages made for them." On the 16th Jim and Adolph were taken to Locke Point and made to swim ashore. Queenie was there, but she had not touched the food left for her. At the Edisens', Lady was eventually reduced to confinement and was sent back across the lake to Conservation Department officials in the upper peninsula.

A day after their release, Jim and Adolph visited a family on a small island in Tobin Harbor where they tore up another laundry. By the time seasonal rangers James Orsborn and Robert Linn got there the wolves had swum to Minong Island, also inhabited. The following day Linn and ranger Ed Kurtz went to the island and fed the wolves some sedative capsules embedded in hamburger. The males began to "stalk" Linn and Kurtz and had to be deterred by gunfire. On the 19th all three wolves were in the vicinity of Rock Harbor Lodge, and a visitor was approached and then followed by a large gray male, most certainly Jim.

Rangers Henderson and Kurtz decided the wolves must be eliminated. Taking the trail eastward toward Scoville Point they found Adolph, who came toward them, "with nothing friendly in his approach." Henderson shot, killing him instantly. Shortly Jim was seen, following Adolph's tracks. Another shot missed. The wolf turned and ran and was not seen again, despite further hunting. As Bob Linn recalls it, Queenie was shot later that same day.

Jim was left as the sole survivor in the wild, and for some years he was something of a legend on the island. As we see it now, with a breeding wolf pack already established, Jim would have been at a great social disadvantage in any attempt to associate with his own kind. It is unlikely that he made any contribution to the genetics of today's population.

The wolf-stocking experiment got the expected unfavorable public reaction.[5] There were indignant letters to the park, the

Figure 1. Natural features and points of interest, Isle Royale National Park.

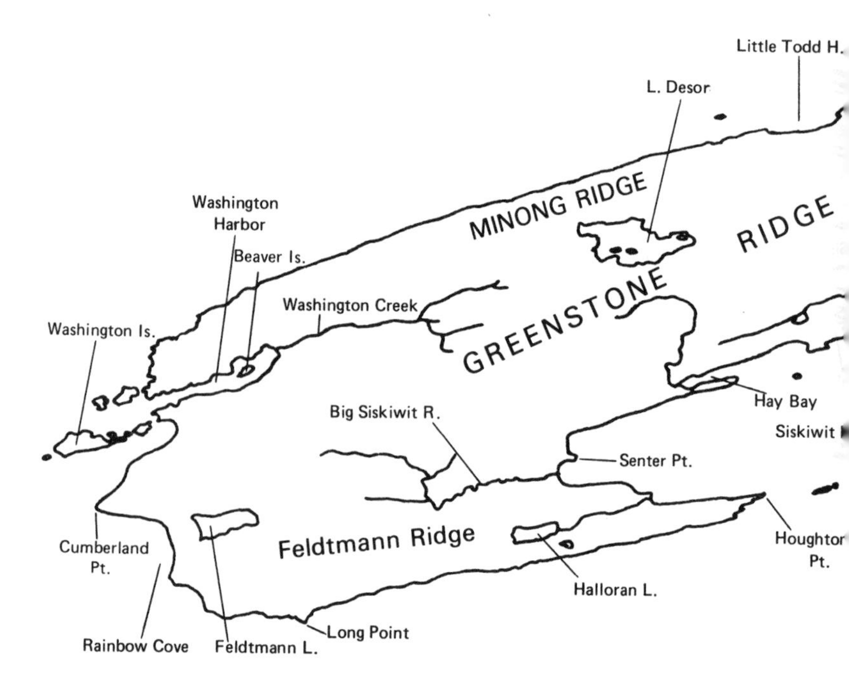

Washington office, and to Congressman John B. Bennett. Comments in the press, featuring letters to the editor, were generally unfavorable. An editorial in the *Daily Mining Gazette* of Houghton, Michigan, for 13 August 1952 quoted local hunters:

> A number of sportsmen were "horrified" and indicated that "that's what we get after trying to build up the island as a tourist area and now the Department of Conservation slaps us in the face."

As one respondent expressed the common view, "It goes without saying that the wolves will in no way attract visitors." Another recalled an "occurrence which took place in Champion about two years ago. Two wolves went on one of their regular jaunts — Result — 30 or 40 deer carcasses strewn over the swamps." The outlook for Isle Royale was conceded to be dim, although the editorial included one comment on a different note

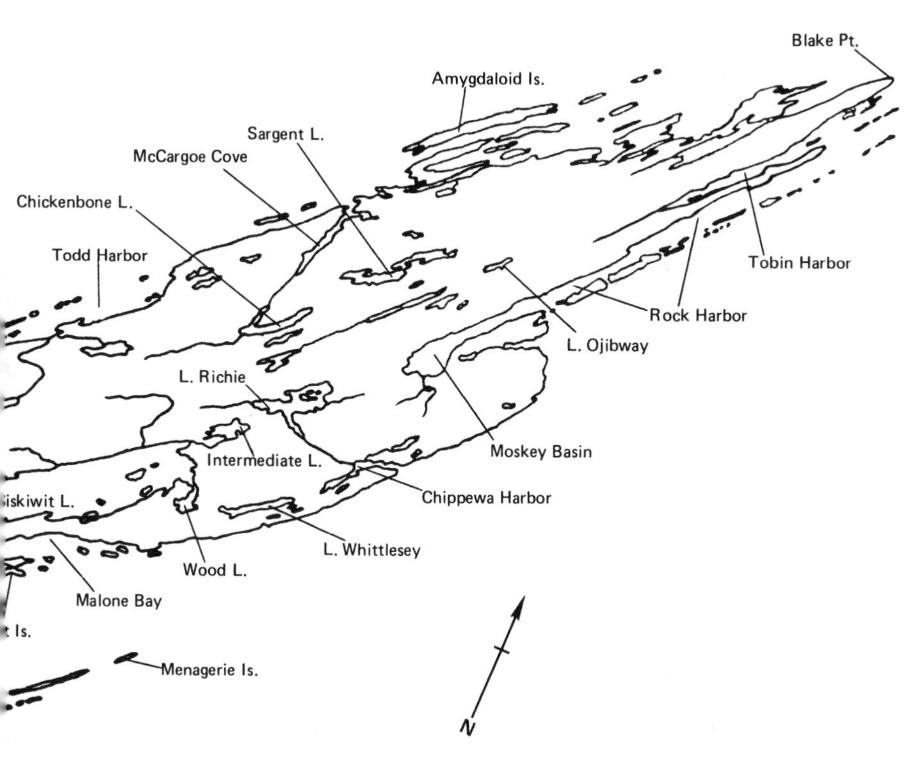

by a resident of Hancock: "I don't believe that they'll be of much harm on the island . . . I sort of believe the department knows what it is doing in this case."

Communications among various offices of the National Park Service indicated a recognition that the plan went awry mainly because the pen at Edisen's was not adequate and the wolves had to be handled roughly in transferring them. Since wild wolves were now on the island, it was considered inadvisable to do anything further with captive-reared animals.

The most significant studies on Isle Royale in the 50s were those of James E. Cole, National Park Service biologist, who was regularly stationed at Rocky Mountain National Park. In late winter of 1952 Cole and ranger Robert Hakala carried out a field investigation of the moose and wolf and confirmed the presence of four wolves in the southwestern end of the island. In 1956 Cole and Linn flew to Halloran Lake and were in the field on snowshoes from 9 February to 8 March. Cole's report summarized their findings:

> This winter evidence strongly suggests there are at least fifteen wolves in the same area. Seven additional wolves were reported from the central part of the island making a total of at least twenty-two. Probably most of the Isle Royale wolves live in winter on the southwestern or central parts of the island, but no extensive census was attempted so doubtlessly the enumeration is incomplete and more than twenty-two wolves now live on Isle Royale.

It appeared that moose were at a low point of numbers and that wolves were building up. Cole noted that all the cows they saw had calves. He also mentioned the recent decrease of beavers on the island, which correlated with the advent of wolves. We will have more to say of this in a future chapter (p. 239). All of the wolf scats collected indicated that moose had been fed upon. The kill of a calf was verified, and the presence of several carcasses of adult moose suggested that they also were being killed.

The following winter Cole was on the island with a pilot and plane from 12 February to 2 March. They flew 31 hours in their 17 days in the field. They made a census of moose in the open burns and estimated the island population at 300. They found eight moose carcasses with evidence that the animals had been killed by wolves.

A point of particular significance appeared in Cole's report of 1957:

No coyotes were observed this February and tracks of one only were noted. Apparently the Isle Royale coyote population has declined substantially the last five years. Off-hand it would appear that the uneaten moose carcasses left by wolves would benefit coyotes in winter. Probably at some time of the year, these smaller predators cannot compete successfully with wolves when food of both is scarce.

This was, in fact, the last record of coyotes on Isle Royale. In early times on western ranges coyotes and wolves were both abundant in the same areas. Wolves lived on buffalo and other big game, and coyotes undoubtedly scavenged in their wake. Coyotes also fed upon jackrabbits and a wide variety of rodents. Under these conditions both species survived. However, on a tight little island in Lake Superior there evidently was too much competition, possibly both social and economic, for both to persist. It was the lesser carnivore that gave ground and disappeared. We suspect that the wolves simply killed them at every opportunity.

Another episode of Cole's winter study has interesting connotations. On the ice of Siskiwit Lake the plane broke a ski strut and was inoperable. The two men had to make a long snowshoe trip back to Siskiwit Camp at the southwest end of the bay. They became aware that a wolf was following them, and Cole had this to say about it:

The actions of the lone wolf which followed the pilot and writer for nearly nine miles across Siskiwit Bay on their mush back to camp after the plane broke down on Siskiwit Lake is probably not typical of the actions of wild wolves. I may be wrong, but I believe this animal was "Old Jim," the Detroit zoo-raised wolf which was liberated on Isle Royale in the summer of 1952. During the six hours it followed us, mostly from a discreet distance of a fourth mile or more, its actions appeared to be those of a friendly but cautious dog. When we rested it lay down, and continued on when we resumed mushing. Of course, we had it thoroughly confused by dragging a trout across our trail. It knew we were humans but certainly we must have smelled like fish.

To our best knowledge, nothing like this has ever happened since on Isle Royale, and Cole may have been correct that this was the last recorded episode in the career of Big Jim.

CHAPTER TWO

Moose Haven

MOOSE BECAME ESTABLISHED on Isle Royale some 40 years before the wolves. The advent of moose was clearly tied to environmental changes along the north shore of Lake Superior in the second half of the nineteenth century.

Primitive forests of that region supported few moose and probably no deer at all. Large stands of pine and great areas of mature white spruce and balsam fir, the type that extends far north into Canada, produce little food for animals like the moose, which are heavily dependent on winter browse.

Browse? It is a word we will need repeatedly. Browsing animals feed on the twigs and buds of woody plants, including conifer twigs with their needles or scalelike (e.g., cedar) leaves. In summer the browsers typically switch over to more succulent plants and the leaves of deciduous shrubs and trees. These foods are available only near the ground. Ordinarily moose reach to about 8 feet.[1]

The most typical hoofed animal of the "climax"[2] spruce-fir forests was the woodland caribou. This race did not form large migratory herds as did the barren-ground caribou of arctic tundras, although it trekked about in small bands. The woodland caribou did some browsing but fed mainly on leafy foliage, herbs, grasses, and lichens, including the tree lichen *Usnea* that festoons the forest edge along humid shores. The caribou — and also what few moose there were — made much use of lowlands, which were extensive in the original forests. These habitats included lake edges, beaver floodings, bogs (undrained marshes), muskegs, and cedar swamps. Characteristic lowland trees were the northern white cedar, black spruce, and tamarack, and there were many shrubs and water plants that made such sites seasonally attractive as feeding areas.

Before the turn of the century, caribou were plentiful on the Sibley Peninsula, directly northwest of Isle Royale. J. Cross, a lifelong resident, estimated that there were 500 caribou in 110 square miles (65). The largest herd recorded for that area was 59. Before the period of settlement, caribou were found as far south as Michigan's lower peninsula, although they may not have been plentiful. Henry Schoolcraft, Indian agent at Sault Sainte Marie and Michilimackinac for 20 years beginning in 1822, mentioned in his journal, "The American species of rein-

Early in the century a few moose reached the island, presumably by swimming, and then built up rapidly on abundant foods supplied by successional forests. This patch of aspen typified excellent browse conditions.

deer, which under the name of cariboo, inhabits the country around the foot of Lake Superior . . ."[3]

The boreal (northern) spruce-fir forest gave way southward on more fertile lands to the lake states hardwoods, principally sugar maple but in different parts of the region variously mixed with basswood, beech, red oak, yellow birch, and hemlock. Light sandy soils commonly supported nearly pure stands of white or red pine. The most sterile "sand plains" were in jack pine, which had little value for timber. In all mature forests the dense shade prevents much development of the understory that could provide moose or deer food. When the canopy is destroyed and sunlight strikes the ground, low-growing vegetation immediately responds.

After 1850 great changes came about with the logging of premium stands of pine. Heedless methods of the time left vast areas of tindery slash to burn in dry summers. The fires swept on across uncut conifer forests. Muskegs and swamps dried out and they too burned, the peat fires smoldering for long periods in years of little rainfall.

Along the north shore the country took on a new aspect. Old-growth forests were gone, and it might take one to two centuries of orderly development through predictable stages of plant succession for them to be re-established. The forest that develops after "disturbance" is a transient type characteristic of burns, blowdowns, and cuttings throughout the region. The two most common trees are paper birch and quaking aspen. But the type includes other wind-seeded, root-sprouting, or fruit-bearing trees and shrubs — willow, cherry, mountain ash, hazelnut, blueberry, and many more, according to local conditions of soil, water, and topography.

This early phase in forest development does not reproduce well in its own shade, and it will gradually be taken over by invasion and spread of the more tolerant conifers. While it lasts, the birch-aspen association is highly productive of wildlife, offering an abundance of winter browse and summer foods for members of the deer family. It is good habitat for the snowshoe hare and predatory species that depend on it, notably the lynx, fox, and various birds of prey.

Destruction of the mature forests greatly reduced woodland caribou. Moose multiplied and spread, and deer invaded new country far north of their primitive ranges in Michigan, Wiscon-

Top: On a rocky slope above Ojibway Lake, birches and aspens slowly cover the rocks where organic soil was burned away in the 1800s.
Middle: Where the glacier dropped a deposit of mineral soil, a forest fire was followed by quick invasion of the wind-seeded birch.
Bottom: Climax spruce-fir forest along the north shore near Huginnin Cove. After a fire it may require from one to two centuries for this stabilized stage to be re-established.

sin, Minnesota, and Ontario. Early in this century caribou were disappearing along the Canadian border and southward.[4] Today we have an ample background of both historical and ecological knowledge to draw upon in understanding what happened in the Lake Superior country. However, a few understood it while the changes were in progress, as attested by a succinct statement in *National Geographic* in 1921. The author, George Shiras III, was a former congressman from Pennsylvania, a famous outdoorsman and pioneer wildlife photographer:

> About 1885 a steady movement of the moose westerly from Quebec was observed and a slower easterly migration from northern Minnesota. Eventually these animals conmingled and took possession of the entire shore, later extending into the interior until they reached the water flowing into Hudson Bay. Following the moose came the white-tail deer and many timber-wolves, when the caribou began yielding the possession of centuries.
>
> After the construction of the railroad, extensive lumbering and many forest fires changed the face of the country, large clearings and a mixed vegetation succeeding dense evergreen forests, and to this change may be principally attributed the influx of new animals and birds.

Historically, there were a few caribou on Isle Royale. Some may have lived there permanently; others moved to and from the island across the ice of the north channel in winter. The Biennial Report of the Michigan Department of Conservation for 1921–22 said, "Various estimates as to their numbers have been placed at from 225 to 300." The last of the caribou were seen about 1925, according to Peter Edisen, commercial fisherman, who came to the island in 1916 and has been there every summer (and several winters) since.

The systematic study of Isle Royale plants and animals began with the work of the University of Michigan expedition headed by Charles C. Adams in 1904–05. They reported no moose and — of later significance to this account — no beavers. It is likely that the moose never were there and that beavers had been nearly, but not entirely, wiped out by trappers.

Moose probably reached the island about the time of the Adams expedition. It is even possible that the party saw moose sign that was attributed to caribou. In various reports it has been assumed that moose crossed the ice in winter, but our acquaintance with this oversized member of the deer family casts doubt on such an explanation. These animals do not venture far out onto open ice. In studies since 1958 we probably have never seen a moose a mile from shore, and half a mile would be the usual limit for such forays. It is much more likely that the initial breeders colonizing the island swam across the channel. The distance could have been as little as 13 miles from the nearest Canadian islands (Spar or Victoria), but it is likely to have been considerably farther. The western third of the Isle Royale coastline, nearest to Canada, features long stretches of cliffs and abrupt rocky slopes. A moose approaching its landfall in that area might have to lie-to in deep water and pick its harborage carefully to make a success of the operation.

Pertinent to this is an observation described to me in 1961 by James M. Godbold, who was then director of photography for *National Geographic*. In May 1956 he employed a bush pilot in Grand Marais, Minnesota, to fly him around Isle Royale in a Seabee. Ten miles out from shore they saw four moose swimming toward the island. The pilot said he had seen this more than once in the month of May.

The temperature of Lake Superior is around 54 degrees F. This level of exposure is quickly lethal to human beings, but the moose is a large animal. It has a high ratio of volume to body surface and conserves heat efficiently, as we will have occasion to point out in other connections. It is at home in the water, and the swim from Canada or Minnesota is well within its capabilities.

Although the build-up of moose evidently stems from the first decade of the century, there were earlier indications of their presence, and animals reaching the island could have been eliminated by hunters.[5] Murie quoted reports that more moose arrived in 1912–13, and the population increased rapidly after that.

With extensive successional forests, induced by some cutting and much burning in previous decades, Isle Royale was an ideal habitat for moose. In addition to foods produced by brush and early tree stages, the old growth had an understory of American yew, or "ground hemlock," a shade-tolerant evergreen shrub, once common in forests of Michigan's upper peninsula. Although poisonous to livestock, it is one of the most palatable, and presumably nutritious, winter foods of both moose and deer. The great proliferation of deer in northern Michigan practically eliminated American yew, to the point where few people now recognize it. Moose so nearly wiped it out on Isle Royale that the finding of sprigs here and there today is always a matter of surprise. An interesting check on these conditions is found on Passage Island, the location of a lighthouse, 3.5 miles off the northeast tip (Blake Point) of Isle Royale. Possibly protected by a strong current, this island was never colonized by moose, and it has an abundant growth of American yew. It is reasonable to suppose that when we look at the vegetation of Passage Island, we are seeing how much of Isle Royale looked in primitive times.

Nowhere do animals prosper so well, live so long, and reproduce so abundantly as where a breeding stock is expanding to fill a vacant and favorable range.[6] For at least 10 years this was the situation of moose on Isle Royale. It can be considered "abnormal" that no natural predator was present to act as a check on this expansion. In its native life community every creature has effective mechanisms for population control, and among herbivores the influence of predators is likely to be important in the scheme. For the smallest prey species, which tend to be short-lived and highly prolific, the meat eaters usually are a secondary factor; weather conditions, food depletion, and population density effects playing basic and closely interrelated roles.

Environmental and behavioral conditions also may be important as we go up the scale of size and longevity in prey species. Some grazing and browsing animals have developed specialized behaviorisms — involving the activities of territorial males — to limit breeding and space out the population over its range. Such adaptations are seen dramatically in certain African antelopes, most notably in the Uganda kob.[7] Caribou of the American Arctic make long migrations; on the move they use their slow-growing tundra range lightly, allowing long intervals for recovery.

For hoofed animals of the world, population "control" means most significantly the limitation of numbers to a level where the food supply is maintained by regrowth. Almost universally, predators play a part in this, and in some cases they are the critical factor. Realistically, there is no "delicate balance" in the strict sense. It is more likely a seesaw of constantly changing weather conditions, food supply, prey productivity, and predator numbers that somewhere strike a mean that can be determined only over long periods of observation.

Among American "big game" species behavioral checks alone are not sufficiently effective to provide a high degree of range protection. Hence, such carnivores as wolves and mountain lions have an important role. If they are killed off by man or are absent for other reasons, as on Isle Royale, the prey population increases to the point where food scarcity becomes the crucial limitation. The onset of such conditions is evidenced by a heavy die-off of animals, which may be followed by a recurring cycle of build-up and reduction. Eventually the population finds relative stability at a low level supported tenuously on a chronically depleted food supply.

During the past half century we have had examples of this progression of events affecting the deer herds of many states, and it has happened to other species, notably elk in Yellowstone and Rocky Mountain national parks and moose on the Kenai National Moose Range of Alaska.

In both North America and Eurasia the wolf probably was the most significant natural enemy with which the moose and its ancestors evolved over millions of years, and doubtless important interdependencies developed. This logic suggests that the wolf is the one predator that might have limited the increase of moose on Isle Royale. Without the predation influence many moose must have lived out their time and died of the debilities of old age. Although coyotes reached the island early in the century and became plentiful, their killing capabilities would have been limited to young calves — those left unprotected by incapable mothers. However, as moose built up, the annual turnover of the herd in the form of carrion should have kept the larder well stocked for scavenging coyotes.[8]

In the typical fashion of uncontrolled populations, the moose doubled and redoubled until, in the 20s, the island was the best-known moose range in the world. The progress of this increase was reflected in estimates of moose numbers mentioned in the biennial reports of the State Game, Fish and Forest Fire Department. The report for 1915–16 stated, "On Isle Royale Michigan possesses upwards of two hundred head of moose. These moose are now roaming at large and an annual increase of the herd is evident." The next two reports gave a figure of 300, but that of 1921-22 (of the renamed Department of Conservation) said investigators on the island "report the presence of moose numbering in round figures about 1000." A picture of a moose in the issue of 1925-26 bore a caption that "Michigan's Isle Royale holds upward of two thousand moose." This is the last actual population figure mentioned in the state reports. Obviously, these were estimates subject to large error.

In 1929 the Michigan legislature directed and modestly funded a survey of Isle Royale resources by the University of Michigan. Adolph Murie was assigned the work on mammals and spent a total of five months in the field during the summers of 1929 and 1930. Paul Hickie assisted him during the first summer and thus got his initial experience on the island. Murie's museum bulletin of 1934 and Hickie's Game Division bulletin of 1943 — based mainly on his later work — are the primary sources of information on the most critical period in Isle Royale wildlife history.

Murie's description of the vegetation depicts an extreme state of deterioration. Water plants were largely cleaned out of the shallow-water feeding grounds in lakes and bays. Ground hemlock — used both summer and winter — was nearly gone. Other species preferred by moose (Appendix II, pp. 431-432) were heavily suppressed, and barking of trees was common. There was little question that moose had reached a density that could not continue to be supported. As for actual numbers, Murie cautiously estimated at least a thousand and considered it more likely that there were 2 or 3 thousand. At that time he was aware of a tendency that has impressed many biologists — it is our common error to underestimate the numbers of wild animals.

Murie proposed the only possible management expedient on this depleted island range: drastic reduction of the moose. But he had misgivings about throwing it open to hunting, and he recognized that transplanting live animals could not be done on a scale large enough to reduce the population substantially. Also, he was realistic about prospects for the future:

> It is plain that even though all the moose were to be suddenly removed, a period of some years would elapse before the vegetation could return to normal, so that for some time it would be necessary to reduce and hold the herd below the level at which it could afterward be carried. For the immediate present the numbers should be reduced enough to prevent further overbrowsing and to permit a recovery of the vegeta-

tion now overbrowsed. If the population is not reduced, the rate at which the vegetation is destroyed will rapidly increase, and, in the near future, the moose will begin to be eliminated by disease and starvation.

Another statement from this far-sighted report contained something of prophecy: "Were it known if and to what extent our larger predators such as the bear, cougar, or timber wolf prey on moose, a possible solution to the over-population on the island would be to introduce an effective predator." This was before Murie's famous Alaskan studies, in the course of which he learned a great deal more about the habits of wolves (194).

It is unlikely that Murie expected his management recommendation to be followed. He knew that, then as now, people do not understand the logic, attributed to biologists, of "killing animals to preserve them." In fact, this catchy bit of sophistry is used frequently by the leadership of that growing school of popular thought that considers any kind of killing by man an outrage against the natural world, a viewpoint that deserves some additional comment.

Individuals with strong leanings of this kind are commonly dubbed "protectionists," which is not a very good term. All of us need to be protectionists in the sense of preserving species and populations of species. That includes the biologist, but he faces squarely the fact that the long-term welfare of wildlife depends on maintaining its habitat in a healthy and productive state. Keeping the population within the carrying capacity of the food supply — thus preventing its destruction — may require killing excess animals.

Since this kind of expedient is offered in the name of "management," it has tended to bring the latter term into disrepute. Other aspects of the situation will be considered later. Here it is enough to say that killing animals to contain a population is not a good long-term solution, and few responsible authorities regard it as such. More often it is a next-best, hold-the-line (keep-the-habitat) solution until we can get the natural system operating.

That natural system is what the biologist prefers — the kind of prosperity shown by an animal population in a richly productive environment. An abundance of habitat resources implies a high rate of reproduction, but this reproduction must be matched by high mortality; otherwise the resources will not hold up. Of course, no species lives alone, and large annual "losses" are not really losses; they are plowed back into the biosystem as a food supply. They are part of the rapid turnover of nutrients that takes place in every vigorous community of living things.

In the natural world it is the fate of the individual to be born, to live as long as possible, then to die for the cause (survival of the ecosystem). In human terms the dying may be too soon, unjust, or unsightly; but in the biological process there is no such standard. It is of interest that well over a century ago the inimitable Thoreau had an intuitive comprehension of the proliferation and wastrel dynamics of life communities. He wrote in *Walden*:

> I love to see that Nature is so rife with life that myriads can be afforded to be sacrificed and suffered to prey on one another; that tender organizations can be so serenely squashed out of existence like pulp — tadpoles which herons gobble up, and tortoises and toads run over in the road; and that sometimes it has rained flesh and blood! With the liability to accident, we must see how little account is to be made of it. The impression made on a wise man is that of universal innocence.

Here was a man who saw the fantastically adapted realm of reproducing organisms for what it is, a system that casually uses up an unending sacrificial march of individuals in order that the species may go on. The pleadings of compassion, he said, must not be stereotyped. In a real sense there are places where the well-meant protectionism of the protectionist is not protection at all. It is escapism that shuns a solution hoping the problem will go away. Since the public often does not understand the kind of forthright action recommended by Murie for Isle Royale, it follows that their representatives in legislatures regard the issue as "controversial" and therefore eligible to be ignored.

That there was controversy cannot be doubted. It need not be detailed, but I borrow from Hickie's bulletin two quotations from writings by island residents. These have a familiar ring to anyone who has followed the wrangles over deer irruptions[9] in such states as Pennsylvania, West Virginia, Michigan, Wisconsin, and California. The Isle Royale observers were adamant in their denial of Murie's observations and their opposition to his recommendations. As usual, the only meat-eating animal around came under indictment. A resident of more than 20 summers testified:

The statement that moose are starving to death on Isle Royale because of lack of browse, is a misstatement of fact. I was in the interior of Isle Royale on two different trips (of two weeks each) this summer . . . There was no evidence of heavy and serious overbrowsing. The principal browse of moose in northern Minnesota and north into Ontario is the balsam or fir. During both of my inland trips on Isle Royale this summer, I did not see one balsam which had sustained "heavy" or "serious" overbrowsing . . . There are not as many moose on Isle Royale today as there were two years ago. The reason, to me, is obvious. The brush wolves (coyotes) have learned to band together and kill the moose. At first, they killed the calves but during the last two winters they have been killing the grown moose as well.

Another resident had an alternative recommendation to make concerning the prospering coyotes:

I think that we are all of the uniform belief that with the wolves [coyotes] exterminated, there will be no question whatsoever regarding the moose both as to living and multiplying; and those of us who are familiar with the island know, absolutely know, that there is an overabundance of forage there for many times the number of moose on the island.

As part of the University of Michigan survey, botanist Clair A. Brown made a field reconnaissance of ferns and flowering plants in the summer of 1930.[10] He took impressive notes on the near-disappearance of water plants and the widespread evidence of heavy moose damage to palatable upland species. Hickie quoted a Game Division report by J. H. Stephenson, who went to the island accompanied by I. H. Bartlett (both deer experts) for an appraisal of the browse situation. Stephenson fully backed the findings of Murie:

It is a somewhat startling fact that *hardly a tree of suitable size to provide browse of the varieties listed bears any available browse.* Either the trees are stripped of all twigs or browse or are killed from overbrowsing or girdling. *And there is no evidence of recent reproduction.* As soon as seedlings appear they are eaten out and destroyed.

It is evident that there is an over-stocking of moose which is becoming more serious each year. To one accustomed to the sight of overworked and eaten-out deer yards, the Isle Royale condition presents a most alarming situation and is deserving of the most serious and early consideration.

The moose in Michigan had been protected by legislative act, and further legislation would have been necessary to change this status. However the matter may have been judged, nothing was done about the problem on the remote island in the north.

In 1934 various reports of dead moose were reaching the Game Division in Lansing. It had been a winter of extremely deep snow, and after surveying the evidence, Hickie thought it probable that several hundred moose had died. Summer observations by people familiar with the island in former years indicated a general reduction of moose numbers and a particularly small calf crop (106, 107).

The chief of the Game Division, H. D. Ruhl, and his staff considered all available choices in what might be done to counter the distress of Isle Royale moose. Neither hunting nor the introduction of predators was feasible — as of 1934 either might "finish off" the moose. That, at least, would be the popular view. The island population could not be fed by any-

Against the background of a May snowfall, the differential browsing of balsam (center) and spruce is evident. Heavy feeding on young firs by moose is converting many forest areas into relatively pure stands of spruce.

thing less than a massive browse-cutting operation, which would be, at best, a temporary expedient. It was not possible to transplant enough animals to relieve the population pressure (236).[11]

However, the transplant idea caught on for a particular reason: Moose had been nearly extirpated from Michigan's upper peninsula, presumably as a result of year-round subsistence killing. It seemed a good idea to remove some of the excess animals from Isle Royale and release them in northern Michigan. It would be fairly expensive, but it might help bring back the moose, and there was no great opposition to the plan.

It got under way quickly. In November Hickie and Ellsworth St. Germain found themselves at Chippewa Harbor on Isle Royale to begin learning how to trap moose and keep them safely in confinement. In the first winter they were assisted by fishermen Holger Johnson, Otto Olson, and Jack Bangsund. The learning process was fraught with problems, but this small crew succeeded in sending 11 crated moose across the lake to Houghton in April — requiring two trips of *Patrol Number One,* the Department of Conservation 65-foot boat, commanded by Captain C. J. Allers.

It was a creditable showing, and the Conservation Commission gave a vote of confidence by providing for another effort the following year. Hickie spent much of his time in Lansing coordinating operations. St. Germain and five men went back to the island to trap moose. For two months in the fall a Civilian Conservation Corps camp was set up on Siskiwit Bay at the site of ongoing timber cutting by Mead Timber Operations. They built traps, corrals, and other facilities for the winter program. There were mortalities among captured animals, but with the help of the experienced Holger Johnson at Chippewa Harbor, a total of 30 moose were shipped back to northern Michigan in the spring of 1936.

The development of Isle Royale as a national park was well under way by the mid-30s, and in 1936 the National Park Service sponsored a CCC camp to build docks, buildings, and other facilities. The CCCs found other employment for part of that summer, which was one of the driest on record for much of North America. Fire broke out in the Siskiwit Swamp slashings on July 23 and spread to the northeast before strong winds. Crews came to the island from all directions in the weeks to follow; Hickie said that at one time 1600 men were fighting fire

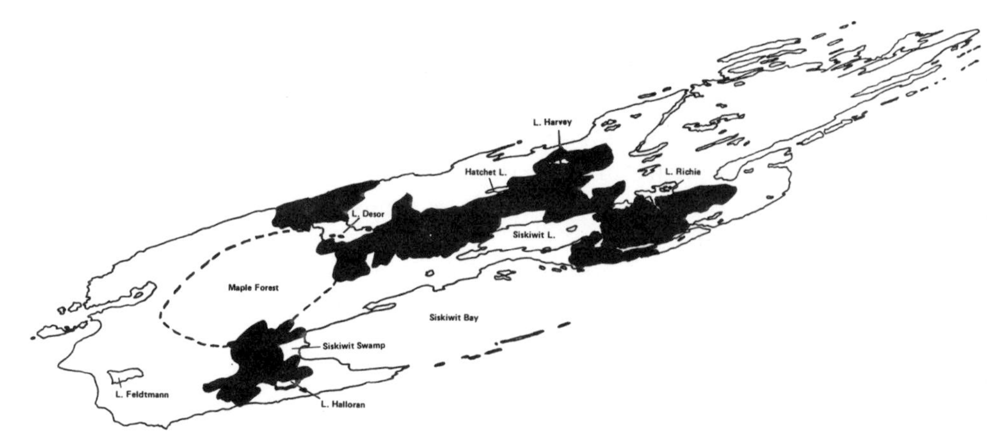

Figure 2. From July to September, 1936, the most extensive fire of this century burned nearly a quarter of the Isle Royale forest. For more than 20 years the resulting brush-stage vegetation was prime range for a moose herd recovering from the die-off of the early 30s.

on Isle Royale. By mid-September, when rains came and the conflagration was finally suppressed, about 35,000 acres of forest, nearly a quarter of the island, had been burned over.

A third winter of trapping was carried out by Hickie and Dale Fay, with the help of five men and the CCCs. They shipped 28 more moose. It was a winter of deep snow — Hickie mentions measurements of 4 to 5 feet — and moose were impressively scarce or absent from the higher parts of the island. There had been an obvious movement to coniferous cover along low shores. This is the first recording of a pattern of moose behavior that has proved highly significant in our recent studies.

A total of 71 moose were stocked in an attempt to re-establish them in Michigan's upper peninsula. It was a good try, but unsuccessful. The moose did breed, reports of calves being fairly frequent in the first years. But there were illegal kills in the hunting season and perhaps others unrecorded. Along with whatever residue of moose that may have survived from early times, the species dwindled and disappeared from upper Michigan.

It is not difficult to reconstruct conditions on the great middle portion of Isle Royale in years following 1936. There are written records, and much can be interpreted from what we see today. Great sentinel pines that had dominated the skyline were now blackened boles that would topple in decades to come. Often beneath their spreading roots the shallow organic soil had burned away, leaving stumps elevated a foot or more above the rock. The sun-baked southeast exposures of ridges would remain bare far into the future; 40 years later the steep basaltic faces are only partially obscured by the creeping mat of lichens, mosses, herbs, and pioneer shrubs.

In growing seasons after the fire the first flush of healing greenery was mainly herbs and grasses, colorful meadows of fireweed, everlasting, and other annuals. After these the perennials spread rapidly — sod grasses, asters, goldenrod, fields of bracken, and early shrub invaders. The latter were mostly species bearing fleshy fruits, whose seeds are scattered by birds: red raspberry, thimbleberry, blueberry, red elder, juneberry, red-osier dogwood, squawbush, wintergreen, and the rock-crawling bearberry. Distributed similarly by birds and the fruit-loving foxes, certain trees were among the pioneer woody plants, notably fire cherry, choke cherry, and mountain ash.

The fire was in two main areas, separated by a stretch of hardwood forest east of the Island Mine Trail (Figure 2). Also, islands of forest on wet ground escaped the fire; these are evi-

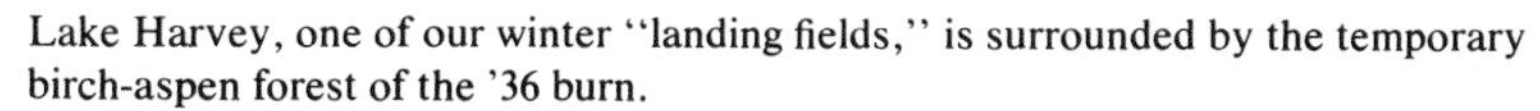

Lake Harvey, one of our winter "landing fields," is surrounded by the temporary birch-aspen forest of the '36 burn.

dent today northwest of Siskiwit Lake and east of Lake Desor. On unburned edges, birches shed their seeds during winter, and these were blown in windrows across the snow to sprout in summer wherever they lodged on exposed mineral soil. Far out in the burn, where old birches were killed by the fire, life quickened in the heavy root masses, and clusters of sprouts came up around the rotting stumps. The sprouts would produce another generation of seed trees for propagation on any ground that remained open some years later.

Birches predominate in the successional forest that took over the '36 burn. There had been many old birches in the original stand, and also overmature aspens that clung to life in sunny breaks among the spruces and firs. The aspens were surrounded by extensive systems of shallow underground stems. After the fire, which destroyed parent trees, a great proliferation of sprouts emerged from the soil.[12] In these thickets the stems that survived moose browsing and competition for light would grow up to form nearly solid clumps or groves of aspen. A shade-tolerant shrub widely prevalent in the original forest was beaked hazel, which also sprouts vigorously from roots; and after the fire it appeared as a thick understory on large areas of south slopes and ridgetops. On such sites it has become clear that red-osier dogwood and red maple likewise are root-sprouters and reclaimers of open land.

Eventually, balsam fir and white spruce — mixed with black spruce on cool northern lake fronts — will largely replace the transient species and restore a relatively stable forest. For these conifers the process of spreading follows a course similar to that of the birch. They are wind-seeded after cones ripen in the fall. The same can be said of northern white cedar, a tree of wet ground found in scattered clumps far up on ridges, where dips in the impervious bedrock form catch basins for soil and water.

Notes left in the park files by Clifford Presnall indicate the condition of the '36 burn eight years after the fire.[13] He said hardwood (deciduous) browse was coming along well and was being extensively used by the moose. He noted that red-osier dogwood, aspen, and birch appeared to be preferred to some kinds of willow. Young spruces were doing well where seed trees had survived on low ground, but balsams were severely browsed. These observations were made at the east end of the burn near Lake Richie.

A year later at the same season, Shaler Aldous described conditions on Feldtmann Ridge in the western unit of the burn. He said the area was growing up to brush, but moose were allowing little of it to get more than 10 feet high. He found ground hemlock and balsam fir to be much in evidence below snow level, but above that they were suppressed. In 1972, 36 years after the fire, I examined this area and found many birches, aspens, and spruces that in places were forming a new forest canopy. However, there was still much brushy growth at browse level. This was especially true along the trail, where it was being cropped back annually. Since Aldous's observations, the ground hemlock has been reduced to a vestige and balsams of any size are a rarity.

On 20 May 1945, Aldous and Laurits Krefting rowed north across Siskiwit Lake and checked browse conditions near the center of the burn on the rocky slopes above the lake. They found moose foods to be abundant and heavily used. Said Aldous, "White birch was the dominant vegetation but nowhere has it grown higher than 5 feet due to intensive moose browsing." In October 1973 I made notes in this same area during a three-day pack trip across the burn. I followed a ridge at about the 800-foot contour, which averages approximately half a mile up the slope from Siskiwit Lake. The young forest was now well beyond reach of moose; and birches — with aspens in clumps here and there — were 3 to 6 inches in diameter breast height (dbh). These stands have grown up in depressions between the exposures of bare rock on the ridges, and often they have a thick ground cover of thimbleberry, a low-grade moose food. I found almost no sign of winter use, confirming what we have regularly observed in the January–March period of our field work.

In the mid-40s, with summer rains again falling on the region, the brush-stage burn was a prime moose feeding area. In their paper of 1946, Aldous and Krefting remarked that it was producing "more browse than the remainder of the island combined." At that time it would have been excellent habitat for snowshoe hares and also for the sharptailed grouse, which is the only game bird found on the island. However, the mid-40s were a period of decline of these and allied species across the continent. It was the most drastic low of the northern "ten-year game cycle" since the beginning of the century.[14] Since Aldous and Krefting saw only a few signs of hares, we can assume that the island population was reflecting the general scarcity of the species at that time. This is also consistent with a remark made in a memo of 1952 relative to Isle Royale by the assistant regional director of the National Park Service. He said that the hare "has been at a low point in its population cycle for many years and is now starting to build up again."

All the professionals who have looked at browse conditions on Isle Royale over the years have been impressed with differences in the propagation and regrowth of balsam and spruce. Not only was balsam subjected to extreme overbrowsing, in the 30s it was further devastated by attacks of the spruce budworm, an infestation that may have been associated with years of continuing drouth. The spruces are rarely taken by moose, and they evidently were not hit hard by the budworms. They have spread thriftily into large areas with little competition from the firs that normally should be at least equally abundant in the climax forest. With the passage of time this trend becomes increasingly impressive. Where individual balsams have got beyond reach of the moose and grown up with the spruces, the two can be distinguished readily at a distance because balsams have been "high skirted," while the unpalatable spruces have branches to the ground. It is obvious that in the future the island will support many almost-pure stands of spruce. A species that has changed greatly in status since the early land survey (1840s) is tamarack. Once plentiful in swamps and bogs, it has been reduced to rarity by ravages of the larch sawfly.

The 1936 burn undoubtedly helped to bring moose back from the low point of the die-off. Aldous and Krefting thought there might have been less than 200 survivors at the time of the fire. The burn was highly productive of moose foods for about 10 years, and thereafter under heavy browsing and with trees growing beyond reach, it became less favorable, to the point where it was getting much less winter use in the 60s. The western half of Isle Royale, where the last glacier dumped substantial deposits of till, supports more moose and sustains greater browsing pressure than the eastern half. This does not apply to the western high ridges, occupied by the hard maple–yellow birch forest, which is practically a moose desert.

It is perhaps natural that most people judge the status of wild animals by how many there are. If a species is abundant, it is doing well, and vice versa. Every big game biologist learns that there is a more significant approach to this question. His criterion of prosperity is what the population is doing to its range. Almost irrespective of how many animals he may see in the field, his judgment of what to expect in the immediate future will be based on the condition of the most important food plants and whether or not they are holding their own in annual regrowth.

An experienced individual can judge this fairly well, but it takes many years to develop such know-how, and imperfections of the system become evident when experts disagree. There is need for some method of measuring the available food supply and how much of it is being used on an annual basis.

Shaler Aldous developed such a technique for deer-browse inventories, and with some modification he and Krefting applied it to the moose problem on Isle Royale. The Aldous survey

The southwestern — higher and warmer — parts of the island support a climax forest dominated by hard maple and yellow birch. Here a stand of white pine on a rocky slope has long been protected from fire by the surrounding hardwoods.

method has much to recommend it, even though it depends heavily on the good judgment of the observer. It makes possible a rapid and relatively simplified sampling of conditions representing a wide and confusing range of variability (4).

It is worth noting here that almost every kind of measurement we attempt to take in natural life communities must depend for its degree of accuracy on the knowledge of the field investigator. For example, we cannot measure the browse available on every acre of Isle Royale. So we must try to measure it on representative acres and extend the results. The big problem is getting our sample distributed so it truly represents the entire range. One never can be sure, so we try to take as big a sample as possible to compensate for errors of judgment. If the results are consistent — that is, if they show a low degree of variance — the statistician will say you have pulled it off. The field referees of a scientific journal will find your paper acceptable because you have put numbers on everything. You base your extrapolation on a neat cover map that blocks off the different types of vegetation, or the different "strata" of moose density, or whatever. But the more you walk over the land, with cover and terrain changing constantly before your eyes, the less that map looks like the natural scene. Did you sample the whole island? Or did you sample what was accessible and possible to work on at specific seasons? Do your consistent results represent realism for only a quarter or half the island? We can have confidence in one thing: As time goes on someone else will apply something different and better. It will show how well you did.

This does not call to question the findings of Aldous and Krefting; there is no reason to doubt that they demonstrated the major trends in food supply and moose on Isle Royale in the 40s. Krefting reported on changing conditions in browse and browse use at the North American Wildlife Conference in 1951. It was obvious that moose had responded to improved food conditions and built up substantially in the decade after the '36 burn. Vegetation surveys in the late 40s showed increased pressure on the available browse. There was more barking of trees, and more use of less palatable species.

In February 1945, Aldous and park ranger Karl Gilbert flew strips on Isle Royale in a biplane (not best for the purpose, but all they had) and counted 122 moose. They estimated their coverage at about 30 percent and extrapolated to a population figure of 510 for the island, "a conservative estimate." That winter they found most of the moose on ridges and in the open burn, which was a favorable situation for counting. Another survey in February two years later gave a minimum estimate of 600 moose, according to Krefting. He considered that to be the peak of the increase, because the winter of 1948–49 was a hard one for moose — conversely, we have assumed that it was an easy one for a wolf pack arriving from Canada.

During three weeks in the spring of 1949, Krefting accounted for 19 carcasses of winter-killed moose, young and adults,[15] mostly in the big burn. On a field trip a year later he saw eight. Krefting thought the numbers of moose reached a peak in 1948, which was followed by a decline. "It appears that the herd decreased by approximately one third, which would place the 1950 spring estimate around 500. These estimates are open to question but without any doubt they show the rise and fall of the moose herd" (143).

Trends of the late 40s posed the definite prospect that Isle Royale was to be another of those big game ranges, degraded by overpopulation, where the animals hang on at the limit of numbers permitted by their food supply, to be decimated periodically by hard winters that concentrate them into starvation areas.

The National Park Service had seen much of this in the western parks, and the situation clearly stemmed from the general campaign of extermination that was carried out against the wolf, mountain lion, and other predators early in the century. It is understandable if administrators were in a mood to try something different on Isle Royale. As we have seen, it did not work out as planned, but natural events took care of the matter. By the time of another moose emergency in 1950, the wolf had become a part of the island fauna, and the way was open to new developments in the complex of plant-animal relationships.

CHAPTER THREE

Island Laboratory

A PARTICULAR ADVANTAGE of islands for research purposes is that their wildlife populations are confined. In counting animals you can be reasonably sure they have not wandered off the tract and also that immigrants have not arrived to confuse further your already sufficient confusion.

Thus by studying a population of mammals or birds on an island of appropriate size the researcher can reduce or eliminate a common source of error. However, he must take into account the fact that animal numbers on islands may well be higher than they are in mainland habitats of the same size and quality.

I first realized this in witnessing the extremely high density achieved by pheasants on Pelee Island (Ontario) in Lake Erie. On this 10,085-acre area of prime lake-bed habitat, Allen Stokes found fall densities up to 378 pheasants per 100 acres. I was already acquainted with pheasant studies by Charles Shick and others on Michigan's 8400-acre Prairie Farm, a comparable area of fertile, farmed, lake-bed soils. There, in what was considered an excellent range, pre-hunting densities were about 48 birds per 100 acres. While there probably were significant conditions favoring high production on Pelee, the record pheasant concentration must be ascribed in part to its characteristics as an island.

We may theorize profitably on the way this works: In a mainland unit of high quality habitat, the animals are productive and build to a density where social pressures and competition mount rapidly, at an even greater rate than population. But the tensions are dissipated in part by the spreading of animals into bordering ranges of lower quality. We may assume that these areas are already populated to the level of their carrying capacity. It seems to be true of most small animals that when the "annual shuffle" of subadults takes place in late summer and fall, many individuals wander into marginal and submarginal habitats, where their mortality rate increases. As a result, the population of the best range, while remaining high, does not reach the maximum tolerable to the species.

On the other hand, there is no spreading from a similar, high quality island range. The population increases on abundant environmental resources and is likely to be limited eventually by what we may call high-density stress. Per the diagram in Figure 3, we can visualize the island as supporting a "stacked" population, in which all kinds of competition and social pressures are built up. Under these conditions, one can expect the potentially high reproductive capacity of so many animals to be offset by

Figure 3. A conceptual diagram of the size (density) relationships of an animal population on an island range and one on a mainland range of equal quality. Confined populations on islands tend to build up and endure a high level of competition and stress as compared with those on mainlands, which can spread into less favorable habitats where their numbers are reduced by increased mortality rates.

ISLAND RANGE

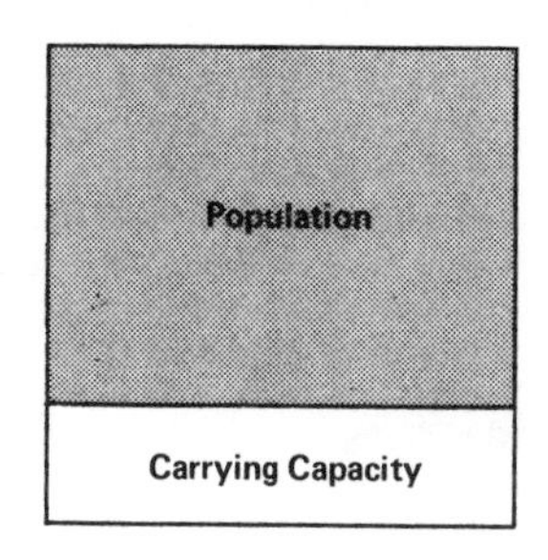

MAINLAND RANGE

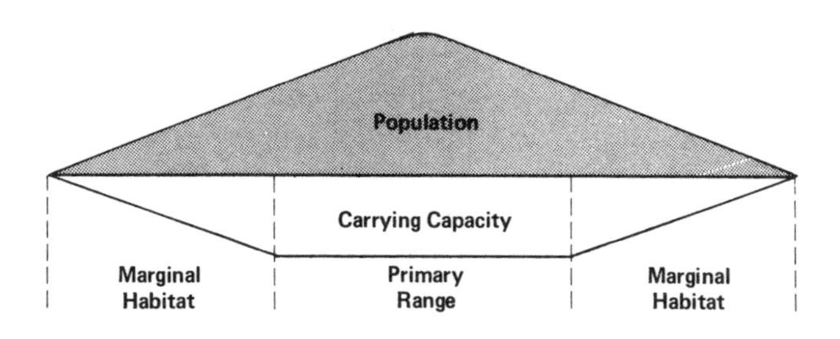

Top: The Isle Royale red squirrel is recognized as a distinct subspecies.
Bottom: Red foxes appeared on the island in the 20s and have been much at home since then.

such things as behavioral disorganization in the breeding season and low survival of young.

Originally we had no idea whether any of the above trends might be observed in the moose, wolves, or other creatures of Isle Royale. But it was evident even when we started our research in 1958 that this study area offers attributes of another kind that probably help to reduce the problems of fact-finding. A number of mammal and bird species common on the mainland 20 miles to the north have never reached the island, and others have disappeared, thus reducing somewhat the complexity of the community. The missing species are most notably the black bear, coyote, skunk, raccoon, porcupine, white-tailed deer, and several species of small mammals, including mice, ground squirrels, and shrews. The ruffed grouse and spruce grouse are absent also. Reportedly, martens and lynx were native to the island, but both were easily trapped and they disappeared, the marten before 1910. It would have been a natural predator of the Isle Royale red squirrel, recognized as a separate subspecies. Relative to the lynx, Mech was told by island residents that it was present until about 1930. This cat of the north would have lived primarily on snowshoe hares. It may be that widespread destruction of ground cover by the overpopulation of moose in the 20s reduced hares and contributed to the demise of the lynx. (In the North, old-timers invariably speak of this animal in the singular as a "link.") There are reports that a dozen deer were introduced to Isle Royale about 1906, but I have been unable to find any official record of it or to verify what happened to them. As noted earlier, the coyote is another former resident that is no longer present.

While there is no such thing as a simple ecosystem, the community of life on the island certainly is less complex, and therefore easier to work with, than what exists in similar habitats in Minnesota or Ontario. Aside from the moose and wolf, the most conspicuous or common mammals are the red fox, beaver, snowshoe hare, red squirrel, and woodland deermouse. It has been difficult to believe that this large island supports only one species of mouse, but widespread trapping by Wendel Johnson and our summer student helpers has failed to turn up anything but this ubiquitous seed-eating animal of northern forests.

Isle Royale's one species of grouse, the sharptail, is a bird of prairie-edge brushlands that moved eastward in the lake states in the wake of lumbering and fires. Sharptails are capable of fairly long flights, and these birds probably came from the mainland by ice-hopping in winter. A small population hangs on around ridge-top openings. Fires and moose browsing have created and maintained the sharptail habitat.

Since our work started in 1958 we have added three species to the known Isle Royale mammal list. David Mech found tracks he judged to be otter in the summer of 1961, and the following winter we definitely identified otter tracks in the snow of Rock Harbor and its western extremity, Moskey Basin. Since then we have often seen tracks or the animals themselves from the air. Otters are now well distributed around the shores and in streams, and one was caught in a fisherman's net.

For several years in the 60s there were reports from the trail crew and other sources of a cat being seen. This too was verified on 8 March 1970 when pilot Don Murray and ranger Zeb V. McKinney (who was our park man at the time) saw either a bobcat or a lynx on the ice of Lake Richie. The next day Don and I measured the tracks, which were so large the cat was almost certainly a lynx. It is possible that we have a small, scattered population that will continue to build up. Our third new species for Isle Royale is the red bat.

Isle Royale has no extensive marshes, so muskrats are largely limited to beaver ponds and lake margins. They are thinly distributed over the island, as is their principal predator, the mink. Both long-tailed and short-tailed (ermine) weasels were reported by the Adams expedition, but in his more recent study of mammal records, Shelton decided that we could be sure of only the ermine, characteristic weasel of northern forests (123). Its tracks are fairly common in winter, at which season it is white. It probably lives on woodmice, squirrels, hares, and birds, although we have learned nothing specific about it. Bats are seen frequently in summer, and four species have been identified.

The earliest historic record of Isle Royale wildlife that I have found mentions some of the species that were being taken there in fur-trade days. The account is in John Tanner's narrative of his 30 years captivity among the Ojibwas. Tanner was born about 1780 and was captured at the age of 9. He was 13 (about 1793) when he and his 17-year-old Indian "brother" were living and hunting in the vicinity of Grand Portage (118).

We had been but a few days at the Portage when another man of the . . . Muskegoes, invited us to go to a large island in Lake Superior, where, he said, were plenty of caribou and sturgeon, and where, he had no doubt, he could provide all that would be necessary for our support. We went with him accordingly; and starting at the earliest appearance of dawn, we reached the island somewhat before night, though there was a light wind ahead. In the low rocky points about this island, we found more gull's eggs than we were able to take away. We also took with spears, two or three sturgeons immediately on our arrival; so that our want of food was supplied. On the next day, Wa-ge-mah-wub, whom we called our brother-in-law, and who was, in some remote degree, related to Net-no-gua, went to hunt, and returned at evening, having killed two caribou. On this island is a large lake, which it took us about a day to reach from the shore; and into this lake runs a small river. Here we found beaver, otter and other game; and as long as we remained on the island, we had an abundant supply of provisions.

It would appear that the first landing of this party was at Washington Harbor and they took a trip overland to Siskiwit Lake. It is of interest that otters were there at that time. Possibly a few survived the era of heavy trapping and became the nucleus of today's population. The same might be said of the beaver. That gulls were nesting on rocky points implies the absence of foxes. Modern records indicate that they first appeared in the 20s. No doubt the formerly plentiful sturgeon ran up Washington and Grace creeks and could be taken easily from a canoe in the shallows.

Relative to what may be found in mainland habitats, reptiles and amphibians are not well represented on Isle Royale. There are only two kinds of snakes, the common garter snake and the earthworm-size red-bellied snake. On warm days in May the hiker of trails who knows the songs of frogs can easily identify them. Spring peepers are common in woodland pools and flooded shallows in all parts of the island. Their birdlike note is joined by the rising trill of the western chorus frog around many bays and inlets and the larger lakes. In wooded swamps and beaver flowages one frequently hears the outburst of hoarse clucking that means wood frogs. Usually a bit later, as waters warm in May and June, the high trill of the American toad is heard from inland waters, including grassy pools left by the snow melt. In early summer, green frogs are gunking from the

Outer reaches of Washington Harbor, our winter base. Washington Island is at upper left, and the tip of Beaver Island appears at lower right.

edges of lakes and beaver impoundments. The similar mink frog has been collected. Frogs conspicuously missing from the island are the bullfrog, leopard frog, and tree frog. Three species of salamanders have been identified (123).

As might be expected, neither migratory nor resident bird life is so limited. At the proper place we will have more to say of our seasonal observations on birds that have particular interest in these studies.

In considering the presence or absence of various species on the island, one must perceive today's conditions as a stage in developments, still in progress, that began some 9000 years ago. Then the Isle Royale bedrocks, which form the northern edge of the deep Lake Superior basin, were unburdened of perhaps a mile of glacial ice. In response, they have been "rebounding," or rising slowly, ever since. The present rate is about 6 inches a century.

These events came at the end of the fourth major Pleistocene glacial epoch, and those affecting Isle Royale have been studied most recently by N. King Huber of the U.S. Geological Survey. In its final retreat, the front of the continental ice sheet moved northeastward across the present area of Lake Superior, undoubtedly "calving" off huge icebergs into a highwater lake that occupied the western portion of the basin. Called Lake Duluth by today's geologists, it drained south and west through the St. Croix and Mississippi River valleys. It was impounded between the ice mass of the glacier on the east, the coastal ridges of Minnesota on the northwest, and the high ground of Wisconsin and Michigan on the south.

With further retreat of the ice front, the southwest end of Isle Royale was exposed. By that time Lake Duluth had been succeeded by a lower stage known as Lake Beaver Bay, with an elevation of about 650 feet above present sea level. Under these conditions Feldtmann Ridge would have been a separate island to the south.

In subsequent developments, ice melted out of lower drainageways, first to the south and then east of the Lake Superior basin. On Isle Royale old beach lines at various levels, some quite conspicuous, tell the story of lake subsidence and island emergence to the present time. Shorelines shaped by the surf of Lake Beaver Bay are the highest and oldest, up to 200 feet

Top: Basement rocks — basaltic lava flows perhaps a billion years old — at the head of Moskey Basin. This area was burned over at some time in the nineteenth century. Birches and aspens could not invade on bare rock, and we see a slow reoccupation by the conifers as organic soil collects in depressions.
Bottom: Old beach lines, representing the shores of early glacial lake stages, are common on the island.

above the present lake level. Lake Superior is now 602 feet above the sea, so the western end of the island has been raised about 150 feet since the strands of Lake Beaver Bay were bare and wave-washed. The highest point on the island, Mount Desor, peaks at 1394 feet.

After glacial ice had left the entire Lake Superior basin, a stage known as Lake Minong was formed at only 450 feet above sea level. Still later the lake went down even farther to 375 feet, about 225 feet below the present surface of Lake Superior, but it regained today's level when post-glacial rebound raised the bedrock of the outlet at Sault Sainte Marie. Throughout these fluctuations, continued uplift of the island dominated the scene, resulting in its increased exposure and growth in area.

The highest and hardest ridges of the island are Greenstone Ridge, which extends more than 40 miles and literally forms the backbone of Isle Royale, and Minong Ridge, which is about a mile from the north shore and is lower and somewhat shorter. These ridges have abrupt northern escarpments in many places. They slope more gently to the southeast, for they are the edges of strata that dip down under the big lake and reappear 50 miles south as the Keweenaw Peninsula. The main ridges are the solidified magma, basaltic lava flows, that came out of the earth's crust more than a billion years ago. They are interspersed and overlain, especially on the south half of Isle Royale, by consolidated accumulations of erosion rubble that formed sandstone and conglomerate.

Although the island was heavily scoured by glaciers several times in the last million years, the latest working-over obliterated all signs of earlier glaciations. The bedrocks are gouged and scratched in parallel lines where ice-embedded rocks were planed across them. On eastern parts of the island these marks follow its main axis in a northeast-southwest direction, but at the high west end the ice moved directly westward at an angle to the ridges. Glacial debris is relatively scarce on the eastern half, forming thin deposits here and there between the ridges and in transverse fractures. But as Huber remarks (114):

> Till is abundant on the west end of the island, effectively mantling most of the bedrock and subduing the landforms . . . Large erratics [boulders] also are abundant . . . Outcrops are limited mainly to small but prominent knobs and the steep north faces of some ridges . . . It

is on this west end of the island that the abandoned shorelines of higher lake levels are so well developed.

The drift dropped by melting ice was favorable parent material for development of the soils that now support the extensive hardwood (maple-birch) forests on the western third of Isle Royale.

At some time in its post-glacial history, the island must have been an eastern outlier of a belt of vegetation that occupied a climatic zone stretching far to the west. This probability is indicated by the presence of plant species more characteristic of the west than of country to the east. The most famous example is devil's club, which is found near the eastern tip of Isle Royale and on two adjacent islands. The nearest other range of this prickly plant is in the Rocky Mountains. Thimbleberry also is a widely distributed western shrub, with a continuous range from Alaska to California. On some of Isle Royale's dry eastern

Figure 4. Surface features of Isle Royale, a topographic map by A. J. Burgess, courtesy of N. King Huber, U.S. Geological Survey Bulletin 1309.

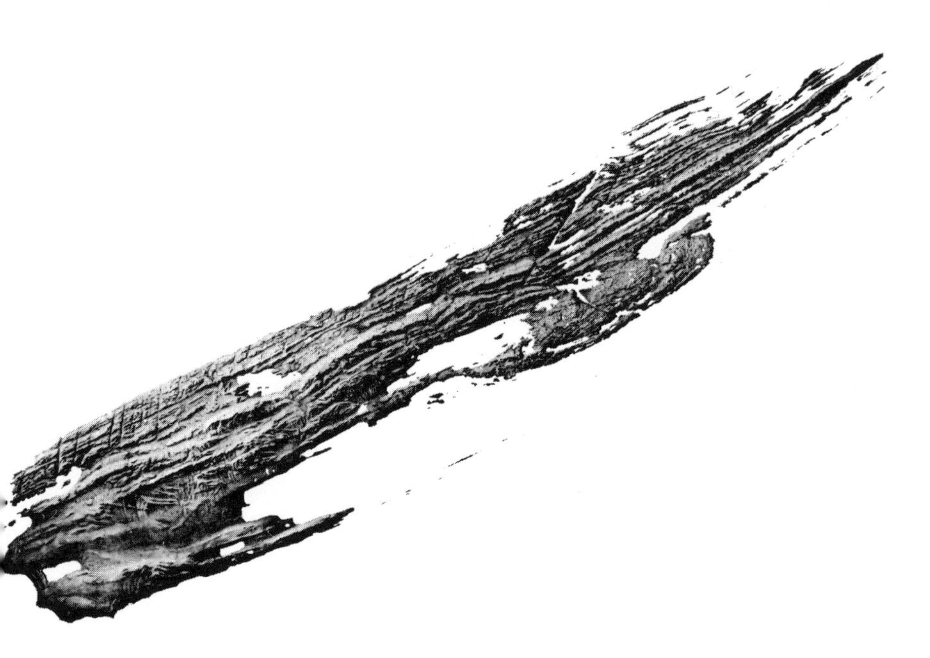

ridges there are stands of buffaloberry, a shrub common on prairies of North Dakota and westward into the Rocky Mountain parks of Canada. Many of the island's heath-type plants — bearberry, pipsissewa, the cranberries, Labrador tea, blueberry, the pyrolas, creeping snowberry, wintergreen — are common to a broad range in boreal forests across the continent.

Except for the abundance of birch, the successional forest of Isle Royale might be considered a counterpart of the great aspen parkland that borders the northern prairies of western Canada and forms a transition zone to the spruce-fir forests farther north. Many of the trees, shrubs, and herbs are identical or closely similar species. Examples are quaking aspen, paper birch, balsam poplar, pin cherry, red-osier dogwood, highbush cranberry, beaked hazel, bunchberry, juneberry, red raspberry, rose, sarsaparilla, baneberry, and wild lily-of-the-valley (31).

By similar logic, the community of lichens, mosses, heaths, and dwarf shrubs that pioneers on the rocks of the island is in some degree similar to arctic and alpine tundras. Its wet phase is seen in bogs and low, moist woodlands. Today we undoubtedly have less of this muskeg type than was present a century ago when it helped support the caribou.

After the leaf fall and under winter snow, geological features show up well from the air — ridges and depressions, and most impressively the fractures where strata have been cracked in lines across the island by differential stresses of rebound. In the forest, green conifers are conspicuous, and bare slopes, burns, and marshes seem open and unencumbered. Many of these sites look very different when you get into them on snowshoes, and they have no resemblance at all to the thickets and swamps that impede a cross-country trip in summer.

Isle Royale is actually an archipelago consisting of the main island and about a hundred smaller islands surrounding it or included in the larger lakes. There are at least an additional hundred major rock clusters and exposed reefs, and even more can be seen rising ominously from the depths when one flies over the adjacent waters on a quiet day. These have been a navigation hazard since early times, and many wrecks have been found and explored by skin divers. Restrictions on salvage activities are now in effect, since the National Park Service

Minong Ridge at a point northeast of McCargoe Cove.

considers wrecks to be a historic feature that should not be exploited.

Because of its impervious basement rocks, subsurface drainage is poor on Isle Royale, and every depression between the ridges and knobs tends to hold water. As a result there are about 30 named lakes, the largest of which is Siskiwit. Situated near the center of the island, it is 7 miles long, 142 feet deep, and 57 feet above Lake Superior. Its outlet, flowing into Malone Bay, is one of the few fast-water streams. The next-largest inland water is Lake Desor, nearly 3 miles long, 55 feet deep, and the highest of the big lakes, being 252 feet above Superior.

Many of the smaller lakes, such as Feldtmann and Halloran, are relatively shallow, with sand and mud bottoms and weed beds that are much sought and depleted by feeding moose. Between ridges, and especially at the heads of shallow bays, slow-moving waterways have been extensively dammed by beavers, creating habitats attractive to moose and waterfowl. The long isolation of the island and some of its lakes has resulted in a restricted fish fauna, only 49 species being recorded from both inland and surrounding waters. These conditions have also produced a number of subspecies peculiar to Isle Royale lakes (113, 152).

Commercial fishing began around Isle Royale at an early date. In putting together this history, Lawrence Rakestraw surmised that the Northwest Fur Company may have been taking fish from the north side of the island before 1800. The first fishing station was established in 1837 at what is now the Belle Isle campground. In the next two years the American Fur Company had 33 fishermen working at seven locations. In 1839 they shipped 5000 barrels of whitefish and lake trout.

Since that period independent fishermen have operated from every suitable site (and some not so suitable) on the bays and islands. Often in unexpected places one encounters the old clearings with moose-chewed apple trees, remnants of docks and fish houses, crumbling cabin foundations, and even the ruin of a long-abandoned boat. During these studies only five small-scale fishing enterprises have been in operation (Figure 5). The Stanley Sivertson fishery, an outpost of Sivertson Brothers in Duluth, is located on Washington Island, the westernmost land of the archipelago. The much-visited base of Sam Rude and his wife, Elaine, is at Fisherman's Home on the Houghton Peninsula; Mrs. Edwin C. Holte maintains a small fishery from Wright Island in Malone Bay; Peter Edisen runs a few demonstration nets for the park from his home of the last 60 years on Rock Harbor; and Milford and Myrtle Johnson are on Crystal Cove at the eastern end of Amygdaloid Island. Three times a week the 63-foot Sivertson service boat *Voyageur II*, piloted with famous precision by Roy Oberg, makes the rounds of the island from Grand Portage. It brings supplies, transports visitors, and picks up barrels of salted fish. The fishing families (including the late Laura Edisen, Ed Holte, and Sam Rude) have been our friends and benefactors since the research work began. Fishing has been a cultural feature of the park, and all who know the island regret that it is passing from the scene.

In his study of "Historic mining on Isle Royale," Rakestraw assembled the record of futile pioneering endeavor that lies

Figure 5. Sites of commercial fishing operations that were active during most of the study period, 1958-76. The Mott Island headquarters, Rock Harbor Lodge, and Windigo ranger station, and location of the winter camp, also are shown.

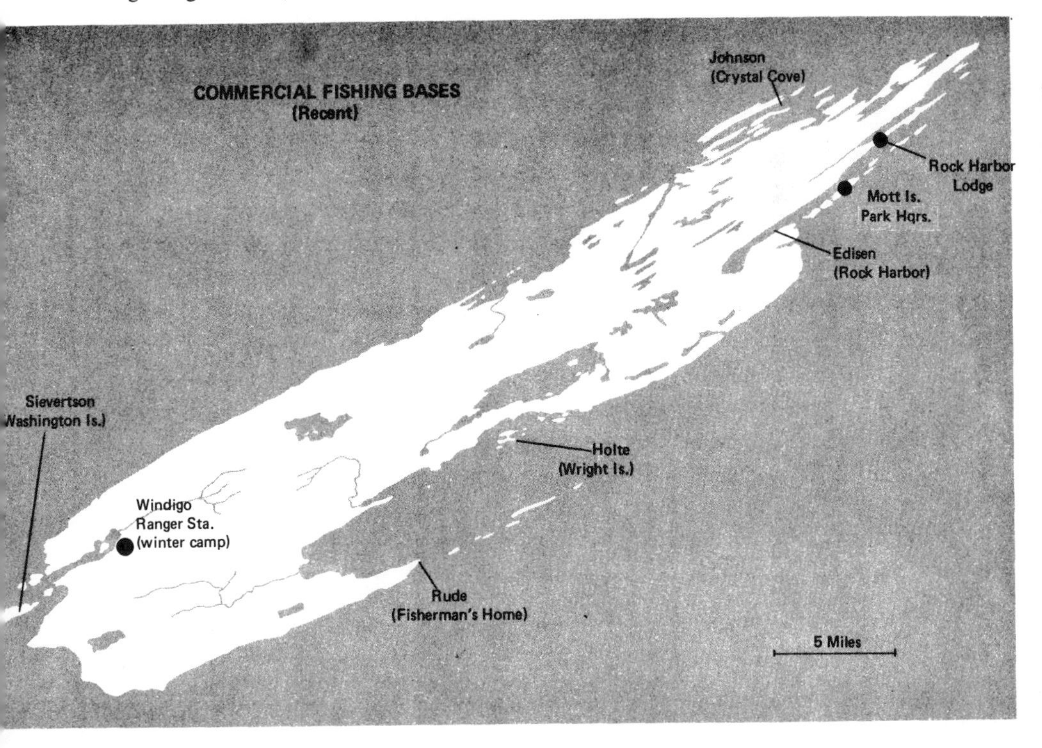

behind the old shafts and pits, rock piles, buildings, and junked machinery that today's park visitor encounters as he hikes over the 174 miles of foot trails (Fig. 16, p. 410). "There was an early boom-and-bust period lasting from 1843 to 1855; a lull, and then a revival of mining activity from 1889 to 1893, ending with a shift to tourism and commercial fishing" (229).

The reference to tourism recalls the development of modestly famous resorts at Belle Isle, Washington Harbor, and other locations. Most of the buildings and docks have vanished, through the industry of park clean-up crews.

Visions of a rich strike in native copper probably were behind the government acquisition of Isle Royale from the Ojibwas in 1843, for exploration began immediately. In the 50-year period of interrupted development, several mines and numerous test sites were worked. Some of them are readily accessible by trail and boat.

Figure 6. The most intensive mining operation of a century ago was the Minong Mine on McCargoe Cove. From Washington Harbor a wagon road extended to Lake Desor and Siskiwit Bay.

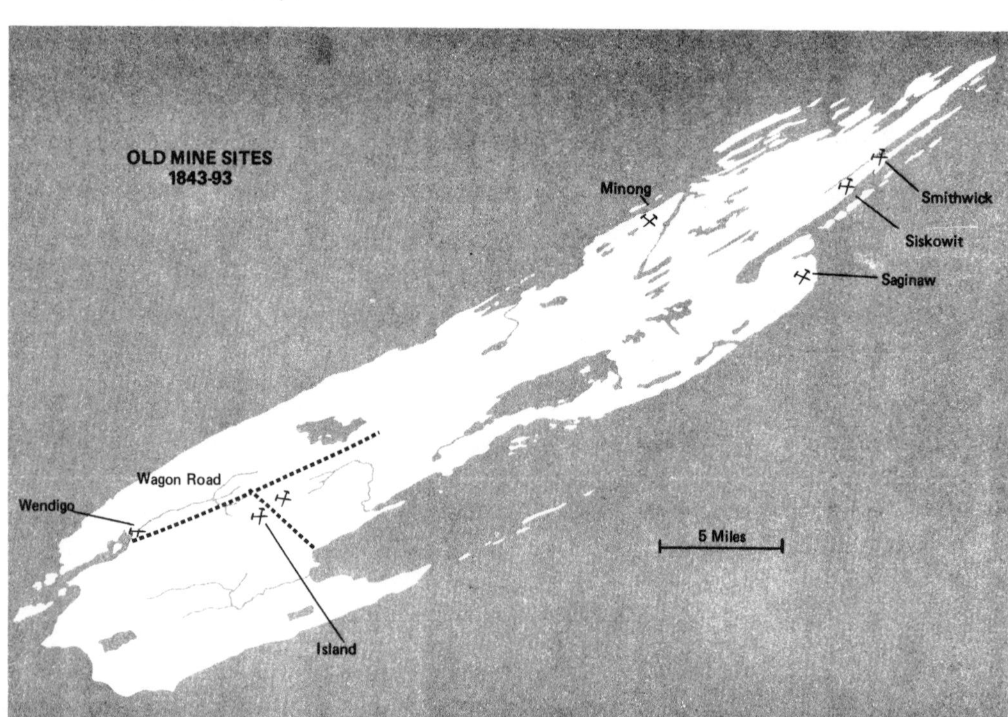

Ranger III at the Rock Harbor dock. This 165-foot National Park Service boat is the main means of transportation from Houghton to Isle Royale.

As the National Park Service motor vessel *Ranger III* (165 feet, master Woodrow R. Bugge) approaches the island on its due-north course from Houghton, it enters Rock Harbor through the Middle Islands Passage. One and one-half miles out from that narrow entrance the citizen newcomer to the park will hear the PA announcement that the headland on the left is Saginaw Point. Nearby is the site of the old Saginaw Mine. The square vertical shaft into the granitic bedrock is now nearly filled with water, and the surroundings are overgrown by forest.

Some of the old mine shafts have proved hazardous to moose, which fall into them. Hikers discover the animals, dead or too damaged for salvage, and word gets to our field researchers, who thereby face another problem. With aid from the trail crew or other willing help, necessary tackle is rigged and the animal is hauled out. Then comes the messy job of a warm-weather autopsy in search of what may be learned of the health and condition of the unfortunate specimen. All is grist . . .

After the boat has turned northeast in Rock Harbor, it proceeds a mile and a half to dock briefly at Mott Island, the administration and maintenance headquarters of the park. On the shore directly across the narrow harbor (here only about 400 yards) is a heap of "poor rock" excavated from the Siskowit mine, once the largest dig in the harbor.

Four miles farther up Rock Harbor, *Ranger III* unloads at Rock Harbor Lodge and Ranger Station, where it ties up for the night. From the lodge, before his dinner in the dining room, the visitor can take a short walk to the site of the Smithwick mine.

If he were at the opposite (southwest) end of the island, at Windigo Ranger Station on Washington Harbor, he probably would have arrived on the 64-foot Sivertson boat *Wenonak* from Grand Portage. There his hike to the nearest mine would be a bit longer. From the ranger station, or the campground on Washington Creek, a 1.5-mile trek on a pleasant trail (a former wagon road) leads to the cavelike hole in the hill that once was the Wendigo mine. Yes, they spelled it that way.

Having referred to transportation facilities, I should add that a daily boat, the 60-foot *Isle Royale Queen II,* crosses Lake Superior by the shortest route, from Copper Harbor, Michigan, to Rock Harbor Lodge. From Houghton, the twin-engine seaplane service is available to Rock Harbor, Mott Island, or Windigo. All services are subject to delay or cancelation because of bad weather.

Today there are no roads on the island, only short vehicle trails a few hundred yards long around the two lodges. However, in the early 90s a road from Windigo extended up Greenstone Ridge about 10 miles to Lake Desor. Beyond Sugar Mountain, a bit more than halfway to the lake, it was joined by another road that led southeast to the Island mine, a major operation,

The dock and headquarters building at Mott Island.

from which the road went on downslope to the northwest corner of Siskiwit Bay. These old roads, on which some of the rotting logs of corduroy swamp crossings can still be seen, are routes followed by park hiking trails.

The largest mine development is the impressive Minong mine near the campground on the McCargoe Cove, where *Voyageur II* stops frequently by arrangement. On a tributary of Chickenbone Creek the skillfully fitted log dam, 12 feet high, is still functional after a hundred years. Actually the impoundment that once supplied water to the stamping mill is silted-in nearly to the height of the dam, but beavers have added their own new dam on top of the old one, backing up another 4 to 5 feet of water. This combination is serving well while an overmature stand of aspens is logged on the slopes around the pond. However, as one stands in the bottom of the rocky stream bed looking up at some 18 feet of man and beaver engineering, he reflects that the law of gravity must inevitably prevail. The time could come when a magnificent avalanche of water, mud, logs, and beavers will sweep down the gully into McCargoe Cove.

At the mine, the wreckage of an old smithy, narrow-gauge tracks and ore cars, cool damp tunnels, and huge piles of excavated rubble (on which white pines are beginning to grow), all testify to the major effort that was made here in the 1870s. The site had given false promises in the form of several large masses of native copper, one weighing 3 tons.

In this area too was found the greatest concentration of ancient Indian mine pits, some dating back more than 4000 years. Anthropologists tell us that many of the copper hatchet heads, knives, projectile points, and other artifacts found at prehistoric Indian sites in the Midwest were made from metal pounded out of the rocks of Isle Royale (93). Hundreds of the beach-cobble hammerstones have been recovered from the pits, most of which are less than 10 feet deep. These depressions are full of debris and obscured by vegetation. Even along ridge trails they would not be noticed by the passer-by except where labeled by the park staff.

Obviously, Isle Royale beaches and rock exposures have been combed for decades by collectors, not only for copper but for gemstones of chlorastrolite ("greenstone"), prehnite, and agate. On ridge trails where attractive amygdules of these minerals were half embedded in the durable pavements, they have almost invariably been shattered by the efforts of someone who tried to break them out of the rocks.

The accumulated abundance of surface minerals that must have stimulated the imagination of early travelers in the region is indicated by a quotation from the Jesuit father Dablon:

> But farther toward the West, on the same North side, is found the Island which is most famous for Copper, and is called Minong [Isle Royale]; this is the one in which, as the Savages have told many people the metal exists in abundance, and in many places . . . Pieces of Copper, mingled with the stones, are found at the water's edge almost all around the Island . . . In the water even is seen Copper sand as it were; and from it may be dipped up with ladles grains as large as a nut and other ones reduced to sand.[1]

Figure 7. Route taken by *Ranger III* from Houghton, the mainland headquarters, on the 70-mile trip to the island park.

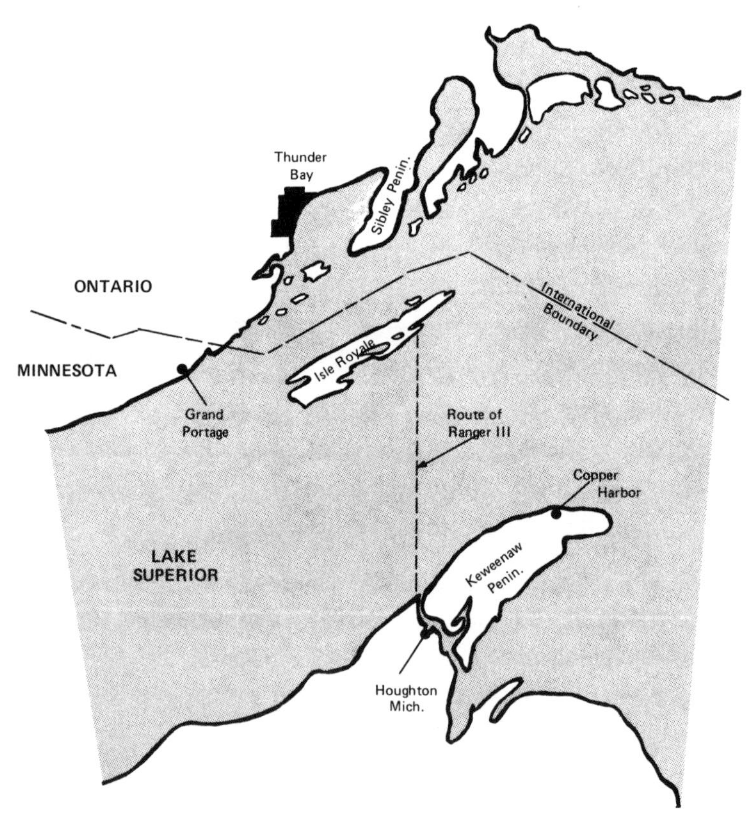

Evidently even taller tales were told by the Indians, and backed up with impressive samples. Pierre LeSueur, entrepreneur of the fur trade, became convinced that wealth lay in the occurrence of copper around the shores of Lake Superior. In 1698 he went to Paris and sought authority to develop mines in the area. The proposal was summarily refused by the Intendant of New France. However, in 1710 this personage made a report on resources of the colony and stated that on the borders of the lake:

> There are found in the sands pieces of this metal, which the savages make into daggers for their own use. Verdigris rolls from the crevices and clefts of the rocks along the shores, and into the rivers which fall into the lake. They claim that the island Minong [Royale] and small islets in the lake are entirely of copper.[2]

It is perhaps understandable if later prospectors expected Isle Royale to become a great copper-producing center. Actually, mines in these same strata on the Keweenaw Peninsula to the south paid off much better. We may speculate that veins of metal lying deep under Lake Superior will rest in peace for yet a while.

Although the old mine sites have historic interest for the visitor to Isle Royale, the half century of mining activity has greatest present significance for its effects on vegetation and animal life. The prospectors had no regard for timber or other natural values. They set fire to the forest to get rid of it, so miles of bedrock could be inspected easily. The result is that, with the exception of the maple-birch forest, most of the island gives evidence of being burned since the early 1840s. Clair Brown said, "In fact there is no place on Isle Royale visited by the writer that does not show some evidence of fire, either by the presence of charred stumps and logs or of plants that usually follow fire."

Fire may logically be regarded as an attribute of climate, for it is true the world over that periodic burnings are an important influence in modifying vegetation on drouthy sites and in dry years. Fire is a concomitant of low rainfall, and natural grasslands on all continents would be far less extensive without it. Man has been setting vegetation fires for at least half a million years, but lightning was effective in this function long before the human influence. One of the conditions to which various forest types are adapted is the frequency and type of burning that takes place in different regions. This is particularly true of pines and certain other conifers, some of which must be aided by fire if they are to make out in competition with the hardwoods.

The last major period of fires in the upper lakes region was the devastating series of drouth years of the 30s, which produced the famous '36 burn on Isle Royale. Hansen, Krefting, and Kurmis summarized the park's fire records since 1940 and found that the total area burned through 1965 was less than 1500 acres, nearly all of which occurred in 1948. That year a tract southeast of Lake Desor, within the area of the '36 fire, burned again. Up to 1965 the park had an active fire-control program. Since then most summers have been above average in moisture, there have been winters of heavy snow, and burning has been inconsequential. Of 48 recorded fires since 1940, 26 were started by lightning. In other words, human influence probably accounts for roughly half the fires on the island.

Like other land-management agencies, the National Park Service has recognized that, up to a point, fires are natural and a part of the primitive dynamics of wilderness. Heavy-handed fire control is in the same category as the one-time program of clearing out predators; it has been reconsidered. The policy of today on Isle Royale is to let "natural" fires burn, as long as they are not a threat to human life or property. This policy is right and proper, but it will obviously mean a war of nerves for unfortunate superintendents in the future. Predicting the exigencies of weather may be involved in any decision to stop a fire while you can. If you don't have a good reason for stopping it, you let it go. Then it is predictable that for months after the event members of the press will be publishing photographs of the blackened ruin that was a forest in order to activate whatever public interest there might be in the situation. If you put a fire out while it is small, the proponents of good wilderness management will say you betrayed your trust. *C'est la guerre.*

On Isle Royale, extremes of heat in summer and cold in winter are tempered by the surrounding expanse of open water. Summers are cooler and winters warmer than on the mainland. Even in July and August, boaters have good use for a down jacket. Cold, blustery weather can occur at any time, and in

September and October it is a sure thing. September appears to be the month of greatest moisture, with 3.99 inches of rainfall. No adequate year-round weather record is available for the island, and annual precipitation is known for only two years. At Mott Island it was 30.69 inches in 1941 and 35.68 inches in 1942. Isle Royale evidently gets 5 to 10 inches more moisture than Grand Marais.

Linn's study of zonal climates showed clearly that lower parts of the island, nearest shorelines, are moist and cool. Hence their forest climax of spruce and fir. The higher elevations are inland and less affected by the lake. They are dryer and warmer, and their northern hardwood forest climax is comparable to that found in similar climates elsewhere, especially to the south and west in the lake states.[3]

To state with certainty what the long-term, average snow accumulation is in midwinter would require more standardized measurements than we have for the 18 years. Not until the winter of 1972 did Rolf Peterson begin to systematize the study of snow depth and other characteristics. His reasons and results will be explained later. In general, we have come to regard 25 to 30 inches of snow to be in the "average" range. During the first three winters of the study we did not realize that wolf and moose activities were being conditioned by relatively light snow accumulations. In his thesis, later published as *The wolves of Isle Royale,* David Mech summarized snow depths as "16 to 24 inches in 1959, 12 to 16 inches in 1960, and 20 to 26 inches in 1961." Deeper snows in the mid-60s also caught us somewhat unawares. We saw new conditions in the field and new relationships between moose and wolf that seemed "exceptional" in terms of our previous experience. It was in two winters, with drastic extremes in snow conditions, that nature bludgeoned us with truths we had not suspected. The years 1968 and 1969 provided those extremes for which the biologist must frequently be thankful; they bring enlightenment that does not show through convincingly in small year-to-year trends. More information on weather is given in Chapter 15 and in Appendix IV.

As our work should demonstrate, this island park has nearly unique possibilities for certain kinds of biological research. Yet there are limitations, and we concur that they should be maintained. Our use of the aircraft in winter is no liability to either wolf or man, but this would not be so in summer. A wilderness hiker is entitled to aloneness, and he cannot enjoy this with a low-level plane overhead. The field marking of animals is another issue, and there are exasperations to which a wildlife photographer should not be subjected: *exempli gratia,* a moose with ribbons in its ears!

A superintendent must have total knowledge and control of things done in his park. On occasion he should say no, and we have met with this. Researchers do better in having their own feel for the out-of-doors and the public interest. Their credit is best when they place strictures upon themselves. They will not be told no when they do not ask.

CHAPTER FOUR

Our Winters North

THE FLIGHT FROM THE Eveleth-Virginia Airport to the easternmost point of Minnesota and on to Isle Royale is about 180 miles. Depending on wind, we usually made it in less than an hour and a half in the Cessna 180 piloted by Bill Martila. It took Don Murray well over two hours in his Aeronca Champion, a tandem two-seater weighing only 750 pounds empty. This was our field plane, and Don and the Champ would be with us the entire seven weeks of our winter work.

On the morning of 26 January 1971, I had called Don and Bill at 7:00 and learned that the weather was "go"; we had waited a day in the motel for it to clear up. Rolf Peterson, Jim Dietz, and I had driven north from Lafayette, Indiana, and met Bill Dohrn, who had come to Virginia from Houghton, Michigan, mainland headquarters of Isle Royale National Park. This was Rolf's first winter on the project, and, because he was tied up on campus with course work, he could be with us for only 10 days between semesters. Dietz, a graduate student in wildlife physiology, had not yet taken on a problem and he had accepted my offer of a temporary assistantship on the island to get field experience. This winter I would be on the job for the full period, rather than my usual four or five weeks, and Jim's help would be needed. As electronics technician, Dohrn was commonly the first park man out, since it was practically routine to have trouble with the generator or radio equipment.

For nearly three months there had been no human inhabitants on Isle Royale. At the end of October *Ranger III* had made its last trip, this time to both Mott Island and Windigo, and carried the staff back across Lake Superior to Houghton for the winter. Perhaps a week later, the few remaining fisherman families had boarded *Voyageur* and gone back to their winter homes in Minnesota. Summer residents, who still hold life leases on a few properties in Tobin Harbor, had mostly left about Labor Day. After late November or early December, the island is icebound and inaccessible to boats. When ice becomes firm in January a plane usually can land in protected bays or inland lakes. Thus there are a couple of midwinter months when our field work is practicable. By mid-March frequent thaws make flying uncertain, and it is time to get out. The break-up comes at the end of April, and the first park boat trip to the island is commonly in the first few days of May.

Usually our winter expedition involved a party of four: One man would go with Don, and Martila would bring the other two in the Cessna — back seat out and loaded to the roof. At about weekly intervals during our stay on the island Bill would fly out to exchange members of the park staff, and the incoming man would bring a supply of fresh produce. These trips were a welcome diversion for Bill. He was an FAA inspector, and his regular business was rebuilding and maintaining aircraft at his marine base on Crooked Lake near the airport.

This year we had five men, so I had chartered another Cessna to pick me up, along with our large load of groceries. Rolf was

The Cessna lands at Windigo to exchange park staff men and bring fresh supplies.

the extra man, but he needed to get the feel of our winter program, even for a few days, because he would have major responsibility for the field operation in years ahead.

It was one of the best days we ever had for flying to Isle Royale: clear and 18 below. Many times, at the opening or on exchange flights, we had been delayed by bad weather, and it was necessary for one or more of us to wait as many as seven days until Bill got a satisfactory weather report and said we could fly. Don Murray was equally careful. It was the way they kept us out of trouble.

Today by the time our plane was over Devil's Track Lake at Grand Marais we could see Isle Royale about 60 miles to the east. Beyond Grand Portage we headed out across the lake and were soon circling over Washington Harbor at the west end of the island. The harbor was well iced in, but other more exposed bays that we could see were open water — the combined effect of warm weather and winds.

It was evident that our plane was first to arrive; the others had gone farther east to cross the channel. The center of Washington Harbor was wind-drifted and rough. So, as we let down, my pilot picked a stretch of smooth snow in the southeast lee of Beaver Island, which is the big convenient windbreak to the west of our landing area on the inner harbor. Coming down by Beaver Island would put us more than a quarter-mile from the dock, and we would need to taxi over the ice to get there for unloading.

The skis settled softly onto the snow, then broke through the surface and began to drag. I knew we were in slush and expected a burst of power to get us onto the firmer ice ahead. Instead, we came to a stop in a foot of water and wet snow that lay over the ice. We were effectively stuck.

The mechanics of this situation are of some interest. If the north country has cold weather in early winter before any heavy fall of snow, the lakes are solidly frozen over and conditions are generally good for flying. However, if deep snow comes before thick ice forms it insulates the lakes and reduces the rate of freezing. Heavy snow also weighs down the ice, causing cracks to form, and water wells up to spread over the ice and form a layer of slush. Prolonged cold will freeze this, adding to the thickness of the ice — a process that commonly builds up 18 to 20 inches on the lakes. However, if the weather stays abnormally warm, the slush becomes a hazard to aircraft operation. Areas of deep snow near protected shorelines are especially likely to be a problem. As the metal skis of the plane contact the slush, ice instantly forms on them, and few light planes have power enough to take off again from this kind of surface.

Before long both Martila and Murray were overhead, and they saw what had happened. They knew these conditions, and they came down on the rougher but more solid ice near the dock. Now we had the problem of getting the stuck plane unloaded and onto a wind-blown area of exposed ice that, fortunately, was only 20 yards away. Our war-surplus track vehicle, the "weasel," had a dead battery and there was no spare. However, we got the snow machine started and took it, with its small trailer, out onto the harbor. We waded water and wet snow to our knees, unloading the plane and carrying supplies to the trailer. From there we hauled them to a grand pile on the ice near the dock. All of us fell a time or two and were thoroughly encrusted with ice.

When the plane was empty, we attached ropes to the skis and, with everyone pulling and the pilot giving it full power, we got it moving and up onto dry ice. From there he could take off. When Martila knew we were out of trouble, he left us too.

As was my custom on these occasions, I went into the ranger station and got a pair of snowshoes. Our cabin was 300 yards up in the woods, and the snow machine could not break trail uphill in the loose snow, of which we had some 20 inches. Someone needed to shovel out the back door and start the two oil stoves as soon as possible.

I was ready to start when an uproar broke out near the end of the dock. Rolf and Bill were waving their arms and chasing a fox, who was carrying a large roast off our food pile and making heroic efforts to get his bonanza to the dock. In fact he did make it and took our 6-pound chunk of beef somewhere under the dock.

The rest of that story may as well be told. Several days later I was near the dock and saw the fox heading out across the ice toward Beaver Island, carrying that same well-wrapped piece of meat. Being unencumbered I had the advantage and ran him down until he abandoned his burden. I found the meat frozen hard and only a few corners gnawed away. That was a roast we particularly enjoyed.

From the generator shack several hundred yards back in the woods, we dragged a toboggan with a spare battery to the dock and got the weasel started. We loaded it with what we hoped were still-unfrozen fresh vegetables, unbroken eggs, and other valuables. Then Don took it carefully up the hill to our cabin. Don was our expert in every kind of mechanical operation. He had run a motor pool for a variety of vehicles in the Korean War. He knew how anything worked or what it needed if it did not. In a couple of trips he had packed a trail the snow machine could negotiate. Bill was working in the generator shack, and — it was one of our lucky years — soon the lights came on. I melted a kettle of snow, and the coffee pot began to heat on the electric stove.

This may sound like an easy and orderly process, but we had seen hard times on many an opening day — occasions when our ancient weasel would not start at all, when the heads of an old and balky generator (the early 60s) had to be heated with torches at 20 below; when recent ice storms had wrecked the power lines, the gasoline lanterns (all four of them) would not operate, and we had to get supper on the heating stove and eat by candlelight — as someone remarked, without the party napkins, yet.

So 1971 was a good year. Often, when Don unpacked the Champ he brought forth a great baking dish of some wonderful

Water hole detail, Murray and Shelton.

casserole sent along by Helen, his wife. This was one of those times, and it was soon in the oven. Reluctantly, the shack began to thaw out from its winter freeze; this was a slow process, and it would be several days before flies were coming out of the walls.

Milk cans were loaded into the snow machine trailer and taken to the harbor, where a water hole was opened. As we pulled the lid off that first can of superb Lake Superior water, we extracted some ice chunks. These served their purpose at supper time, when a bottle appeared and we sat around the table enjoying our first "social hour" of the season. It was a standing rule that we never had a social hour unless there was some notable accomplishment to discuss and celebrate, and there were times when this taxed our creativity somewhat. It tended to keep the crew in an optimistic frame of mind. Ice from the milk cans was definitely a compromise. After things got settled down, a fringe of icicles would be hanging from the eaves, and this was the regulation source of supply for our once-a-day libation.

That night, as our cold beds gradually became more bearable, we knew we had made it again and our program was under way. The radio was alive, and we had even tried to report to Houghton. We could not raise anyone, and that showed how much confidence they had in us.

Usually it required about a week to get established in the winter camp and have our work routine operating. That is, if you can call anything routine that depends from day to day on the weather. Our weather reports by radio were the most important news we got from the outside world. These helped considerably even though Isle Royale does not conform to conditions on the mainland. The lowest temperature we have recorded is 35 degrees below zero F., whereas 40 or more below is not a great event in either Minnesota or northern Wisconsin. In receiving daily reports from Thunder Bay (formerly Port Arthur and Fort William), Ontario, we note that our temperatures frequently are 5 or 6 degrees higher.

From open water to the south, low morning clouds often drift over the island, and it is midmorning before we can start flying. Lake Superior never actually freezes over, although rare periods of calm weather permit a thin, gently undulating skim of ice to form, which quickly disappears before any kind of breeze. On these occasions newspapers carry headlines about how the big lake has finally frozen over "for the first time in history." One of the early jobs of each winter period was to measure snow depth and get the weather station operating.

Soon after our arrival at the island, Don took steps for the safety and operation of his airplane. With the weasel he packed down a parking area beside the dock and smoothed out another strip at the end of the dock, on which our cache of aviation gas was left last fall in 17 barrels. Depending on ice conditions, the weasel might also be used to pack down slushy snow, which would then freeze into a convenient runway for takeoffs and landings. In the parking area Don froze anchors into the ice for tying down the plane. At night he plugged a cord into the hot line on the dock and placed a large light bulb under the cowling of the motor. Then the nose was wrapped in old blankets and a canvas. Almost irrespective of cold, the motor would start in the morning. On occasions when the generator stopped at night, there would be a delay before takeoff while oil was drained from the motor, heated, and poured back in.

Our landings on rough ice around the island have cost Don a lot of off-season maintenance on his plane, including replacement of the skis. But there has never been major damage during the winter operation and never an emergency. This record is standing tribute to his meticulous care on every point of concern, his skill as a mechanic, and his consistent avoidance of taking chances. A number of times I have been with him when we did a job with a narrow margin for error, but Don knew what his plane could do, and he never misjudged.

There are some things we don't like to think about. Like a year early in the study when our gasoline was being delivered in old rusty barrels that obviously had been long without a cleaning. When Don took the Champ back to Virginia for its 100-hour check, they found a chunk of ice in the wing tank that melted into more than a quart of water!

Don replaced his aircraft twice, but he stuck to the same model, which will throttle down to less than 70 mph and maneuver readily in circling over animals or points of interest. It is light enough so that one man can lift the tail or a ski. For a number of years he had a thermometer taped to the wing strut in such a position that it could be read from inside the plane.

To me this seemed a convenience, and on occasion I would mark down the temperature while we were in flight. After this had gone on for a couple of years, he and I were untying the aircraft one morning and I glanced at the thermometer. Then it occurred to me that every time I read it, the mercury registered 10 degrees above zero. As a matter of curiosity, I asked Don about this. He nodded and got a faraway look in his eye.

"Yes, that's right," he said. "I guess it's one of the few things a fellow can depend on."

Long ago in these studies we learned that good weather should not be wasted. Thus, even if many basic things were not yet done, when conditions were right for flying, the plane would be in the air on the second morning. Our first objectives were to find the wolves, survey the island for tracks, and locate old kills. We always took notes on moose, foxes, and anything unusual.

In early years of the project, writing notes in the air was a demanding chore. It took time away from observation, the unheated plane was "airy," to say the least, and stiff fingers wrought hieroglyphics that sometimes were difficult to interpret when the job was done over in the evening. After trying a tape recorder with two reels, I decided that the bulk and technical difficulties were not worth it.

Then I discovered a tape recorder with cassettes and a numbering system by which any spot on a tape could be identified and relocated. The wonderful machine could lie behind one's feet on the floor and be managed entirely from a separate microphone — which it was good to shield from the roar of the motor.

In the winter of 1971 the records of my field work improved immeasurably. At the worktable at night I could replay my observations and write up notes as the job ought to be done. In the following winter, when Rolf first came to the island for the full seven weeks, he brought another recorder and we continued to use them. When there was action, we could now keep our eyes on what was happening and dictate the whole story — with gloves on.

Much of our flying has been at 300 to 500 feet, and each winter our initial exploratory work involved following shorelines around the island perimeter and also the inland lakes. Often

Chores at a kill site.

there was good ice below, where a landing could be made if need be. Most flights lasted about two hours before we came in to gas up, and many points of interest the length of the island could be checked in that time. A flight might be prolonged by one or more stops. If a kill was found near a landing place, and no wolves were present, we were likely to stop and examine it at once. For this purpose the plane was equipped with snowshoes for two, a hatchet, and large and small plastic bags. It also carried survival equipment in the tail — boxed rations and a couple of sleeping bags from the park fire-fighting cache.

The commonest thing we looked for was wolf tracks. It is nearly impossible to tell which way the animals are going, without landing for a closer check. In one direction we might follow the pack backward to an old kill. Such a site usually is indicated by a complex of tracks leading inland, and there may be a moose bone or two, or a scattering of hair, where a wolf has lain out in the open and chewed at length on a leg or something else removed from a carcass. We would circle close in over the deep-worn paths leading inland, for they might point directly to the kill, and if one or more ravens flushed out of a clump of conifers, it usually meant we had found the place. Even then, a kill under the spruces, balsams, or cedars could be difficult to see, and we would circle at length until, usually, Don spotted it. There have been times when the kill was in such thick cover we could not see it at all but knew approximately where it was

Moose jaws, as cleaned up by the wolves.

by the radiating wolf tracks. Then it took a landing and an investigation on foot to verify the find.

Almost any kill the wolves have fed upon will have the viscera removed — these soft parts probably are eaten primarily by the dominant animals of the pack — and the skeleton is likely to be dismembered or in the process. The contents of the digestive tract will be intact, frozen, but with meaty parts peeled off. Usually the carcass is lying in a bed of sheared-off hair, or it may have been dragged away from the kill spot. The jaw will be clean and ready to be picked up. By the extent of its tooth wear an old animal probably can be aged within an accuracy of two years. But a better job can be done with the upper molars (p. 98), so we knock off one of the tooth rows; it comes away neatly with a piece of the maxillary bone when hit properly with the hatchet. Beginning in 1971 we also took a metatarsus — the lower section of the hind leg, including the hock. This was to be cleaned up and measured for purposes to be described later. Finally, we placed in a small plastic bag several handfuls (frozen chunks) of the coarsest contents of the rumen, which would provide information on what the animal had been eating — instructive when compared with notes on what was growing at the site.

All these specimens we joyfully carried back and piled into the plane. Often the passenger had part of it in his lap on the way back to camp, especially if we also brought in a sacrum or skull showing something of interest.

It happened sometimes that, from the air, we would see many tracks, signs of a chase, and blood on the snow, without pinpointing the kill. A light snow fall might help to obscure events of a day or two before. This could mean that a moose was injured and got away, but we always landed to investigate if possible. I well recall that Don and I checked such a site on the edge of a small island a short distance from the south shore of Siskiwit Lake. Two wolves had been there since the snow, and they had fed on something, leaving moose hair. Then, under overhanging cedars we found the snow-covered contents of the intestine. Where wolf tracks crossed the ice and entered the woods, I picked up the center section of the leg bone of a calf; both ends had been gnawed off, and ravens or whiskeyjacks had picked marrow out of the cavity. That was all we ever found, even after following wolf trails into the woods.

In this case we knew that a calf had been killed, probably by two wolves (the big pack had been elsewhere). The main thing lacking was the sex of the victim. We have learned that calf kills are frequently cleaned up in short order and the remains carried off. On occasion the only evidence is a jaw, a hoof, or the small size of the frozen stomach contents, which usually retain the shape of the organ.

Every dead animal examined, of whatever species, gets the latest number in our autopsy file, and the card bearing that number is the recording place for all information regarding it. However, before examining a specimen on the ground, we might have seen a kill from the air — often far from a landing place — and kept watch on it for a week or more. Such kills were spotted on a map taped to the wall of our shack. The kill was given a number that begins with the year. Thus, 71-5 was the fifth kill we located in the winter of 1971; and after specimens were picked up it also got autopsy number 540, which means that since the beginning of our studies in 1958, it was the 540th animal so recorded. All but a few of these numbers represent moose, and that number has long since been doubled. As might be expected, the autopsy number goes on each specimen (teeth, rumen contents, bones, etc.) collected from a particular animal. The results of laboratory work are expected to be entered on the autopsy card. That statement hedges a bit because sometimes, inevitably, it does not get done. When I find that I, personally, have failed to do something so simple, I cannot imagine why.

When, in our aerial tracking, we followed a traveling wolf pack, it was often possible to catch up with them. In the distance one of us would see them, mere specks on the white of a frozen lake, and we would approach and circle, counting carefully. Or we might encounter one or two wolves following the trail of the others; then we flew on to intercept the main pack and, again, make the best possible count.

In our first day of flying we liked to cover the island as thoroughly as possible to get a reasonably good idea of how many wolves were present and in what combinations, although it might be many days or weeks before we could put together what was considered a reliable inventory of the entire population. Good counts early in the season were highly desirable, because they could then be checked repeatedly during the remainder of the study.

We tallied every moose seen, and in recent years we recorded whether it was standing or lying down. Standing or walking moose usually are browsing, while a moose lying down has stopped browsing and is ruminating — chewing its cud. Gathering and processing food is a constant preoccupation for this large animal, and our accumulated observations have helped us understand the daily rhythms of its activities. Later in each winter period some time was set aside for the close-in checking of more than a hundred moose for age (adult or calf) and sex for comparison with counts from the previous fall and other years. These counts are indices to the annual productivity of the herd. A good, unbiased sex-age inventory in winter requires determining the status of every animal seen, and in our daily flying we did not have time to do the circling and flushing that are necessary when a moose is in thick cover. Simply recording the "easy" ones — such as a cow and calf in the open — would give a distorted (biased) statistic on the herd, although in all cases our notes included such information.

On marginal flying days, especially those with enough wind to make air near the ground bumpy, some of us have had a physiological problem: air sickness. Close circling over a kill site is a particular trial. As the plane continues to bank, you have the dizzy feeling that your stomach is sailing off into orbit.

Any continuation of the maneuvering means the loss of one's latest meal and possibly a feeling of malaise that may last the rest of the day. Some people are prone to air sickness and others are not. Through repeated flying you can outgrow it in degree, but probably not completely. I have always had problems of this kind, and Don knew my tolerances. When I tapped him on the shoulder after some rough flying, he knew it was time to straighten out and land on the first favorable lake for a few minutes of recovery. Dave Mech and Phil Shelton also had such troubles. Pete Jordan, Mike Wolfe, and Rolf Peterson were practically immune. In the first two winters of the project, 1959 and 1960, Dave was going it alone: he had no other professional help. He and Don also put in longer hours in the air than anyone has ever done since. Though sometimes he was unspeakably ill, he stayed with the job in a manner that anyone who knows what it is like could only regard as heroic.

Beginning in 1961, with the elimination of my spring teaching, I participated every winter, usually for at least a month. But one flight a day was my normal quota, when I did any flying, and the great bulk of this and other work in field and laboratory has been done by my students and associates. I have had the advantage of long tenure in the program and the important privilege of profiting from the work of the others. None of us is likely to forget that every time something was observed from the air Don Murray was there. Thus, he has been far more than a pilot. As an observer and tracker, we regard him as the world's best.

By the third or fourth day of our winter program we were getting settled. For one thing, we had dug out the entrance to the root cellar behind Windigo Inn and removed our canned goods — some left over from last winter, plus a new order purchased for us in Houghton and sent out via *Ranger III*. With these additions our cooking became more substantial. The sourdough starter, carefully dried in flakes and put away last winter, was activated, and the yeasty-smelling crock would occupy a shelf behind the kitchen oil stove for the remainder of my own stay on the island. The pancakes, bread, and rolls that came out of it represented a part of my self-appointed function in the camp, as was most of the cooking. My interest in these arts developed around the campfire under the instruction of one of the greatest of scoutmasters, G. M. Wilson of my home town, Fort Wayne, Indiana. Since those early days of hike chowder, wild greens in spring, and biscuits baked in a tin can, I have regarded cooking as a relaxing and creative pastime. My sourdough starter has ancient and distinguished origins, being a crossbreed of cultures from Alaska, Yellowstone, and Grand Portage. Like all living things of noble lineage, when neglected or mistreated it can get temperamental, but when properly fed with unbleached flour and handled with consideration it reproduces mightily and yields comestibles of rare virtue.[1]

Each winter when we went to the island we had decided on a moose autopsy program involving one or more animals to be collected and examined inside and out. We seldom have had an opportunity to inspect a fresh kill — the first thing wolves do is open the body cavity and pull out the viscera. Also, as will be discussed later, the wolf kills commonly were restricted to the young and old age classes. So, to get information on the health and breeding status of "prime age" and other select moose, we took time to collect one or two each winter. In addition, we might finish off and examine, if conditions allowed, any moose found injured and abandoned by the wolves or otherwise disabled.

An "autopsy day" was one when we were grounded by weather and nothing else demanded our time. Then I was likely to be found on snowshoes scouting the shorelines of Washington Harbor and Beaver Island looking for a moose of the sex and age group we considered preferable as a specimen. Often this was largely guesswork, but we never failed to profit by an autopsy. I would be carrying the only gun in camp, a .30-06 rifle contributed to the program by one of my fellow members of the Boone and Crockett Club.

I spent many a half day amid the falling snow looking in vain for the kind of moose we wanted. When I did find one, certain conditions had to be met. It needed to be killed near the shore; four men do not drag a moose very far. In addition, it was desirable to kill the animal with a neck shot to avoid mutilating internal organs. Then a report to the shack via pocket radio, and the crew would bring the weasel loaded with our weighing gear and autopsy tools. Once out on the shore, the moose would be hooked to the weasel and dragged over the ice to a point across the harbor from the dock. Then, after weighing, we took it apart, preserving such tissues and samples of organs as we, or someone else, had use for. The carcass was left as a bait about half a mile away and in full view of the ranger station. If the wolves came through the harbor they were almost certain to stop and feed, sometimes staying a day or two. Then we kept away from the waterfront, and our student investigator spent the day with a spotting scope at an open upper window of the ranger station learning all he could about the social behavior of wolves. Rolf used this method to special advantage and learned to recognize several members of the west pack by individual characteristics.[2]

Our winter cabin was a portable building originally constructed by the Civilian Conservation Corps when they had a camp at Windigo during the 30s. It was T-shaped, with three large rooms, the middle one being our kitchen, eating, and work room. The top of the T was divided into a heated bunkroom on the east and an unheated storage and "cold room" on the west. The only plumbing operational in winter was the drain under our kitchen sink.

In summer these quarters were used by seasonal help, and the park administration had winterized them for our benefit. This especially meant wallboard and ceilings. There was a time at the beginning of our work when the north (downhill) side of the shack was perched on 4-foot pillars, and the wind swept under it unimpeded. Anything placed for very long on the floor — feet, for example — would freeze, and we learned by experience how far above this frigid substrate to store our fresh vegetables and other produce to keep them at a desired temperature. Things that could be kept frozen simply went into the cold room.

In addition to the cabin, we inherited from the CCCs a lot of heavy-duty kitchenware, dishes, and implements, including a huge oil-burning kitchen range that might have seen early service in a lumber camp. These practical facilities served our purposes well, and it saddened us somewhat to see them progressively disappear over the years. Some of the stuff probably was thrown out by unimaginative rangers. Our cookie sheets, of just the handy size (20 by 30 inches), now catch oil under motors in the generator shack. Some items went to the lodge kitchen, and I have encountered familiar ceramics of the old days in various trail cabins and fire towers around the west end of the island. Fortunately, we have had a source of replenishment in the outworn dishes and kitchenware relegated to "surplus" from dormitories and the student union at Purdue University.

One gets attached to the old things — in midwinter on Isle Royale it takes about a week. I recount these details more in regret than bitterness, for we have been adaptable and largely uncomplaining. The truth is, of course, that because of our own personnel turnover every three years or so, Don Murray and I are the only ones who have known the full range of conditions. No doubt we get a bit wistful in telling newcomers how it used to be. Several times I have heard Don describe how "Doc spilled a bucket of water on the floor, and before he could reach the mop it had turned to ice." This was true enough. Between the back door and the bunkroom door there was a difference of four inches in elevation. In cold weather, water running north would freeze before it reached the hole we had bored in the floor.

The man who brought greatest disruption to this homey scene was superintendent Henry G. Schmidt. Hank spent nearly a week with us in the winter of 1962, and we enjoyed his company little suspecting what was going on in his progressive mind.

The following winter, even from a distance, one could see that things were not the same. The foundation of our shack had been completely boarded up, cutting out both wind and snow. Inside, the impact of change was even greater, for our venerable kitchen range had disappeared and in its place were an oil-burning heater and a propane cook stove — the latter would give way to a modern electric stove, since we had an excess of power. In the bunkroom, metal clothes lockers had been set up between our cots, and some reed easy chairs and a couch were arranged suggestively around the stove. All these amenities had been scrounged from some public institution that went out of business in Houghton or Hancock. As a final touch, we discovered that someone had gathered up the ancient army blankets and sent them to the laundry.

We stood about somewhat embarrassed by it all. Then, amid colorful expletives the thought came out. "Things are getting pretty soft around here!"

According to the park-drafted operating procedure for the winter study, one of the first jobs to be done at the opening was

Hay Bay and Hay Peninsula at a time of open water. Beyond, at right, is Spruce Point. Siskiwit Lake, showing Ryan Island, is in the distance.

to shovel snow off the roof, a sensible provision. Three feet of snow could collapse an old frame building. Often this chore fell heavily on the park man, since the rest of us were preoccupied.

At some time, nearly every winter, the shoveling would be neglected and new snow piled up on the roof. Under this insulation the roofing warmed and the snow began to melt. Water ran down onto the eaves, where, without heat from below, it would freeze and build up a ridge, impounding a pool of water that backed up under the roofing. In the middle of the night, the water came down through the ceiling onto somebody's bed. As chance would have it, Don and I occupied cots in the cold east corners of the bunkroom, and the water never bothered us there. But it was somewhat disturbing to hear the others scurrying around and water drip-dripping into half a dozen pans.

Usually, by the time our winter period was half over, the finding of old and new kills was well in hand. We would have picked up specimens from most of them where landing conditions were favorable. As of mid-February we nearly always had a good or better count of the wolf population.

Then came a major assignment involving long hours of flying drudgery by Don and our research man: the winter moose census. More of this later, but it has been one of our most perplexing problems (as in nearly every wildlife research project) to count animals with the best possible accuracy and *then be able to prove it.* I escaped this entirely because it needed to be done, critically and consistently, on days representing in reasonable degree a given set of weather conditions, by one man, working with Don. The counts were made on sample areas plotted over the island in a design that, we hoped, was representative of the entire area. The object was to fly parallel lines that covered the census plots completely, doing enough circling over dense cover to flush out or see every moose.

The winter count was not our only source of information on moose numbers, but it has been the most important and probably the most accurate. The work usually carried over into March and was complicated by periods of warm weather, rain, and slush, which conditions caused us, in recent years, to move the entire program up a week and plan to close camp by mid-March. The close-down chores dominated every activity during the last week. Bones had to be cleaned and all specimens packed for the trip back to the campus. Any accessible kills — those that could be reached in half a day of snowshoeing from a landing place — were picked up. Other kills were carefully spotted on a map so they could be found via cross-country trips in the spring. Locating such sites on foot was seldom easy, but where this was likely to be especially difficult a long streamer of brightly colored plastic flagging was dropped to hang up in the trees and serve as a marker. Sometimes it helped.

Closing also involved cleaning up the place, returning canned goods to the root cellar, and caching our permanent winter gear in chests and boxes in a corner of the storage room. When the date for leaving was imminent, our main interest in life became the weather reports. Would we make it, or not? Not until the plane was on its way would we shut down the generator.

There is excruciating frustration in having everything ready, and waiting out the last hours in uncertainty. Then, in midafternoon you finally get the big letdown, a radio message from Houghton. They have been informed by telephone that it is snowing in Minnesota and the flight for today has been canceled. There is little incentive to do much unpacking, and everything is on a standby basis. I recall one of Don's appropriate remarks on such an occasion:

"We can't sit around and tell stories about how the wolves attacked us, because no one here would believe it."

CHAPTER FIVE

Wolves and Their Hunting: Prey Selection

WHEN DAVE MECH AND I went to Isle Royale in late June 1958 for our first field trip, there were questions in our minds on which we expected to get solid information. The abundant popular literature on wolves described a creature so destructive and wasteful of its food supply that it was difficult to see how the modern animal and its ancestral forms could have survived to the present. Indeed, it has done more than survive. Before men took over total administration of earthly affairs, the wolf in its various races occupied a huge range in Eurasia and North America, extending to practically every region where there were productive populations of ungulates (hoofed animals) to feed upon.

It was evident that, like every other successful creature, the wolf had to play some useful part in the operation of its ecosystem, but we did not understand how that worked.[1] Obviously, tales retold often become embellished and exaggerated. But it is intellectual arrogance to discount offhand what someone says he has seen, especially when many of the stories were accepted and repeated by people who were sincere and usually careful in their reporting.

Did wolves kill for fun in great orgies of slaughter when prey was easily available? Did they habitually cut out and kill "the best animal in the herd" for their nourishment? Did they pursue and hamstring their victim, then pull it down the easy way? Did they prefer to devour their meals alive? This chapter and the one that follows will provide some clues.

Needless to say, testimony has been conflicting on many points. Today re-evaluations are proper because we are sometimes in a better position to interpret early accounts than were the original observers. Which is to say, our accumulated hindsight counts for something. The eminent C. Hart Merriam has already been quoted on wolves in relation to deer. As was usual in his time, he considered wolves to be a threat to the deer and a liability to human interests (p. 11). A similar view was expressed by George Shiras III, a good woodsman well acquainted with the Lake Superior country, as were his father and grandfather. In an article in *National Geographic* in 1921 he included a picture of four wolves hanging by the hind feet — poisoned on the lake where his camp was located, near Marquette in Michigan's upper peninsula. The caption said:

> Next to man, these animals are the most destructive foe of the white-tail deer in the upper lake region. Fourteen years ago, during a severe winter, when deep snows were crusted for several weeks, the wolves destroyed nearly all the deer within a ten-mile radius of Whitefish Lake. From the carcasses found it was estimated that over two thousand deer were killed in this limited area.

Anyone acquainted with the history of deer and hard winters in Michigan and other northern states may have misgivings about this report. The situation would be easy to misinterpret. In winters of exceptionally deep snow the deer concentrate and form "yards" in the best of their winter range, usually cedar swamps. If this condition persists for very long the browse supply is reduced and malnutrition results. Young animals are at a particular disadvantage because they cannot reach as high as adults. In the past 30 years many thousands of deer have died in Michigan yards where there were no wolves, especially in the lower peninsula.

Wolves present under these conditions will have easy living on the weakened and concentrated deer. But what they are doing is not necessarily destructive; it could be part of a workable long-term design. Often something we see happening may appear meaningless, or it can have the earmarks of disaster, but it might help preserve the entire ecosystem, on which each of the associated living things must depend.

Such hitched-together events as deep snow, concentration, malnutrition, and excessive predation may thin out deer in the yards soon enough to prevent total destruction of the browse. Several hard winters could reduce the deer considerably. Several "open" winters in succession would permit them to remain scattered over the range (as in summer), feed better, and build up again. When deer are scarce, brush-stage food plants make a comeback; then deer can do likewise. In cases like this, where processes beget their own controls, the scientist says he is witnessing a feedback.

In the operation of ecosystems we see repeatedly that calamitous extremes of almost any kind are countered by compensations. The high-to-low fluctuations tend to even out in a pattern that spells survival over long time periods for all the kinds of life that fit the local organization. Actually, that "delicate balance" we hear so much about is not especially delicate, nor is it a precise balance. It is tough and durable, an eternal teetering about a midpoint.

The Merriam and Shiras accounts of wolves on a jamboree of slaughter had impact on an impressionable public, but they did not represent usual conditions. In 1938 Sigurd Olson brought together information from the best observers he could find in the Superior National Forest of northeastern Minnesota. Counts of deer kills by wolves on two areas of about 60 square miles each indicated "the general conclusion that nowhere is the kill

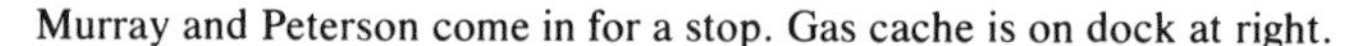

Murray and Peterson come in for a stop. Gas cache is on dock at right.

more than one deer per square mile and a quarter." Within the limits of possible accuracy, this indicated a toll the herd could sustain, and Olson pointed out that the deer were holding up.

From 1946 through 1953, Milton Stenlund carried out a wolf study in the same region. His conclusion was that wolves were killing 1.5 deer per section annually out of a population of 9 per section, or 16 percent of the herd. On the average, hunters killed fewer than one deer per square mile; thus the drain on the herd from these two sources of mortality was near 24 percent, which Stenlund considered approximately equal to the annual increment in years of nonextreme weather. Available information indicated that the wolves themselves were under a yearly toll of about 41 percent under the bounty system.[2]

More recently, in Algonquin Park, Ontario, George Kolenosky found that over the winter period a pack of eight wolves killed about 10 percent of the deer in their range, which indicated a consumption rate of one deer per wolf every 17.6 days. In this case an "open" winter and a winter of exceptionally deep snow were averaged, of which we will have more to say in a later chapter.

Graphic accounts of mayhem in the woods have spread the idea, cherished by many, that the peace-loving vegetable eaters are at the mercy of their "natural enemies" (a term I use advisedly). On the other side of this card is the simplistic claim of casual naturalists that predators kill only the diseased and weak. Both generalizations are faulty.

Obviously, in a population of prey animals — meaning practically all species that feed on plants — some individuals are more vulnerable than others. Plenty of evidence in scientific literature indicates that the predation odds are against creatures handicapped in any way: physically, mentally, behaviorally, or ecologically.[3]

On the other hand, meat eaters do not always select weak individuals out of the prey population. Healthy animals can be taken under many conditions, especially when they are overnumerous because of unusually successful reproduction, when they are moving extensively, or when some catastrophe has lowered the carrying capacity of the habitat. Characteristically, predation is a "density-dependent" mortality factor. The success rate of the hunter is high when the density of prey is high, and vice versa.

Dave and I felt reasonably sure that these were reliable generalities, but how do they apply specifically to the wolf and its prey, especially when that prey is the moose? In 1958 the most definitive piece of research we could draw upon was Adolph Murie's study of the wolf and Dall sheep in Mount McKinley National Park. Murie made a large collection of skulls or fragments of skulls that could be aged accurately by annular rings on the horns, and also by tooth wear. In sheep you have the advantage of correlating in many individuals the degree of tooth wear with known age on the basis of horns, something not possible in deer and moose.

Dividing the material into two classes, Murie found that 221 of his specimens had died in the five years just past, 1937–41. Of these, 69 percent had been nine years old or older. Significantly also, an additional 9 percent of younger sheep showed evidence of disease, a severe jaw necrosis often accompanied by malformation or loss of teeth. With less expectation of accuracy, he found the same trend in a series of 608 older skulls

Temporary ice west of Long Point. A few days of little wind permit such formations, which sometimes are suitable landing places for picking up kills.

representing sheep that had died before 1941. Including lamb remains (these disintegrate rapidly and are under-represented in the sample), Murie summed up the situation this way:

> In the absence of predation we would expect the mortality to be distributed among the weak, namely the old, diseased, and the young, but in the presence of a strong predator like the wolf, known to be preying extensively on the sheep, it is interesting that so few animals in their prime are represented. In the recent material 211 skulls, or 95 percent, were from the weak classes in the population and only 10 skulls, or 5 percent, were from sheep in their prime which were [presumably] healthy.

Corresponding figures from the earlier period were 88 percent and 12 percent, respectively. It was evident that, whatever proportion of these sheep had been killed by wolves, a strong selection for the weak was involved. Murie also had convincing observations on the chasing of caribou bands by the wolves. Calves or other animals that were unable to keep up fell behind and became wolf meat. This was selection in action.

As a personal matter, I can say that I studied the wolves and moose of Isle Royale because I came along a century late to study wolves and buffalo on the plains. But many early field observers mentioned what they saw, and the culling of substandard bison by the abundant wolves is explicit or implicit in their notes. One of the earliest statements of this kind was in the journal of William Clark when he and Meriwether Lewis were conducting their famous expedition up the Missouri on the way to "Oregon." The date was 20 October 1804 (44):

> I observe, near all gangues of Buffalow, wolves and when the buffalow move those animals follow, and feed on those that are too pore or fat to keep up with the gangue.

The following spring an entry in "Captain" Lewis's journal (22 April) gives a detail of the same kind:

> Capt. Clark informed me that he saw a large drove of buffaloe pursued by wolves today, that they at length caught a calf which was unable to keep up with the herd. The cows only defended their young so long as they are able to keep up with the herd, and seldom return any distance in search of them.

A young British adventurer and journalist, George Frederick Ruxton, who recorded accurately a wide variety of information

For every hour spent in the field, a biologist (Peterson) must expect to spend at least an hour in the office and laboratory to complete his work.

on life in the plains and mountains during the 1830s, described a scene that was familiar to him:

Dense masses of buffalo still continued to darken the plains, and numerous bands of wolves hovered round the outskirts of the vast herds, singling out the sick and wounded animals, and preying upon the calves whom the rifles and arrows of the hunters had bereaved of their mothers. The white wolf is the invariable attendant upon the buffalo; and when one of these persevering animals is seen, it is certain sign that buffalo are not far distant.

Under some conditions, at least, there was not much doubt about the wolf's propensity for doing its hunting the easy way. Elsewhere, Ruxton made another pertinent observation on wolf habits (223):

If a deer or antelope is wounded, they immediately pursue it, and not unfrequently pull the animal down in time for the hunter to come up and secure it from their ravenous clutches. However, they appear to know at once the nature of the wound, for if but slightly touched they never exert themselves to follow a deer, chasing only those which have received a mortal blow.

The last point made by Ruxton has particular significance as an impartial report of what wolves were seen to do on the plains a century and a half ago. It is fortified by a similar observation on the part of Horace Greeley (hardly a wilderness sophisticate) in the course of his historic crossing of the continent in 1859. He saw the great grassland and its creatures immediately before the building of the railroads and the rapid changes that were to follow. In western Kansas he wrote of the wolf:

> His liveliest hope, however, is that of finding a buffalo whom some hunter has wounded, so that he cannot keep up with the herd, especially should it be stampeded. Let him once get such a one by himself, and a few snaps at his hamstrings, taking excellent care to keep out of the way of his horns, insures the victim will have ceased to be a buffalo, and become mere wolf-meat before another morning.

In a country where literate observers were a rarity, Greeley's descriptions, like Ruxton's, had outstanding value. There is no reason to doubt their general accuracy. However the mention of hamstringing could mean that the great editor had fallen into a common assumption. The thought that wolves hamstring their prey has always appealed to authors of accounts for popular reading. It is part of the ravin-depravity syndrome that characterizes the wolf in fiction. It was handed down in idle backwoods palaver as a regular technique of canine killing, and it found its way into early scientific writing, that of Vernon Bailey in his *Mammals and life zones of Oregon* being a case in point:

> Usually they catch their prey by the ham or flank, tearing out the flesh and sinews, often hamstringing the larger animals and rendering them helpless to be torn and eaten at leisure. Bloodthirsty in their hunting methods and terribly destructive to game and livestock, they are nevertheless among the most intelligent of our native animals and most difficult to trap or hunt successfully.

It appears that Bailey got some of his information from trappers and ranchers, many of whom believed through hearsay in the prevalence of hamstringing. Without making any judgment in the matter, we must have reservations as to whether the above statement was based on personal experience.

On the other hand, there is no reason to question the faithful recording of observations on many points of natural history by the late Lois Crisler in her book *Arctic wild*. It did not involve wolves, but she described how their dog Tootch brought down a caribou by grabbing its hind leg. She ended with the statement that Tootch had hamstrung him. On meeting her at one of her lectures[4] I queried her on this point, and she thought she might have used the term loosely. She was not certain that the Achilles tendon of the caribou had been severed, which is the usual connotation of "hamstringing."

In his studies of wolves in the Rocky Mountain national parks of Canada, Ian McT. Cowan summed up his description of killing methods with the statement, "No instance of hamstringing has yet been seen or reported to me."

Although our chapter and verse are yet to come, it probably is appropriate to state here that in the Isle Royale work we have examined some hundreds of wolf-killed moose that were fresh enough for us to determine whether the hock tendons were intact. Although wolves were seen to seize a hind leg on occasion, we do not know of a case of hamstringing. This does not mean it never occurs; if I made any such claim, it would be only a question of time until someone proved me wrong. Wolves have been watched so little that we will go on learning new things far into the future. Guggisberg evidently knew of authentic cases of hamstringing by both lions and tigers when these big cats were dealing with such large prey as buffalo and elephants.

A point of great significance emerges from chance accounts in the literature and most convincingly from our own work of 18 years. It is the keen sensitivity of the wolf to the condition of its prey. In judging what he can handle and what should prudently be left alone, this carnivore brings to his daily work sophisticated skills completely beyond our human ken. These are made possible by inborn capacities effectively tuned and developed in the young animal through an apprenticeship that only the capable survive.

Our island wolf history divides logically into two periods, the first 10 years and the second 8. The initial decade brings us through the winter of 1967-68, Mike Wolfe's second year on the wolf-moose project, and into the third year of Wendel Johnson's study of smaller mammals. Here it might be good to turn back and review the tenure and overlap sequence of the major investigators as given on pp. xix-xx, since contributions of each will be cited. Actually, for some purposes we will draw upon the entire span, from '58 to '76, including Rolf Peterson's long stint that began in the summer of 1970 and goes on, we hope, well into the future.

In our suppositions of what Dave might find in his field work, we were not totally dependent on what had been written. In that first field trip of 1958, we saw many large scats[5] on the trail, and Dave began to collect and dry them for further analysis. The summer's collection showed that moose was the predominant food, and most of the hair and bones were from calves. Beavers were the secondary item, and that was about all the wolves had fed upon. In later years we would find a trace of snowshoe hare fur sometimes, but the slight evidence indicated the killing of a hare to be largely a chance occurrence. Wolves were not fruit eaters, as foxes most obviously were.

In addition, we knew what Jim Cole had found in his third winter of work for the park. In 1957 he had a plane and pilot and recorded eight wolf-killed moose. He thought there were at least 15 wolves on the island and possibly as many as 25 (51).

In March 1959, when Dave returned to the campus and we discussed his first seven weeks of winter work, we knew he had

A comparison of tooth wear in two moose killed by wolves in the winter of 1968. The teeth of a yearling, below, show high sharp crowns. Above are the worn teeth of a 16.5-year-old, three premolars at right, three molars at left. Typically, the first molar of an old moose shows the heaviest wear.

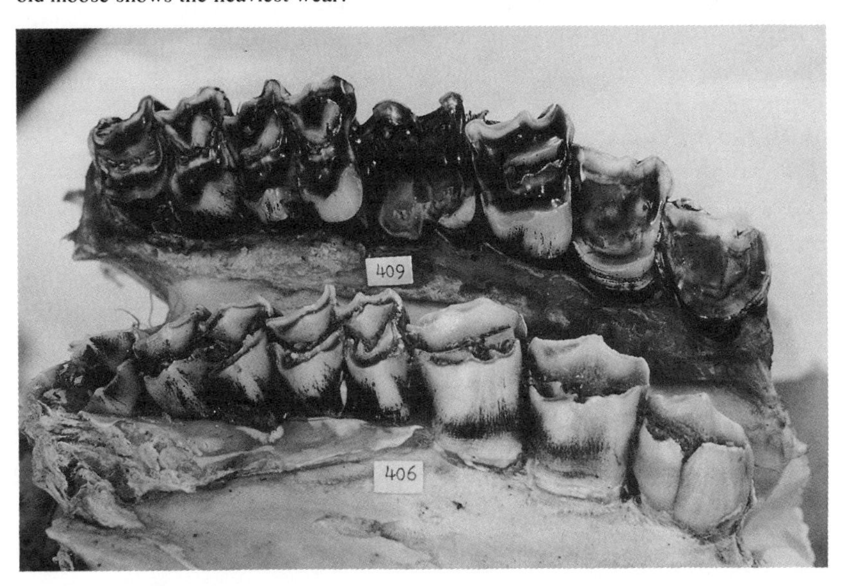

established a firm basis for the future. In Don Murray we had a pilot of great skill who had offered his services as a regular member of the team.[6] That first winter the wolves were not afraid of the aircraft and could be approached at close range when necessary. The observers could circle over them and watch their hunting and killing without affecting their activities. As Dave had demonstrated, it was even possible to snowshoe in on the wolves at a fresh kill. Far from being belligerent, they simply spooked and ran away, usually to come back within a day or two.

Mech and his first-year pilots had flown 110 hours on 27 days between 3 February and 14 March. Later we would appreciate that they had exceptionally favorable weather. The nucleus of the wolf population was a breeding pack of 16, which included a trailing wolf that brought up the rear. When sighted by the others this wolf was chased and picked upon. I was largely responsible for a theory at the time that this wolf was trying to join the pack — that it might even be the alien Big Jim. This idea turned out to be wrong.

Dave examined 11 moose kills on the ground. The mandibles were collected and aged (in the laboratory) by the tooth-wear method.[7] The age distribution was of great interest: Four were calves about 10 months old, two were 6 and 8 years, and the other five were all over 10 years old. It recalled Murie's findings on the sheep, but obviously our sample was small.

That first winter Dave noted that a group of five split off from the main pack and hunted by themselves for a period, and three of these were especially light colored and puplike in appearance. With no previous experience to go on, he decided that this was a family group of the spring of 1958. His judgment probably was correct, for five wolves continued to drop out and go their way occasionally during the next two years. The population estimate of the first year was 19 for sure, and possibly 20, comprising the breeding pack of 16 (15 + 1), a group of 3, and a possible loner. It could have been one of the three.

We had no evidence that the Isle Royale wolves produced any surviving pups during the next two years, and the population appeared to be stable. Later Dave decided that there was, indeed, a loner at the east end, in addition to a duo and the nonbreeding group of three. Thus, total wolves on the island probably numbered from a minimum of 20 to a maximum of 22 in all three years. However, in Table 1 (p. 276) the figure for each year is given as it was calculated at the time.

In the third winter the "big pack" was 15, without any trailing wolf. For two years this hanger-on had been familiarly designated as "Homer," and it was a notable event when he failed to show up in 1961. I heard much about this, for it was my own first winter in the field.

As later observations were to show, it probably is usual for a breeding wolf pack to have a "trailing" member, and Homer was quite typical. As Dave and Don observed, he frequently was a mile or more behind the pack, following in their tracks and cleaning up what he could at kills. On other occasions he was within sight of the other wolves and was able to associate amicably with two or three of them; then he would be chased away, presumably by the more dominant animals. Ordinarily Homer would cower, tuck his tail, and show other manifestations of deference, but on rare occasions he was able to join the pack and for a time would be indistinguishable from the others. He obviously was of low social status. Over the years it has appeared to be the ultimate fate of such a wolf that he drops farther behind and eventually becomes a wandering loner having no direct association with the pack. It became evident, however, that two or more loners might pool their interests, thus forming the nonbreeding duos, trios, or even a group of four that we continued to observe.

Without certainty, I believe that these nonbreeding groups are composed of animals of the same sex, and I never identified a female among them.[8] We have positive evidence that females leave the breeding packs for at least short periods, but whether they become true loners and live out their days in isolation is an open question. In recent Minnesota telemetry work, Mech has known young females to travel long distances alone, but this seems comparable to the wanderings of adolescents of all species before they settle down to normal social adjustments in a home range (181). In general it would appear to be manifest destiny for wandering females to find a mate.

From the foregoing it will be clear what is meant by a trailing wolf, which we sometimes lump in as a pack member, especially by use of the device, 15 + 1. For many years on Isle Royale, nearly all groups other than the "big pack" were nonbreeders.

*

The details of changes in wolf numbers and packs will be left to a later chapter, as will our observations and conclusions on relationships of wolves to people and human activities. However, one aspect of the latter subject needs recording here.

In the first two winters, the tolerance of the big pack for close approaches by the research aircraft — on open lakes and bays, where a landing could be made in case of a low-level motor failure — was a major advantage. There is no substitute for watching the animals behave naturally at close range. However, in the third winter (1961), when Don and I first circled the large pack on Hay Bay they showed an obvious uneasiness, finally scattering toward shore and into the woods. Don was nonplussed by this; after several days he and Dave agreed that the behavior of the wolves had changed.

The change was far more evident in the two winters to follow, when Shelton and I were on the island. Many times the pack would head for cover when the plane came within a quarter-mile. This avoidance of the plane placed Phil under a handicap in making observations of hunting and social behavior during the two winters of his work, 1962 and 1963. Since wolves were being hunted and bountied in both Ontario and Minnesota, I conceived a personal theory that the island wolves were being disturbed, and possibly shot at, each winter before our arrival.

For once, my theorizing seems to have been correct. In March 1963 Don and ranger Bob Peterson saw in the park an unauthorized aircraft that took evasive action. The number was reported to the FAA, and it proved to be an unlicensed plane from Beaver Bay, Minnesota. That aircraft probably did not cause further problems, for when Pete Jordan and Don saw the big pack for the first time in 1964, there was no evidence of fright. This tolerance continued with fair consistency through the years. However, for reasons we may not know about, the Isle Royale wolves never seemed to regain the level of unconcern for aircraft approaches that was so impressive in the first two winters.

Hunting habits and killing selectivity were big issues in Mech's thesis research, and the work of winters 1959–61 clearly exposed the trends. He found that on the average the big pack killed a moose every three days in winter, and they did not kill just any moose. They approached and held at bay or chased — in other words, "tested," as Dave aptly described it — about a dozen animals for every one killed. The pack was seen to detect 131 moose, of which 77 were tested and 6 were killed. In two winters Shelton saw 14 "encounters," 8 of which involved the big pack. He concluded that 9 moose had been tested, of which one was killed immediately, and another may have been wounded and killed later. Peterson summarized statistics for the winters of 1972–74. He and Don recorded the detection of 49 moose by various packs, the testing of 38, and the killing of one.

It is evident that kills actually witnessed have been few. I have personally seen at least a part of the action in four, all of which were calves. Sometimes it has been necessary to leave a chase because darkness was approaching or gas was needed. The fact that the wolves start at once to feed and many kill sites are not near a good landing place means that we have examined few intact carcasses, usually one or two a year, and some of these were moose that died after the wolves had wounded and abandoned them. Often a partially or totally frozen animal is useful only for external examination, until the bones can be gathered in spring.

Except in Canada's Wood Buffalo Park, the largest prey animal now hunted and killed by wolves is the moose. It is relatively long-lived, and discrimination among vulnerable and nonvulnerable individuals should be most clear in this species. Not only is it clear, but in the initial years of this project, before 1965, prey selection was so rigid it seemed almost unbelievable. Later we would see that this feature of wolf-moose relationships is greatly dependent on the depth and condition of winter snow, a subject to be dealt with in Chapter 15. It is evident now that the winters in which Mech and Shelton did their work were ideal for establishing what we may call "baseline" data. They demonstrated wolf selection of prey under weather conditions generally favorable to both species. On Isle Royale this means in particular a snow depth for most of the winter of less than 30 inches. The decade we are discussing, 1959–68, began and ended with a year characterized by snow depth of little more than a foot.

My first winter on the island (1961) was a characteristic baseline winter of nonextreme conditions. It turned cold in January, and by the time work began in early February the channel between Isle Royale and the mainland was almost solidly frozen

Of moose chased or brought to bay, about one in a dozen was killed (photo by Rolf O. Peterson).

over; it firmed up quickly and stayed that way. We had ice on all the Isle Royale bays and a shelf of ice around most shorelines; these are landing places and also open travelways for both wolf and moose. During that winter snow was 20 to 26 inches deep.

Every winter there is some concentration of moose in lowland areas; they seek the conifer cover — cedars, spruces, and balsams — on the edges of lakes and bays and in stream bottoms. They do not stay there and "yard up," as deer do frequently, but move widely through the forest, especially depending on

weather conditions. In 1961, and it was true also of the two preceding winters, we sometimes found moose high on Greenstone Ridge, and there were tracks showing fairly extensive use of the '36 burn north of Siskiwit Lake and elsewhere.

Wolves traveled widely also, and for their long treks they would often be seen strung out for a mile or more along the shelf ice and frozen shorelines, or crossing bays or inland lakes. Here the snow was more solid because of exposure to sun and wind, and traveling was easier. As Dave had observed earlier, a dominant wolf in the big pack often carried his tail elevated, and in February he regularly ran at the shoulder of his mate, whom we (correctly, no doubt) took to be the dominant female of the pack. In 1962 there were signs that an additional wolf had high social privileges in the pack; it was a male that also carried its tail high at times.

It is a fairly frequent observation that before wolves start off on new activities, such as leaving a kill or proceeding after a rest stop or a moose check, they come together for a huddle, nose to nose with a great wagging of tails and the appearance of communicating. This greeting ceremony is regularly seen in organized packs, the dominant male being the center of such behavior.

The conditions described were about what we found each winter through 1964. During this period calves accounted for about a third of the total winter kills. We had no kills in age groups 1 through 5, even though these young animals obviously were the most plentiful moose in the population. The average adult killed by the wolves was 9 to 10 years old. As moose grew older and were under an increasing threat of being killed, they had a higher incidence of jaw necrosis, bone deformities that could generally be classed as "arthritis," and the accumulation of lung-clogging cysts of the hydatid tapeworm (of which more in Chapter 9). Some of the specimens showed a deficiency of

Arthritic acetabulum (below) compared with a normal hip socket. Among bone deformities seen in old moose, this is one of the more obvious (photo by Rolf O. Peterson).

bone-marrow fat, which meant they were in poor condition from some cause we could not determine with certainty.

Even though the aging of specimens was only approximate, according to tooth wear, it was evident that a rigid selection for calves and for old or unhealthy adults was taking place. The pattern would be confirmed more precisely when Mike Wolfe re-aged the entire collection on the basis of first-molar cementum annulations in 1967 (289).[9]

In the first five years, another kind of selection had become evident. During the winter nearly two thirds of the kills of adults were cows. The higher losses were balanced by increased mortality among bulls at other seasons, since the sex ratio in the population was nearly even. While we still do not have a completely acceptable explanation of the factors involved, it was evident that our understanding was getting better by 1974 when Rolf was reviewing the records of more than 400 adult moose mortalities of which the sex was known. On a year-round basis, we can say that selection by sex in wolf-caused mortality is not particularly significant.

A useful index of body condition in hoofed animals is the fat content of bone marrow. The section of a moose femur at left appears "normal"; in the one from a different individual on the right the marrow is like red jelly, totally fat depleted.

Although our interest commonly centered on the most easily observed group of wolves, the big pack, they did not do all the moose killing. The twos and threes also found a victim they could handle at times, which meant they might live securely for a week or two. This brings to mind a question that is asked frequently, "Can a single wolf kill a moose?"

The answer to that inevitably must involve a counter question: "What moose?" Long before there were any wolves on Isle Royale, moose were being born, living out their time, and dying, just as do humans, elephants, whales, and other species not subject to high rates of predation. The coyotes that survived into the early 50s undoubtedly lived in part as scavengers on dead moose.

Thus it is true that every moose will eventually reach a moribund condition where one wolf could finish it off. A pack that includes competent hunters can kill an animal they appraise as vulnerable when to the human eye it appears to be healthy. We have grown to think of Isle Royale moose as receiving regular "physical examinations," which they either pass or do not. The day of reckoning will come. Some strong, healthy individuals are passed by the wolves with hardly a second glance. Others are attacked and hold out amazingly well. Sometimes it appears to us that the wolves have made a mistake, although we have learned not to make hasty judgments in such matters.

On first consideration, it might seem that calf killing is nonselective, but this is not entirely true. Whether a calf survives wolf attack depends in part on the capabilities of its mother, particularly in the kind of "normal" winters I have described. Cows send their calves ahead and defend the rear, lashing out with bludgeoning hoofs both fore and aft. The wolves obviously respect these weapons, and when charged by an enraged moose they scatter quickly. When the charge is by a cow defending a calf, it may be the opportunity for one or more wolves to cut in and separate the two moose, which may be fatal to the calf. One or two wolves usually can pull a calf down quickly if the cow is diverted.

Defense is especially difficult in those situations where a cow has twin calves. Although we have occasionally seen twin yearlings that had survived the winter, it probably is usual for one of them to be killed, an important factor being snow conditions. Two feet of fluffy snow would favor the moose. Three feet of snow containing a high-level crust would favor the wolves. Under summer conditions it may be easier to protect the calves, after they are well along in development. On the Huginnin Cove Trail north of Windigo in late August 1970, I saw a cow with a large gash in her rump — evidently a wolf-inflicted wound. Further observations revealed that she was marshaling two healthy calves through the brush. At Feldtmann Lake on 27 June 1972, Ron Bell saw a cow with several large open wounds on her thigh and rump areas. I observed another rump-cut cow on the Mount Franklin Trail on the first of May 1974. This one was accompanied by a single calf that was only a month from becoming a yearling. On its birthday that calf would have made it through the crucial year and would then stand a good chance of escaping the wolves for at least five years.

Results of the first 10 years of the Isle Royale work left no doubt about the selectivity of wolves in killing the weakest moose in the population. Until that time, notwithstanding Murie's work on the Dall sheep, many people assumed that in some situations the wolf could kill indiscriminately. There was logic in these suspicions, because killing a large animal that can defend itself, such as the moose, could be a different matter from catching prey that can only run away, such as the deer.

Two reports had given some preliminary support to the possibility that wolf hunting and killing can be indiscriminate. One was the work of Milton Stenlund, previously mentioned. From all available sources, including wardens and biologists, as well as personal field work, Stenlund gathered information on the age and condition of deer killed by wolves in the Superior National Forest. The study covered winters from 1946 through 1953 and was centered in the area of Ely, Minnesota.

Based on records of 113 wolf-killed deer, he concluded that there was neither sex nor age discrimination in the deer brought down by wolves. However he considered deer up to 6.5 years of age to be in the "prime of life," and of 36 specimens aged by tooth wear only two were above that age. The take of fawns (25.7 percent) was similar to the proportion shot by hunters and actually existing in the population. Of 9 kills from 1 to 3 years of age, 7 were taken by the wolves a year after a winter of deep snow and starvation, which could have been significant, based on our own moose work (Chapter 15). It is evident that Stenlund's sample of aged deer was small, and in some respects his conclusions are a matter of interpretation. However, it was the first major effort of its kind on deer, and the entire study was of great value.

The second report referred to was from Alaska, where Bob Burkholder followed by aircraft the doings of a pack of 10 wolves for a period of six weeks in 1958. He found that they made 22 kills of moose and caribou, averaging a kill every 1.7 days. He was able to land at 8 of the kills and judged the animals to be in good condition on a basis of fat under the skin, which was also true of caribou collected on that range. He found no evidence of selective killing relative to age or condition, although all calf kills may not have been found.

The bison probably can be considered an even more difficult prey for the wolf than the moose, and the only study of this type of predation was made by W. A. Fuller in Canada's Wood Buffalo Park. Wolves in the park live primarily on the bison herds, which, incidentally, have been plagued with disease. Of 11 kills examined, 3 were calves, 3 were young adults, and 5 were "very old" animals. Of the calves, one presumably had been orphaned in a slaughtering operation — the wolves did not kill another lone calf nearby. Of the young adults, one had tuberculosis, one had an infected bullet wound, and the other had a broken leg that may have been present at the time of attack. In Fuller's small sample, the wolves seem to have selected weaker bison.

The first large-scale evidence for selection in wolf predation on deer emerged from the research in Algonquin Park, Ontario. In six winters of work during 1958–65, Pimlott, Shannon, and Kolenosky collected the mandibles of 331 deer that appeared to have been killed by wolves. Aging these by wear class, they found that only 42 percent of the animals were under 5 years of age, and the rest were older. They could compare this record with the age distribution of 272 deer killed by cars or collected for purposes of their research. This sample could be assumed to represent the distribution of ages in the population, which was not subject to hunting. As it worked out, the control sample showed 87 percent of the deer to be less than 5 years old and 13 percent more than 5. Obviously wolves were more effective on old animals. The wolves also killed more males, as shown by a ratio of 146 males to 111 females, the opposite of our information on Isle Royale, where the wolves took more female moose than males in winter (220, 222).

The findings in Algonquin Park were borne out by a somewhat similar job in northeastern Minnesota by Mech and Frenzel. A sample of 142 wolf-killed deer averaged 4.7 years in age. In the same general area in a kill of 433 by hunters the average age was 2.6 years. Deer over 5 years old represented 48 percent of the wolf kills and 10 percent of the hunter kills. Old deer showed the same kind of jaw necrosis that Dave had seen in his studies of Isle Royale moose. "The age of 5 years seems to be the beginning of the period of vulnerability for adult deer." It should be kept in mind that in this region of the Superior National Forest hunters were taking a substantial crop of young deer before they got old enough to be killed by wolves, and thus would reduce the proportion of old prey. As another factor, wolves undoubtedly killed young animals wounded during the hunting season. The wounds would not always be evident in an examination of leftovers at a kill. Thus the figures of Mech and Frenzel must be considered conservative as an index of old animals killed.

There is now no reason to doubt that wolves habitually select the weaker individuals from the populations of their large prey species. But later on in this account it will become evident that the vulnerability of adult moose on Isle Royale is not always associated with old age. For the young it will be found that the rules may change with conditions in the field. We will examine this proposition further in Chapter 15.

What seems to be an effective culling of prey by the wolf does not justify a similar assumption for all large predators. However, a certain amount of selection must inevitably occur in many situations because obviously weakened prey will be taken more easily. In his studies of the mountain lion in the Idaho Primitive area, Maurice Hornocker found that the big cats kill elk more often than the more plentiful mule deer. In both an age selection for the young was evident. At least half of all the prey taken showed bone-marrow fat depletion as a result of malnutrition, but on the basis of the rather small sample it was not possible to demonstrate a statistical significance. Among adults older individuals were not favored. Hornocker had evidence that hunting lions tended to avoid the mature antlered bulls, for reasons that probably are understandable.

The cheetah of the African plains and savanna runs down its prey in open chase, after an initial stalk where cover permits. According to Eaton, Schaller (241), and other observers, it favors the smaller and more abundant species of antelope. Eaton stated, "In areas conducive for only open pursuit hunting, where prey are also usually aware of the predator, the cheetah's role is more one of a natural culler of less fit individuals from prey populations. In such areas, cheetahs appear to prey more selectively by making a greater number of hunts per unit of time and kill, and thereby sample prey herds for less fit animals more effectively." He found 5.2 chases per kill.

Only the most wary and fleet of prey are likely to outrun the cheetah, which can exceed 60 miles per hour for a short distance. Hunts usually are by individuals, family groups, or several cooperating adults. The cape hunting dog operates in larger packs and is more successful in its kills. Estes and Goddard said that prey selection from a herd is made after the chase begins and the "leader can overtake the fleetest game [antelope of small or medium size] usually within 2 miles." They reported a success ratio of more than 85 percent. With a speed perhaps half that of the cheetah, the wild dog depends upon endurance and a strict hunting discipline in which 20 or more pack members may be strung out as much as a mile to the rear and yet concentrate on following the lead animal and its intended victim. As the dogs catch up, the exhausted prey is set upon and torn apart from the rear. Selectivity is apparent in this process, but its high success percentage indicates that prey abundance and the exigencies of the chase also are important.

Hugo van Lawick-Goodall drew somewhat similar conclusions from work on the wild dog. Out of 91 observed chases, 39 ended in a kill. He asked the questions that beset us all:

> What are the dogs looking for when they stand watching a herd run past them? And why do the individuals of a normally closely united pack sometimes separate in seeming disorder? The answers are, I believe, the same; both techniques enable the wild dogs to select an individual from the herd that is, in some way, weaker and slower than his fellows.

This is the evolutionary process at work. It deals in trends produced by a mix of relationships involving large numbers ("statistical significance") and long periods of time. The inscrutable regime of predator inspection makes only one distinction, that of vulnerability. In the wild even minor and temporary disabilities can be fatal to the unlucky individual.

Amid the abundance and wide variety of prey species available in certain African ranges, the above probably applies also to the hunting of the lion and the hyena. Schaller observed different rates of success relative to ungulates hunted by the lion. "Once a lion has launched itself into the final rush, its chance of catching a reedbuck or topi is 13 to 14%, a gazelle, zebra, or wildebeest 26 to 32%, and a warthog 47% . . . figures which speak well for the vigilance, agility, and speed of these species." He noted that, irrespective of species, hyenas kill about 32 to 44 percent of the individual animals they attempt to capture. Both lions and hyenas greatly improve their score when members of the pride or pack hunt cooperatively. That ecological conditions sometimes greatly affect prey vulnerability is evident in a record cited by Kruuk from his studies of the spotted hyena. He found that the "wastage" (remains left for other scavengers) of kills by the hyena ordinarily is small. However,

> Once in November 1966, a group of hyenas killed, in one night, more than 110 Thomson's gazelle and maimed many others, eating small parts of only a few victims (13 in a sample of 59). The carnage had taken place on a very dark night with heavy rain, and the hyenas had been able to kill at leisure . . . Here, two points are important in discussing the effects of predators on their prey populations: first, a slaughter like this is a very rare occurrence and, second, because the hyenas ate little of what they had killed, there is probably little or no relation between numbers killed and numbers of predators present.

It is evident also that in such a situation the physical condition of individual animals was a minor matter.

But ordinarily it did matter, as Kruuk pointed out. With a dart gun and tranquilizer drugs he captured and marked antelope and then released them after they appeared to have recovered fully. He found that hyenas were able to spot and run down these marked animals. One wildebeest that had been hit by a dart was taken by the hyenas, and it was discovered that the dart had not discharged the drug! From field observations, it was evident that hyenas would attack, and they probably searched for, any inferior animal that happened to appear in a herd, but Kruuk remarked, "most likely the hyenas' criteria are far more sensitive than mine, because I can rarely see anything different in the behavior of an animal that has been selected by the hyenas." Among three principal prey species he found that different categories were taken. Very young individuals were obviously vulnerable, and also very old wildebeest and probably gazelle. Among zebras it appeared that age was not particularly important once an individual had achieved adulthood. In the aggressive stallion the bands of zebra had a kind of protection the other species did not.

In certain areas of Africa not yet overrun by human proliferation we have a surviving remnant of that great flourishing of mammals that took place on nearly every continent during the Pleistocene. Between the great abundance and variety of prey and the competing predators we probably are witnessing effects of the density factor in greater degree than where relatively few species are present. Selection certainly is there, but the game of chance also is often favorable to the wild hunter.

With fewer species available to sustain him, the wolf must be considered an obligatory specialist keenly appraising the individual animal according to cues routinely revealed to the expert. Whether, in a different set, the wolf is more skilled at this than the hyena we could only guess, but skilled he is. The acuity required for survival is a matter of necessity and practice — and also degree. For with my own crude senses I have perceived clearly the forlorn expression of a limping moose and the utter misery on the face of a fox with an injured leg. One might learn more than all mankind has learned of this and other mysteries if he could see through the eyes of a wolf for a single day.

CHAPTER SIX

Wolves and Their Hunting: Killing Methods

PROBABLY MOST OF US are at least vaguely aware of a universal reality in the natural world. All living things are destined to die and be recycled as part of the flow of energy through the life community. Which is to say, a creature must feed, and sooner or later it will be fed upon.

According to the long-standing ecological plan, some species subsist on plants. They reproduce abundantly, their lives are brief, and the yearly turnover of numbers is massive. The bulk of these typical prey animals have their careers cut short. At an early age they contribute their substance to the direct support of the meat eaters. Eventually it happens to the more durable and older individuals, the ones that live long enough to breed, and to the predators themselves. Anything not eaten at once will disintegrate into the soil by way of various decay and scavenging organisms.

For any creature it is a concern of living to put off the recycling as long as possible. But biological accounting is not to be denied, and even the carnivores are subject to an early weeding out. Thus the young of every kind — chicks, nestlings, fawns, calves, cubs, pups — are a part of the communal food supply and will be drastically reduced in the natal year. Those that survive to produce another generation are the select few that continue the slow evolutionary course toward a more competitive level of adaptation.

The dominant feature of this regime is that nature gives short shrift to the weak, the incompetent, the disadvantaged, and the unlucky. The effect of the plan is plainly evident: It preserves the species. For himself, of course, man has disavowed the system, and we are waiting to see how it works when the individual comes first.

As previously noted, many of the hoofed prey animals associate in bands and herds that can be sorted conveniently for vulnerable individuals by the carnivores they nourish. Typical of the more social ungulates are antelope, bison, caribou, and horses. Likewise, some of the carnivores are organized for action, notably the wolf, wild dog, and hyena. In lesser degree this is true of the lion and cheetah, and we probably should add the coyote. Typical solitary hunters are the leopard, mountain lion (also called panther, painter, puma, and cougar), and lynx.

If any predator could kill its prey at will, one might suppose it would be the tiger. It seeks its victims by stalking in protective cover and kills by rush from a critical distance. The great cat is magnificently specialized for its role, but Schaller indicated that in its hunting the usual predator-prey relationships prevail to some extent:

> . . . although tiger predation undoubtedly culls the less fit animals from the population, leaving the vigorous ones to propagate, a considerable percentage of healthy, prime animals are also killed, having increased their vulnerability by frequenting certain habitats, by being pregnant, and perhaps by other means.

If, under normally prevailing conditions, there is a species of predator that feeds *ad libitum* on totally helpless prey, its relationships have yet to be demonstrated. (Obviously we are not considering plankton feeders like the blue whale, which scoops in krill like a front-end loader. The huge size of this creature, the largest animal that ever lived on earth, may be an adaptation for energy conservation and defense.) If it does exist, we may be sure that population control is effectively built into its social system. On this whole subject it might be informative to know more about the killer whale.

Except for seasonal "family" groups and loose associations of several adults, the moose is solitary. The cow protects her calf, but otherwise there is no mutual defense. Adult moose outweigh adult wolves by roughly 10 to 1. Over geological time, hunting methods of the wolf have adjusted to the protective strategies of its various prey species, which also were undergoing change. So wolves can kill moose of certain categories. The rules are not totally hard and fast, because wolves vary, moose vary, and situations vary. In the hunting methods of Isle Royale wolves we have seen much evidence of skill and competence and possibly an occasional example of malpractice.

It is a fair guess that the pattern of behavior by which a moose avoids immediate recycling is more stereotyped than the battle plans through which the pack harvests its next meal. My logic is that the moose needs a largely inborn mode of reaction to attack because it gets only occasional emergency drills that call for defense. And like the parachute jumper, it must be right the first time. On the other hand, the wolves get regular practice at dealing with their large intractable prey, and they miss frequently.

The predator probably inherits a more general program of physical and behavioral capabilities. This genetic score must then be orchestrated through a regime of training that involves puppish play, example, experience, and quite possibly handed-down pack traditions. In dealing with moose the single wolf must be largely a scavenger, but the pack can kill. The senses, abilities, and techniques it brings to this job have been revealed in part during the years we have watched things happen.

Although the island wolf population has undergone a total turnover of individuals, and new packs under new leadership have been organized in the 18 years, methods of hunting and killing have remained the same. They also conform to observations made by workers in Ontario, Minnesota, Alaska, and elsewhere. Compared with the state of our knowledge 20 years ago, a great deal has been learned in North America about how the wolf detects and handles its prey.

All canids are keen of nose, and we have had demonstrations that scents borne on the wind play an important part in prey seeking by the wolf. As for actual scent tracking, we have seen wolves use it in following one another, but it is less important as a hunting technique in winter. We have no knowledge of the extent to which it is used in summer, when it probably could have greatest value.

Moose break many trails through the snow, and often these are followed by traveling wolves. This is a matter of convenience, since nose-to-the-snow tracking for long distances is not seen; there are easier and more direct ways to find a moose.

I had a clear example of the scenting capabilities of two wolves that were headed east along the shore of Amygdaloid Island on 6 February 1962. Don landed me in a cove at the east end of the island and took off again. The wolves were not yet visible when, wearing a white parka and with my camera and long lens on a gunstock mount, I buried myself in a snowdrift behind a large birch that had fallen into the edge of the water (now ice). I was in the line of travel, and chances for a picture seemed good. Unfortunately, a light breeze was blowing from me directly toward the wolves.

The two animals came into view on the snow-covered ice and approached at a trot. About a quarter-mile away, they slowed down and stopped, exhibiting signs of uneasiness. It appeared that they would come no farther, and Don landed west of them on Amygdaloid Channel, hoping to urge them along the shore. They would have none of this and turned back the way they had come, by-passing the plane and continuing on to the west. A few minutes later, from the air, we saw them curl up in the brush on the edge of Amygdaloid Island. I felt certain the wolves had not seen me, but they evidently detected my scent at a distance of about a quarter-mile.

Some of Dave's early observations indicated that the wolves could scent moose at a much greater distance. In his notes he described how, on the afternoon of 7 February 1960, the pack of 16 stopped their play, gathered together, became alert, and pointed upwind. They began traveling and searching for three quarters of a mile in that direction. They got within 250 yards of a cow and calf, then lost the scent and went a third of a mile into a swamp. Finally they turned toward the two moose as the scent evidently came to them. They chased this cow and calf for more than a mile and killed the young animal. There was only about a foot of snow at this time.

In his book on *The wolf,* Dave said, "Of the fifty-one hunts in which I could tell whether the wolves trailed the moose or scented them, forty-two involved direct scenting . . . Usually the wolves scented moose when within three hundred yards downwind of them, but once they detected a cow and twin calves about 1.5 miles away."[1]

Although wolves live in a world of scents, their eyesight is likewise good, and they become aware of any motion at a long distance. In his exhaustive work on the vertebrate eye, Gordon

Walls noted that the big reason for sight is that things move: "Indeed, if nothing on earth moved, there would never have been such things as eyes."

Anyone working with animals in the wild, for observation, still hunting, or photography, has long since discovered that most creatures have little or no perception of form, but they react quickly to motion. A move toward or away from the animal is less likely to be picked up than one in a lateral direction, for evident reasons. In many situations the best camouflage is to sit in front of the cover rather than to attempt hiding behind it. The stalker's necessary movements should be made with great deliberation — literally, an inch at a time — with full advantage taken of intervals when the observed animal's head is down or turned away, or when it is moving and its head passes *closely* behind a tree. The watcher can change position quickly in the instant when its vision is blocked. The faster an animal is passing through vegetation the less likely it is to see motion at a distance. Anyone can easily check this; his best chance of seeing something is to sit immobile until his eye picks up a movement. When a breeze is fluttering the leaves, slight movements are most likely to be undetected, as are sounds.

Don and I had a chance to check the visual acuity of wolves when we were circling a pack of eight on solidly frozen Grace Harbor less than 4 miles west of our camp. The wolves were lying around sleeping on the ice. That morning they had left a kill on Johns Island 1.5 miles to the west.

After the pack abandoned the kill, a single wolf arrived and rummaged among the bones. As we watched from the plane, we could see this wolf leave Johns Island and follow the trail of the pack over the ice toward the opening between Washington and Booth islands. It was evident that the loner would come into view at a point about three quarters of a mile from the sleeping pack. When it did, the animals on the ice arose at once and started toward it at a rapid trot. The single wolf immediately began running south and west toward Cumberland Point. The pursuit was only half-hearted, and when the loner was a mile ahead the pack turned back into Grace Harbor.

When wolves come through Washington Harbor below our camp, they are likely to travel along the north shore and stop at any bait station (moose remains) we have established there. Often it has been evident that they were aware of someone moving around the ranger station half a mile away on our side of the harbor. In going from the bunkhouse to the ranger station when wolves are on the ice, we normally take the back trail through the woods well out of sight. If they see us on the waterfront, they may get restive and trek off to the west behind Beaver Island.

Wild creatures get accustomed to almost anything that is there all the time and offers no threat to them. They do not recognize what is "natural or unnatural" in their habitat. Thus in winter the moose and wolves wander among the buildings at Rock Harbor Lodge or Mott Island headquarters without concern that these structures are anything different from rocks and trees.

On the other hand, our wolves must often encounter both the sight and smell of hikers during the summer. For one reason or another, the appearance of a human sends them quickly into the brush. There are rare exceptions to this, and why it should be so at all is a good question. So far as we know, no one on the ground on Isle Royale has ever been any threat to the animals. I have suspected that wolves are intelligent enough to recognize the human form at close range, but one can never eliminate the possibility that scent or motion was involved.

As an example, on 12 July 1962 Larry Roop, Mike Long, and Phil Shelton were north of Ojibway tower on the way to Lake Eva. As they started around a pond where a moose was eating yellow water lilies, they were surprised to see a wolf coming toward them on a trail. The wind was from the wolf, and it did not see the party until it was about 20 yards away. Then it whirled and ran back up the trail, stopping once for an instant to look back.

In June the following year, Shelton and Roop watched a wolf walk and trot along an open ridge northwest of Lake Ojibway. They had it in view for about two minutes, and the animal was 65 to 85 yards from them at the closest points. They kept quiet, and it did not discover them.

Wolves surely can see as clearly as dogs, and dogs seem to do fairly well at recognizing shapes. Walls cited research in which 14 police dogs (as dogs go, not far removed from their wolf ancestry) were tested for visual performance. It was found that the "best" individuals could descry moving objects at distances up to nearly 1000 yards, but they also recognized stationary objects at about 650 yards.

As we see them from the air, the dominant pair — here in center, male with tail up — are likely to be in front somewhere (photo by Rolf O. Peterson).

It should be realized that the eyesight of the wolf, and for that matter, of all its mammalian prey, is not aided by color vision. Although fishes, reptiles, and birds usually have excellent color perception, it is only among the primates (mainly the monkeys, apes, and man) that this ability is well developed in mammals. Color vision is an adaptation of animals usually abroad in daylight. As birds well demonstrate, they commonly react to their own gay hues as well as those of fruits, insects, and other things they feed upon. The environment of wolves and most other mammals is seen in shades of gray, largely distinguished by varying degrees of brightness. The majority of mammals are nocturnal, and even man does not see color when he is abroad at night. To the vertebrate eye, colors lose their value in dim light. But even in sunny midday the wolf sees a hiker wearing a red pack in about the same tones as one wearing a poncho in the rain, the latter a not-uncommon occurrence.

When traveling on an icy shoreline or the firm snow of a frozen lake, the wolf pack commonly is scattered out in a long line, with individuals or groups of two or three taking slightly different routes. A worker on foot can count the tracks easily under these conditions. As we follow them from the air, the dominant pair usually is up front somewhere, and the male may be out ahead, evidently choosing the route to be taken. Often they cross the base of a forested peninsula in getting from one bay to another, and here they close ranks and go single file through the soft snow.

In the first winter of the work, Dave and Don had an opportunity to check the traveling speed of the large pack. On 21 February at 4:55 P.M., the pack started north from Houghton Point across Siskiwit Bay. They stopped for about 5 minutes of sexual play at one point, but otherwise they moved at a steady trot to Crow Point, arriving at 6 P.M. The distance is 5 miles, which was covered in an hour. Other checks have shown that this is the rate at which wolves commonly travel when the footing is good. Rolf and Don timed a loner on the ice around Washington Island on 13 February 1974. It went 3.9 miles in 59 minutes. Later they saw a single wolf running all-out over the ice, and it covered 1.9 miles in 7 minutes, a rate of 16.3 mph.

A moose upwind of the pack's line of travel is likely to be scented by the first wolves, who turn and point on rigid alert, tails out behind and noses raised in the direction of the message-bearing wind. As the lagging wolves catch up, there may be a tail-wagging ceremony around the dominant pair, after which the inspired hunters bound away toward their intended prey. On a fresh scent there sometimes is a scattering of groups or individuals heading in the same general direction. More often they labor ahead, nose to tail, on a trail broken by a front-running wolf. This leader may or may not be recognizable as the dominant male.

Frequently wolves approaching from downwind will be discovered by the moose when only a short distance away. Although there is something different in every encounter, we may categorize the moose generally in two classes: those who stand and those who run away. There can be vulnerable or healthy individuals in either class, although the moose that backs up against the thick spruces, lays back its ears, lowers its head, and, literally, invites the wolves to do their worst — to the point of charging and kicking out with its hoofs — usually will be abandoned by the pack in short order.

The other category of moose that do not flee comprises those that cannot for physical reasons. Specific debilities will be discussed later, but the effects of a few are evident. A moose disabled by arthritis or with its lungs clogged by cysts of the hydatid tapeworm is doomed regardless of what it does, for the wolves easily detect its plight.

When a moose is strong and able to run away through deep, soft snow, it will not be followed very far. Struggling through such snow is an exhausting process, and the wolves soon give up. Under these conditions they confine much of their travel and activity to lake ice. They dig up old kills frequently, and at these locations they may not pursue loners or foxes with any real determination. It just seems to be too much work.

The situation described was well exemplified after mid-March in 1963, when two snowfalls of 6 inches each brought the total depth to more than 3 feet. It was definitely hard going. Shelton said his snowshoes were sinking 10 inches, which means you avoid any long trips. Moose were well concentrated in conifer cover in the lowlands, and this was especially evident along the southwest shore. At that time we were just beginning to realize the implications of snow depth.

We saw the same conditions two years later in early February, when there was more than 3 feet of snow on the ground, soft and fluffy. The big pack, now numbering 18, had finished a kill and were hunting as we circled over Grace Harbor watching them. Don and I saw them briefly chase two moose, both of which simply ran away and left them.

The pack went on to work the south shore for several days. Moose were again plentiful in the area, and the wolves were seen to pass by numerous animals that were either undetected or located by scent or sight. Several that were chased by a few of the more persevering wolves easily outdistanced them and did not seem particularly alarmed. It often appears that single wolves and a moose have no common problem. Pete and Don, who made most of the above observations, saw a wolf pass within 10 to 15 feet of a moose, and there was no visible reaction on the part of either.

In a relationship that is almost infinitely variable, it would not be realistic to expect consistent behavior. Yet some things are difficult to explain. At about noon on the first of February 1974, Rolf and Don saw a moose on Eagle Nest Island, an islet near shore in the Malone Bay group. The east pack was coming downwind from the west. The moose detected them and then moved onto the ice toward the Malone Bay shore. A couple of wolves saw the moose and stood looking for a moment. The rest of the pack bolted and ran around the end of the island *away* from the moose!

On 15 March 1963 Phil and Don watched two wolves approach a cow and calf, which were feeding on a low ridge north of Stickleback Lake. As the wolves came near, the lead animal rushed at the cow, who simply stood looking at it. The wolf retreated, and the calf kept on browsing a few yards away. While the wolves stood watching, the two moose walked leisurely away.

It often happens that after a nonviolent encounter a moose will continue placidly browsing, or a bedded individual will not move and will simply go on ruminating. On other occasions there will be definite signs of lasting alarm, like the moose that was chased for about 100 yards on the south side of Siskiwit Lake in February 1974. Rolf said it did not stop until it had gone 2 miles to the west end of the lake.

Where the wolves maintain long pursuit of a moose it is nearly always under conditions where the wolves have good footing. In 1960 there were only 12 to 16 inches of snow on the ground. Near Wood Lake the pack of 16 scented a cow and two calves upwind at about three quarters of a mile. The pack headed toward them, and some soon caught up. Dave and Don said that a wolf ran on each side of the three moose, with which they easily kept up in the open. But through heavy cover, snowdrifts, and blowdown they lost ground. On a favorable stretch the pack overhauled the moose and killed one of the calves in a clump of cedars. In his later studies in Minnesota Dave found that deer were often able to run away from the wolves in 15 inches of snow (173).

The attack on an adult moose is nearly always from the rear, with hard driving bites at the muscles of the upper leg. If a leg is injured, the moose can neither kick nor stand on it effectively, and further attacks are easier. They may be directed at the other hind leg, the flanks, and the anal region. With half a dozen or more wolves hanging on, the victim goes down. Sometimes a moose will be seized by the nose, and Dave reported seeing a large wolf swung around through the air while maintaining a nose grip. After a moose is disabled or down, a throat attack often helps to bring the affair to a finish.

Usually there is some evidence of leadership in the initial onslaught, evidently by a few old hands at the business, who close in from opposite sides. These will stay with a running moose, darting in for a gouging slash of the canines at vulnerable parts. An injured moose will turn at bay and back into heavy cover, especially thick conifers, when possible. It will be harried by the wolves, not allowed to lie down or browse, stiffening and weakening from loss of blood. Some of the pack eat bloody snow, curl up to rest, or just idle about, now and then threatening the moose, evidently to test its condition. Often enough, this is the situation when we must leave for the day, and the kill is made during the night.

When a moose is thrown, the entire pack may pile in and help complete the job. Shelton and Murray witnessed the end of such an episode on 24 February 1963. In the morning the wolves were lying around in the Siskiwit Swamp within a couple of hundred yards of a moose that, from the signs, had fought them off. Later in the day the wolves began to move, but Phil became sick with the close circling and they had to leave for a stop on a lake. When they returned, the wolves had the moose stretched out on the snow. One wolf had it by the nose and several others by the rump. Its hind legs appeared to be out of action, and it could not get to its feet. The struggle went on for about 10 minutes before the moose was dead.

That wolves sometimes inflict severe injury while a moose is still on its feet was indicated in 1963 when Don and I investigated the kill of an old cow near the shore of Wright Island. The carcass was not entirely cleaned up and lay beside a cedar log (autopsy 171). The surrounding brush had been sprayed with (arterial?) blood that froze into a bright red coating.

As mentioned in Chapter 4, when the pack feeds on a carcass, they shear off gobs of the longest hair, which form a mat clearly marking the kill spot. The abdomen is opened and the viscera pulled out. Our autopsies of collected moose show that by age 2.5 to 3.5 years every moose has begun its lifelong accumulation of hydatid tapeworm cysts. Such a young animal may have up to half a dozen small cysts in the lungs (the usual site) and occasionally a cyst or two in the liver. A wolf that eats the lung of an old heavily parasitized moose undoubtedly takes in thousands of the living larvae, insuring that it will maintain a maximum infection of the minute adult tapeworms that line its intestine like velvet. These do not seem to affect the health of a vigorous animal. There is little doubt that all old moose and adult wolves on the island are infected with *Echinococcus granulosus*.

Wolves have no taste for the intestinal contents of a moose. The gut is allowed to lie around and freeze, after which the meaty parts are eaten. After the internal organs, the fleshy carcass is vigorously fed upon, and a few hours after the kill the bulk of the pack will move out onto the ice or to a hillside where they curl up to sleep it off. This is the time when one or more subordinates, who have been waiting patiently at a distance, gather to have their turn at the banquet. When it appears expedient, foxes, ravens, and whiskeyjacks join the festivities.

Ordinarily a pack of 15 to 18 wolves will use up an adult moose in about three days. As we snowshoe around the area of a kill and try to piece together the story, signs may tell with some accuracy what day it was when certain tracks were made. On the first day of feeding the fecal material of the wolves is predominantly black and semi-liquid, often melted into the snow along the trails that are quickly established around a carcass. By the second day more solid parts, including skin and hair, are being eaten, and droppings are of the more typical form that we find on trails. However, most of these will have been picked apart by the ravens. At this stage the remains are being dismembered and dragged around or carried off for private chewing.

Three days after the original countdown, it is evident that conditions have changed. Then, and when scavenger feeding occurs thereafter, wolf droppings contain a high proportion of calcium and bone fragments. By that time all soft parts have been consumed, except for the skin on the lowest joint of the legs. This adheres to the bone and only a hungry loner is likely to peel it off. Of course it is the smaller and softer bones that are eaten or trimmed, including ribs, nose and palatine structures, and spinal processes. Wolves cannot crack the large leg bones of a fully mature moose, but they can eat nearly all of a calf, including the ends of the humerus and femur, as well as most of the skull.

Tooth rows are left unbroken. It is usual for the jaw to be meticulously cleaned and left somewhere around the kill site, and this is a specimen we always collect. The consistency with which one finds the jaw, even when most other bones have been scattered, has been striking, and the teeth are the most durable

A 13-year-old cow killed by four wolves in February 1967. The lower half was frozen into the snow and ice; it may not have been totally utilized until the spring thaw.

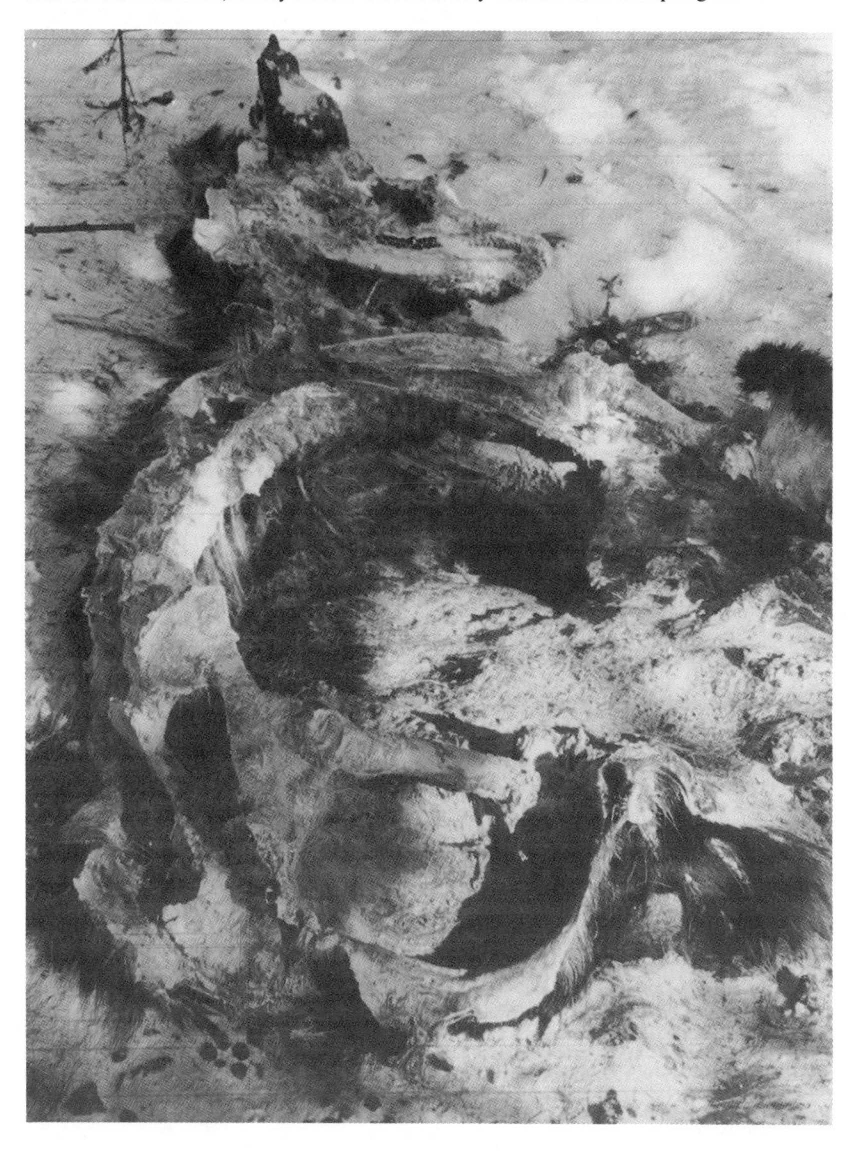

part of the carcass. I have mused that it is easy to understand such a place name as Moose Jaw (in Saskatchewan). Among paleontological specimens the jaw and teeth are the most frequently found and the most useful in studies of extinct animals; this applies to hominids as much as to other groups. It is now possible to age large collections of the teeth of extinct horses and to determine the population structure of animals inhabiting North America thousands of years ago.

Usually, on a day when nothing remains but scattered bones, hair, intestinal contents, wolf droppings, and a pervasive smell, there will be a greeting ceremony and a show of excitement in the pack. The dominant male or female starts away, and the others trail along behind. They are off on their travels, following lakes, streams, shorelines, and ridges where the footing is the best currently available. They may kill another moose within hours, or they may go as many as four or five days without food. It may be 5, 20, or 40 miles to the next vulnerable moose, for in these irregular and unpredictable excursions they may be assumed to be hunting.

It has been shown experimentally that a wolf can eat 20 pounds at one feeding if it has been without food for an extended period.[2] On the average the Isle Royale wolves probably consume 5 to 10 pounds of moose per day in winter. In confinement 2 or 3 pounds is a sufficient ration, and Minnesota wolves that feed on deer probably get by with about 4 pounds per day (174).

This, like many other biological questions, can be complicated with a lot of provisional details. Loners are unlikely to feed as well as pack members. Within the pack high-ranking animals eat first and undoubtedly get the choice parts of a carcass. When we say 10 pounds, how much roughage does that mean? For practical purposes, food is anything the wolf swallows.

It is evident that attacking a fairly vigorous moose involves some hazard, and the wolves do not always escape without penalty. On one of the first flights of 1969, Mike Wolfe and Don saw the carcass of a wolf in a tracked-up area at Chippewa Harbor. They retrieved the still-unfrozen animal, which, as shown by the tracks, had been caught and killed by the main pack. The dead wolf was a mature male, and subsequent autopsy (aut. 436) in the veterinary science laboratory at the campus showed that it had healed rib fractures and scarring of the pleurum (punctures of the lung cavity) on both sides of the back.

In August 1970, near the Lake Richie outlet, John Vanada retrieved the bones of a young wolf (aut. 532) that appeared to have died the previous winter. The right scapula was splintered and adjacent ribs damaged, evidently by a blow. The meager evidence does not tell us what kind of blow, but the kick of a moose must be suspected in such a case. When moose-wolf relationships reach the point of ultimate decision, hoofs are the moose's defensive weapon that can mean survival for yet a while. The experienced old wolves, at least, are aware that the moose can kick or strike with devastating effect.

Mech mentioned in his thesis that he had seen two wolves beaten into the snow, one by a calf, and in both cases the wolf appeared to come out with no injury. On another occasion, early February 1964, Jordan and Murray were flying over Siskiwit Swamp when they saw the large pack (numbering 20) go after a cow and calf. Only two of the wolves actually closed in, and the cow trampled one of them into the loose snow. The wolf got up and immediately returned to the attack.

Some years ago, when Alaska still had a general wolf bounty and hunting from aircraft was legal, Robert A. Rausch made extensive collections of carcasses. He examined more than a thousand skeletons, the only major job of its kind. He found numerous healed fractures of ribs and other bones, including those of the skull. He surmised that such injuries might be the result of preying on moose and noted that a wolf sustaining a square blow in the head probably would not live to provide any evidence. In British Columbia, Stanwell-Fletcher described a dead wolf with many broken bones and tracks that showed it had been trampled by a moose. It would be of great interest to know how many foolish young wolves are weeded out of the population when they are inspired by a fleeing four-legged banquet to dash in for glory and a good meal. Dave Mech has proposed — with good logic, I think — that it is the running-away of the quarry that stimulates the chase. In a measure this must be true of many predators. When the prey moves out, there must be immediate action or there will not be another chance. Much of this is speculation, of course.

We have little real information on why, in a certain winter in a certain pack, five pups of a litter survive until February, while in another winter there is only one or none at all, even though we saw copulation in the previous breeding season. That moose have been kicking wolves for a long time is implied in an observation by district ranger Frank Deckert and a companion on Washington Harbor in the summer of 1972. They were in a canoe and had paddled close to a cow that was grubbing water plants in shallows near the shore. As they sat quietly watching her, one of the men picked up a small pebble from the bottom of the canoe and flipped it at the cow. It struck her in the hind leg, and that hoof instantly lashed out in a sidewise kick.

This is the kind of defense that would be most effective if a moose were running away from the wolves and the attackers were coming up from the rear, as they usually are.[3] Such a reaction to the touch of a pebble suggests that, at least in part, the hind-foot strike has become genetically programmed into the moose nervous system as an automatic function. Since predation is a fact of life for all hoofed creatures, the kick mechanism may be very old indeed.

Many people, not too well informed, who like to idealize the beauties of nature, commonly visualize the predator as making a quick clean kill and then eating the prey. Often this does occur, as in the case of the lioness who breaks the neck of the antelope or the cheetah who throttles its quarry by gripping the trachea in jaws too weak to do much else. In contrast the cape hunting dog literally tears its victim apart, although there usually are enough of them so that the process does not last long.

There would be some realism in saying that about all most meat eaters require of their prey is that it hold still. I have been impressed that prey animals caught by predators often seem to stop struggling and go into something resembling a cataleptic state, calmly submitting to being eaten until death ensues.

An example of this, and not the only one we have seen, occurred on 12 March 1961. There had been no flying for several days, and after this happens it always is in order to find the wolves and backtrack them to learn what they have been doing. Such strategy has led to the finding of many kills and sometimes a wounded moose that has been abandoned and left "in storage."

Dave and Don found eight members of the large pack feeding at a fresh kill on the south shore near Halloran Lake. The backtrail led northeast to Francis Point and then north across Siskiwit Bay. Beyond Hay Bay they saw a single wolf also working its way northward in the trail of the pack. The plane continued backtracking to a point 1.5 miles southeast of Lake Desor. There was a moose, standing bloody and alone, in the midst of a great complex of wolf tracks. The big pack had injured this animal and gone on. At 12:55 P.M. the lone wolf came to the moose and lay down about 20 feet away. Twice in the afternoon Dave and Don flew back to check. When the moose lay down, the wolf would threaten and make it stand again.

On a flight late in the afternoon, Don and chief ranger Ben Zerbey saw the moose down and the wolf feeding on its rump. The moose lay quietly with its head up watching the wolf. On arriving at Windigo, Don took Dave back, but by the time they arrived the moose was dead.

Dave snowshoed to this kill from Lake Desor the next day and performed an autopsy (aut. 91). He found it was a bull, later determined to be 11 years old. Its bone marrow had a normal fat content. The left hind leg was nearly bare of hair and loaded with winter ticks. Most of the viscera were gone, but the lungs remained and they contained at least 35 cysts of the hydatid tapeworm and a "heavy yellow rubbery mucus congesting the bronchi and bronchioles." The upper thighs were heavily damaged, and the anal area had been penetrated into the body cavity. In days following, the wolf returned repeatedly to feed on this carcass, as did at least one fox.

This case recalls a record published by Cahalane, who witnessed the killing of a deer by two coyotes, who also reduced the animal to submission and then proceeded to feed while it watched with head up.[4]

Although wolves can be almost uncannily sensitive to the vulnerability of their prey, sometimes they take on a moose that will not go down but which is cut up in the attack. Here we must remember that infectious pathogens so well known to our own kind are a part of the community of life and that they play their part in the recycling of nutrients in the ecosystem. Even a slight wound may result in a systemic infection. A single tooth penetrating to the coelomic cavity by way of that common point of attack, the thin-walled anal area, could bring on peritonitis. Such a moose would be an easy victim for the pack on its next trip through, or even for that patient social subordinate who trails the pack to take advantage of such happenings.

On several occasions we have recorded that an injured moose was watched by the pack for three or four days before they moved on. Then the moose may have disappeared, not to be seen again. Possibly it recovered and will survive for the present. Or it may have wandered off to die in thick cover, not to be fed upon by wolves or other scavengers until after the snow is gone in spring. Getting anything like the full story on a moose mortality requires a rare degree of good timing, good flying weather, and a favorable location for landing during the winter season. It does not happen regularly.

The manner in which moose predation may stretch out over days and weeks was illustrated in 1962. Phil, Don, and I had been keeping track of a "duo" in the area of Amygdaloid Island and Duncan Bay at the east end. These observations began 14 February, and on the 19th Phil and Don found the two wolves lying near an immobilized moose on the north side of Duncan Bay. No blood was evident from the air. During the next two days the situation was essentially unchanged. Under these circumstances the wolves simply do guard duty, the moose has no opportunity to feed, and usually it is kept standing.

On the 23d both wolves and moose were gone. The wolf track led 5 miles east and was lost in drifting snow. On a return to the site of the siege, as Phil expressed it, they found three ravens sitting about, and for the first time blood was seen in a bed where the moose had rested. Phil and Don landed and went in to inspect the area. As they headed up the bank, the moose got up out of a stand of trees and ran stiffly over a low ridge. It was obviously wounded on the rump and flanks. Don and Phil went back to the plane at once in order to leave no more human scent. From the air they saw the moose move about 200 yards and bed down again. The next day Don and I could see that it had gone another 20 yards, and two bloody beds were evident.

On 25 February, 11 days after the initial attack, the moose was obviously dead. On the 28th the carcass was getting some attention from foxes and ravens. A day later Phil and Don visited it again and "autopsied" (aut. 146) the frozen remains with a pulaski (fire ax). Our eventual aging showed this to be a 12-year-old bull. Phil found a heavy load of hydatid tapeworm cysts in the lungs, some minor jaw necrosis, and normal marrow fat. By 10 March two wolves were at the carcass and evidently had fed on it for several days. They were not seen there again during the winter, although half a dozen ravens were present on 13 March. In late May the bones were scattered, and the head could not be found at all.[5]

What actually killed the moose? Undoubtedly loss of blood — not a great deal in the early stages — shock, and 11 days with little or no food had their effect. Whether infection was involved we cannot say, but it is a probability. Certainly, more than 50 large cysts in the lungs were a burden to this old animal and may have been the reason why it could not run away through 2 feet of snow. This is an example of how time was on the side of the two wolves, who were unable to bring the moose down at first attack.

During the last stages of the above drama, we had a somewhat similar episode in progress at the west end of the island. On 25 February Phil and Don found the large pack, numbering 17 during this period, around an injured moose west of Hugginin Cove on the north shore. There was a large bloody spot behind the moose, which was facing toward a group of 12 wolves that were standing and lying on the ice and snow of the shoreline. Within 25 to 30 yards of the embattled animal four more wolves were curled up in the snow. Nothing much had changed when Don and I flew over before sundown. The area was tracked up, and the moose was looking around, definitely on the alert.

The next afternoon Don and I checked them again, and the moose was in the same position, but the wolves were along the shore 200 yards northeast. Later in the day when Phil was flying, 15 wolves were on the ice near the moose. However, they did not stay, for the next morning they were tracked 6 miles to the northeast, where they made a kill (not found at this time) near the shoreline west of Lake Desor. All 17 wolves were present, and they remained near the site through the following day.

But a day later, 1 March, they were back at the scene of action near Hugginin Cove. Phil said that 4 wolves were on the ice and 7 more were running up the hill behind the flat where the moose had been originally. Shortly after, 4 wolves were seen confronting the moose on the hill, and they were quickly joined by the 7. For about five minutes wolves dashed at the moose from all directions, but it faced them off with no evident contact. Then the pack moved away, and later in the morning they were resting about 100 yards from the moose, which still stood looking toward them. By late afternoon the moose had lain down, and only two wolves could be seen in the area. The rest of the pack had gone south.

For a day the wolves could not be found, but on 3 March they were on the ice three quarters of a mile west of where the injured moose still lay. It was in a bed on the hill 75 yards from where it had been attacked. The 17 wolves went on west to the islands off Washington Harbor. Superintendent Schmidt and I had left Isle Royale on 1 March, and Jack Raftery had come out to the camp. On 4 March he and Phil attempted to find the injured moose for a thorough examination.

They found several moose in the area of the skirmish and a maze of tracks. No obviously injured moose could be located, so the affair ended unsuccessfully, as did many another. It could be that the wolves went back much later and, with keener perception than ours, collected their quarry. For that matter,

For most of the winter, beavers are unavailable. Dam is at left, lodge in the middle.

the moose may be represented by a number in our autopsy file. It may be one of those sets of wolf-gnawed bones picked up by summer assistants with the note that it is an apparent wolf kill. If located within a few months the hair mat would be found where the carcass was fed upon and dismembered.

However it was, there is a hiatus in the record. We shall never really know how it ended.

We tend to assume that the state of health of a moose is the main determinant of the animal's fate when checked by the wolves. However, other factors may be involved. One of these is how long it has been since the pack's last kill.

This is suggested in some observations made in 1964, at a time when the large pack numbered 20. On the evening of 7 February they killed a cow on the north edge of Siskiwit Swamp. On the third day thereafter they finished feeding and left the site about 3 P.M. Half an hour later they were traveling when Don detoured to fly over them in bringing Dick Igo back to Windigo from a trip to Mott Island. He picked up Jordan, and they circled above the pack while it checked 10 different moose that had concentrated within an area of a few acres. None of these animals got much attention, and Pete remarked that the wolves were not especially motivated, which is to say, hungry. Of course they might well be stirred to action by a particularly weak moose.

It is not unusual for the pack to go from two to four days between kills, and after such a fast they seem to take a more responsible interest in the moose they encounter. Especially during the February mating season, their vigor fairly overflows, and woe betide the fox that bounds away over the snow before them. At other times, around a kill after the initial feast, a fox, or even a snowshoe hare, may move around the sleeping wolves with impunity. It is a common observation that prey animals know when the enemy is full fed and do not bother to take evasive action. This has been observed frequently in Africa.

In case the distinction is not clear, wolves and other predators kill *vulnerable* individuals in a prey population, but this does not always mean weak or diseased. Whether a moose can be attacked successfully depends in part on how it takes advantage of deep snow, thick cover, or other habitat conditions. From various reports and a few observations, it appears to be fairly common for them to seek refuge in water. Obviously, the long-legged moose can stand in water at a depth where the wolves must swim. No wolf would be interested in that situation.

On one of my first visits to the island I talked with two campers who had witnessed such an occurrence at the head of Tobin Harbor. They saw a cow and calf wade out into the water and stand there looking into the woods. Three wolves appeared, paced up and down the shore a few times, and then disappeared. In September 1967, Dave Kangas, then in charge of the trail crew, saw a calf escape a wolf by running into the water at the west end of Siskiwit Lake. Dave thought the calf might have been wounded slightly, since it shook itself vigorously. The cow was nearby. A resident of Tobin Harbor told Mech that she had seen a moose prancing and snorting about 20 feet from shore in Gutt Bay, evidently reacting to a wolf that was on the shore and which then left the scene. In June 1974, two young women who worked in the park described to Rolf and Carolyn Peterson how they had seen a cow and calf escape a wolf by retreating to water on the south side of Siskiwit Lake.

An unusual attack involving a water retreat in winter was witnessed by Jordan and Murray on 19 February 1964 near Long Point on the southwest shore. In late afternoon they saw the large pack (then numbering 20 at full complement) flush a cow and calf from their beds. The calf went ahead, and the wolves took the cow to the ground in a clump of balsams. The calf stood for a moment about 30 yards away, then ran off down the shore. In less than half a minute the cow struggled out of the thicket with several wolves hanging on. She immediately plunged into the lake. This wave-pounded strand is rocky, with a shelving gravelly bottom. The wolves disengaged, and the cow swam out for more than 100 yards. Then she came back in to standing depth about 150 yards from her point of entry.

Fourteen wolves watched her as she waded into shallower water. Some were lying down and some were playing. Later the wolves began to wander away, perhaps following the trail of four or five that Pete thought might have gone after the calf. Nearly an hour and a quarter after the cow entered the water, she was standing in the shallows and a single wolf was watching her from shore. The plane had to leave at that point. The temperature was about 22°, and there was a slight offshore breeze.

The following day the cow was no longer at the site, and the pack was feeding on a calf kill half a mile west of Long Point. An adult was browsing 120 yards from the wolves. On the 23d, four days after the attack, the pack killed an 11-year-old cow 200 yards north of the calf remains.

Except for the immediate proximity of the lake, it appeared that the cow would have been killed at first contact, and it is unlikely the calf could have survived on its own. Circumstances suggested that, in the end, the wolves did get them both.

The great preponderance of our records indicate that the wolves seldom kill more than one animal at a time. When they have a carcass to feed upon, they pay little attention to other moose in the area. We have never actually seen a double kill,

Usually these bony parts, plus intestinal contents, are all that remains of a wolf-killed calf.

but it evidently does happen, and in the cases we observed it involved one or more calves.

On 18 February 1959, when Mech, Linn, and Murray were on the island, they saw the large pack come out of the mouth of Washington Creek and proceed westward along the north shore of the harbor. Later the pack was found feeding on a kill in the woods near the west shore of Thompson Island at the mouth of the harbor. The following day, as the plane circled above, Don spotted a second kill only about 50 yards from the first, and it too had been fed upon. Dave examined these kills on the ground on 21 February. They were a cow (age 11, with normal marrow; aut. 6) and a calf. The condition of the kills indicated they could have been made at the same time.

In 1969 a pack of three killed a cow on Lake Mason, northeast of Chippewa Harbor. This probably took place on 11 or 12 February, and 2 days later a calf that appeared to be wounded was seen in the vicinity. When Mike and Don examined remains of the 8-year-old cow (aut. 446) on 18 February, they found what was left of a calf also, again only 50 yards away.

A year later, about 9 February, the west pack killed a calf on the ice about 25 yards offshore in McGinty Cove, which is northwest of Washington Harbor. After a few days we saw a second kill on the shore at the base of a cliff. John Keeler snowshoed to the scene on 26 February and decided that the two calves could have been twins killed at the same time.

Peterson found excellent evidence of a double kill in 1972. Between the south shore and Lake Theresa (a small lake due south of Moskey Basin) he and Don recorded a kill (72-32) that had been made by the new east pack of 10 wolves. This was 5 March, and the remains could not be examined until Rolf and Carolyn searched out the site on 16 May. There they found the well-chewed bones of a cow and a calf, evidently killed at the same time and on the same spot. The cow was a 7-year-old (aut. 665) showing a necrosis of the jaw on both sides.[6]

The strategy of the wolves in attacking a cow and calf is to separate the animals and kill the calf. It is possible that a fleeing calf could be taken by two or three wolves, while the rest of the pack are killing a vulnerable cow. Also, if one of the two is killed, the other is likely to stay in the vicinity and be exposed to later predation.

In the Northwest Territories, during August, Banfield recorded the killing of two caribou calves by wolves in a single chase out of a band of about 50 cows and calves. One calf was largely eaten, and the tongue and throat were taken from the other. Ten days later a wolf returned and ate the second calf. Banfield remarked that "When Caribou are abundant, wolves often kill in excess of their immediate needs." He noted that surplus carcasses can be regarded as caches of food for future use, although they are likely to be consumed by scavengers such as the fox, wolverine, and bear.

On a limited range such as Isle Royale, the chief competitors of the fox for defunct moose are the wolves themselves. Meat keeps well during winter and early spring, but its age and condition are of little concern to the wolves. After the ice is gone and the interior waters warm somewhat, they will drag bloated and reeking carcasses from the water and devour them with the same dispatch that is accorded a fresh kill.

In winter we sometimes see centers of wolf and fox activity that suggest a kill, yet no moose has been chased and there is no blood in evidence. In these cases we have learned to suspect that an old kill is being excavated for further cleaning. On 9 February 1969 Don and I examined such a site on the north side of Tobin Harbor. It was an old bull (aut. 441) with red, gelatinous bone marrow (no fat). The bones were under 3 feet of snow and probably represented a kill of early winter.

A similar specimen, a 5-year-old cow, was retrieved by Jim Dietz from a deep crevasse in the rocks north of Lake Whittlesey on 23 February 1971. It was on the ground under 3 feet of packed snow, and the rear end had been dug out by a couple of wolves that we recognized as the Moskey Basin duo. A week before, we had seen that something there was attracting the local foxes. Jim found the head and front quarters intact. However the cow met her end, it probably happened in late fall, as judged by the odor. Sometimes we find single bones lying about in strange places in winter. The smell and the shrunken condition of the marrow may be the key to a mortality of the previous fall.

On this island where the wolf mainstay is such large and long-lived prey, the great carnivore must work hard for his meat. It is not surprising that he does not require that it be strictly fresh.

CHAPTER SEVEN

Summers in the Field

FOR TWO PERIODS OF THE YEAR we have had to fill in events by indirect evidence. Fall field work must terminate about the end of October, since that is when the park staff close down their operation and *Ranger III* makes the final crossing of the lake to Houghton. Usually the fishermen stay a bit longer; then they board *Voyageur* on its last trip around the island and go back to Grand Portage and on to their winter homes. In earlier times they commonly took their own boats to the mainland in November or December, according to weather.

Shelton took advantage of one of these late seasons in 1962 and continued his field work on beavers, especially profitable in the fall, through the month of November. He made extensive trips in our boat, *The Wolf*, and hiked widely over the eastern two thirds of the island, carrying live traps from pond to pond and getting notes on colonies now ready for the winter. If he had suffered a major mishap on any of these travels, we might still be wondering what became of him.

Phil accompanied the Johnsons and Edisens around the south shore and west end to Sivertsons on Washington Island on 6 December. There they waited out the weather and crossed to Grand Portage on 12 December. It was three below zero, and ice was forming on quiet waters. A light snow was on the ground, hares were turning white, and winter was close ahead. Isle Royale's creatures would have the place to themselves until we landed on Washington Harbor in late January.

The other blank in our firsthand records is the 5 or 6 weeks following closure of the winter camp in March. At that time, with harbors full of unstable ice, the only only access to the island would be by helicopter on balloons, and we never had the money to give it a try. In an average season ice goes out of the more exposed bays in late April and disappears from inland lakes in the first week of May.

As soon as there is open water at Mott Island headquarters, the Isle Royale Seaplane Service is on the job. Pilot Dale Chilson will be setting the Dornier float plane down with four or five of the park staff to begin opening up. When enough ice is out of Rock Harbor, *Ranger III* brings the maintenance crew, and seasonal operations get going in earnest.

Our research requires the earliest possible start in spring because there are several kinds of work that can be done only at that time. For example, one way of measuring winter population density in such animals as deer and moose is by means of a pellet count system. A statistically distributed plot sampling of winter habitats is set up according to (in our case) expected moose density. Then, as early in spring as possible, you count and remove the winter's deposition of pellet groups on each of the plots. Using some base figures on defecation rate, you convert to moose-days of occupancy since leaf fall in the various habitats. Pete Jordan set up such a system and he and Mike Wolfe, with our seasonal help, ran the plots for several years. It was a time-consuming job, subject to uncertainties, and in 1970 we decided that the important spring period had to be spent in other ways. However, there are good possibilities for a more intensive job of this kind in cross-checking the problem of moose counting on Isle Royale.[1]

The principal challenge in spring work of recent years has been to find and examine remains of moose that died during the winter. In 1970 Rolf Peterson began his work in a period of years characterized by increasing wolves and a steady build-up in the number of winter kills, including many that could not be visited during February and March because of bad weather and lack of time. These were spotted on a map and had to be found in cross-country trips before the new growth of ground cover came up to obscure them. Rolf has been meticulous in recording kill locations, and his woodsmanship in finding them — and even directing his assistants Joe Scheidler and Jim Woolington to them — has been phenomenal. A casual statement in his thesis recorded, "Of 141 carcasses located in winter, 1971–74, 136 were examined on the ground." Actually, the ones not found are likely to be calves whose bones were largely con-

Isle Royale's 174 miles of hiking trails have heavy visitor use from May to September.

sumed. Also, we lose a few kills on peripheral shelf ice remote from any landing place; these disappear at the breakup.

In the course of cross-country explorations, field parties find remnants of moose whose demise we knew nothing about, including skeletal parts from former years. If teeth are present for aging or other bones provide any precise information, such a specimen is given a number and card in the autopsy file. A rather unusual finding of this kind occurred when my wife, Dorothy, and I were canoeing on a June evening near our Caribou Island camp. As we glided slowly on the quiet water she had her head out over the gunwale watching the bottom intently. Suddenly she announced, "I see bones down there!" The incomplete skeleton was extensively decalcified by years in the water; but we recovered some teeth for Mike Wolfe to grind, and eventually we knew that an 11-year-old antlerless bull had died near the shore on the ice of Rock Harbor and probably had been fed upon by the wolves.

Another spring assignment is carried out before park visitors begin using the trails in mid-May: that of gathering and analyzing for food content the fall and spring accumulation of wolf scats, which otherwise would soon be trampled and destroyed. Scats of November and December have been well preserved under the snow. They tend to be bleached and usually are not difficult to tell from fresh spring deposits. Fox scats are recorded in the same manner. An early clearing of scats will make it possible to date, at least roughly, the use of trails by wolves later on.

About the second week of May, masses of silvery smelt run up Isle Royale streams to spawn. Then the fish-eating birds — gulls, mergansers, loons, and even goldeneyes (not to mention park employees) — will have a field day on Benson Creek at the Daisy Farm campground. Sucker runs occur at the same season. I saw an impressive run of them from McCargoe Cove up the outlet to Chickenbone Lake. The Indian Portage Trail follows this stream on the west side, and beneath a plank bridge the clear waters of a narrow tributary were crowded with the wavering bodies. Somewhere on a gravel bottom they would lay their eggs and then retreat for another season in the lake around the shores of the island. Lagler and Goldman said that in these cold waters the white sucker may take nine years to reach a length of a foot. The ones I saw were about that size.

As will appear from time to time, there are additional reasons for an inspection of Isle Royale habitats in the weeks immediately following the snow melt. Despite the fact that trails are muddy and weather can be especially cold and disagreeable, it is one of the best seasons to be afield. There are no biting

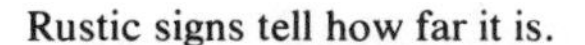
Rustic signs tell how far it is.

insects, visibility in the woods is excellent, signs and animals are undisturbed. It is a time to move quietly and see much. After five days in the field in early May 1976, I realized that I had seen no human footprint but my own.

The biologist has a term for the progress of the seasons; he calls it phenology. It becomes a matter of habit to interpret almost any observation in terms of what has happened and what is going to happen. The present is a moment in a sequence of changes. The basis of phenology, of course, is the climatic cycle through the year.

Isle Royale has a growing season of about five months, May through September. The first flower of spring is the skunk cabbage, whose purple-striped spathe may be seen, even in late April, poking through holes in the ice of wet bottomlands. A few days of warming sun will bring out hepaticas on south slopes, clusters of bloom ranging from white to blue and rose. Marsh marigolds spring from the muck along watercourses. They may be blackened by a delayed frost but will soon revive.

I would not know when the winter wren reaches Isle Royale, for when we arrive in spring it is always there, pouring out its vivacious song from lowland thickets. In sheer exuberance, if

A fall wolf scat of moose hair recently uncovered by the snow melt.

not in melody, its throaty bubblings are matched by only one other species, that elusive avian mite, the ruby-crowned kinglet. Long ago I judged these two the musical nonpareils of the north woods. Later it gave me a feeling of kinship with Audubon to discover his statement: "We shot also a Ruby-crowned Wren [*sic*]; no person who has not heard it would believe that the song of this bird is louder, stronger, and far more melodious than that of the Canary bird."

The two earliest sparrows to make themselves heard are the song sparrow and the whitethroat. Both are common nesters, and the white-throated sparrow's plaintive lilt from brushy coverts is truly the song of the island well through midsummer. Dorothy is enthralled by it; this is her totem creature beyond doubt. Each spring she listens for the first migrants in Indiana as they pass through briefly on their way north. They call soon after sunup, and they must be forgiven their trespasses against tiny sprouts of lettuce in backyard flower beds. That modest croft may have to be replanted. Dorothy has had good ideological company in her admiration for this bird. Said Thoreau of the whitethroats by whom he was awakened in the Maine woods: "What a glorious time they must have in that wilderness, far from mankind and election day!"

Robins are commonly thought of as dooryard birds, as indeed they are. But they likewise nest widely across northern forest lands, and they are especially plentiful during the migration on Isle Royale. They arrive by early May, and they provide the predawn thrush chorus a week or more before other members of the family are heard. Later in the season this function will be taken up mainly by Swainson's thrush. From many camps I have listened for the veery but have heard it only sparingly, which is somewhat strange, since its cheering is common mornings and evenings to the north in Minnesota. Over great areas of North America, in spring and early summer, one can expect some kind of thrush to awaken the woods in the cool darkness before sunrise and again to signalize the end of day in the gloaming. Around my boyhood campfires in Indiana we listened particularly for the fluting of the wood thrush.

From Isle Royale ridges one looks out over far-reaching stands of birch and aspen, which gradually turn from gray to green in the first two weeks of May. A part of the color is in thickets of beaked hazel now bearing inconspicuous greenish

In mid-May new leaves and new ground cover have just begun to sprout. It is a good time to see things.

female flowers. Near at hand there is unlooked-for color where the branching tips of spruce are highlighted by large, rose-red, male catkins.

In this period the last snowbanks disappear from shaded depressions in north slopes. However, masses of ice in wave-beaten clefts and caverns along the north shore may still be available to knowing fishermen until early June.

As woodland pools warm — though there may be a skim of ice on cold mornings — the spring peepers, chorus frogs, and wood frogs awaken to their reproductive duties. On the edges of inland lakes the birdlike call of peepers can be heard at intervals throughout the night. Later in the month, the trilling of toads is a mellow ripple of sound from standing waters, many of which will dry up by early summer.

Mid-May is the time of swelling buds and burgeoning greenery, aloft and on the ground. In successional forests the bigleaf aster pushes tapered shoots through leaves matted by the weight of winter snow. Thimbleberry is leafing out again in coverts that will be rank and shoulder-high in another month. Several kinds of ferns unroll their fiddleheads among the mossy rotting logs of swamp margins. Ground pine (a club moss), wintergreen, creeping snowberry, pipsissewa, bunchberry, horsetail, and a myraid of varicolored mosses and lichens create a fairyland forest in miniature at ankle height. By June there will be clumps of diminutive and fragile Calypso orchids emerging from the moist duff of trailsides.

We hope that about now our summer assistants can be on the job. They have little time for a breaking in, for there are 174 miles of trail to cover, with many excursions through the brush to those kill sites of last winter.

I began sending undergraduates to the island in 1962. It was evident that we must know as much as possible about the entire wildlife community and should look forward to an accessory study (later, by Wendel Johnson) that would involve the fox and its prey. Under an undergraduate training grant from the National Science Foundation, Michael T. Long and Larry J. Roop went to Isle Royale to study the red squirrel and snowshoe hare, respectively. One or two undergraduates, usually wildlife majors, have been there each summer since (Appendix I-B). Their sleeping quarters at our cabin are the two one-room buildings that Jack Bangsund moved across Rock Harbor ice from the old CCC camp at what is now Daisy Farm campground.

Depending on the source of funds, some undergraduates had their own assignments and wrote reports for credit. However, all have worked with the help and surveillance of our graduate investigators. And since they got to many parts of the island, they contributed valuable records to the research on wolves, moose, and beaver.

Among the early special projects was one on the island's birds of prey, carried out for three summers (1963–65) by Erik Stauber, a veterinary student. In his native Germany Erik had been an avid falconer and had a great fund of information and almost mystical lore about the feathered predators. He had no sense of personal hardship, and no swamp was too difficult to cross, no tree too hazardous to climb, if it would put him in position to train his binoculars on a hawk nest. He gave us a valuable survey of the island's avian predators.

Another special job was that of William K. Seitz, an Iowa

State undergraduate who continued the field study and trapping of beavers in 1964 and 1965 after the termination of Shelton's project in 1963.

Following two summers of undergraduate employment on the small-mammal work, Larry J. Roop became a park seasonal naturalist in 1964 and then returned to the same position in 1968 and 1969. In each of these three years we hired him to assist on the wolf project during the fall. John C. Keeler also worked for the park in 1969, after two summers of assisting Jordan, Johnson, and Wolfe. He helped me through the winter of 1970 after the departure of Mike Wolfe in mid-February. Two other men who had three summers of experience and did their bit in our studies were John A. Coble (1966–68) and Ronald L. Bell (1969–72).

As our work schedule in Appendix I-C shows, we had two summers of help from Joseph M. Scheidler, Philip W. Simpson, and James D. Woolington. Those with one summer were Michael J. Doskocil, William A. Knauer, Michael N. Kochert, Timothy C. Lawrence, Steven W. Ruckel, John D. Vanada, and Michael W. Wrighthouse. As mentioned before, James M. Dietz was a graduate student when he assisted through the winter and summer of 1971. Needless to say, without all of this help much less could have been accomplished in our time afield. One of the penalties of my own position was that I had nothing to offer the dozens of other students, both men and women, who sent me inquiries about work on the island. I am sure that among them are names that will be well known in years ahead.

A part of the charm, and also the difficulty, of research in wildland habitats is that your scientific endeavors must be combined with development, in some degree, of pioneer skills. Modern equipment — notably light tenting, packframes, and sleeping bags — help to mitigate the problems, but modern conditions in a well-populated national park also tend to complicate them.

For example, in summer the lean-to shelters at many campgrounds are full, and our people do not compete with visitors for such accommodations. The park has a small trail cabin here and there, but these usually are in use by seasonal rangers and other help. Also, much of our overnighting is done off in the brush somewhere, so it is almost always necessary to carry a tent. Fires are built only at campgrounds, so if you want hot food or need to boil water you take along a little Primus stove

A sleeve tent is adequate for ordinary nights of rain. The pack hangs on a tree in its own cover.

or something comparable. Our researchers commonly used their summer help in groups of two when long or cross-country trips were necessary. This was some advantage, since two people can carry lighter packs than one person operating alone. Most of the time packs were heavier (with bones) coming back than they were going out.

My own field work late in the season was greatly restricted by teaching until 1972. That fall I took a sabbatical leave and continued in the years thereafter on full-time research. This made possible early spring and fall periods on the island that featured leisurely 5-day excursions through areas I had not previously seen on the ground at these times of year. On most trips before and after visitor season I see no one at all and enjoy the use of campgrounds, trail cabins, and fire towers. Or I camp wherever it suits my purpose.

Why five days? It is reasonably easy to carry what you need for that length of time, and by the end of it you are footsore and ready for a shower and change of clothing.

The pioneering side of living and working outdoors has changed greatly in 40 years. In my first sleeping on the ground I had the protection of a cotton blanket and a World War I army blanket. It was routine that we froze after the fire went out and had to get up early to kindle another. Later, in my first summer as a camp counselor in Canada, I bought a pair of 4-point Hudson Bay blankets weighing 12 pounds. I pinned them into a sleeping bag with horse blanket pins and thereafter slept in comfort. Thoreau must have had something similar, as his list of equipment mentioned a "blanket, best gray, seven feet long." Today's sleeping bags are in a different class entirely. The down bag I use spring and fall weighs about five pounds. Depending on weather, its efficacy ranges from luxurious to adequate, it being taken for granted that one keeps his longies on and wears the heavy wool socks he will hike in tomorrow morning (usually changed at noon).

The zipper has revolutionized such things as jackets and sleeping bags, a fact especially evident when they do not work. On a February night, with a tape recorder, I slept out near a kill on the north side of Siskiwit Swamp — vainly, since it snowed 3 inches and the wolves were silent. I had taken a sleeping bag from the fire cache without checking it. On unrolling it inside my tent I discovered that it was completely open with a defective zipper. However, I did fairly well rolled up in it wearing my alpaca flight pants and down jacket.

My favorite kind of camping is from a canoe. I share Sig Olson's personal feelings about this wonderful craft.[2] With it you can carry enough gear to be comfortable, and you have no water problem, although portages can be misery at times. I am a slackwater canoeist entirely and years ago learned respect for Canada's large lakes. Many times Dorothy and I have taken our 16-foot, like-new (restored twice by loving hands), Old Town "Otca" model (long out of print) to the island and used it to explore around the Rock Harbor area.

Yes, you could coast Isle Royale, but only if you are in excellent physical condition, wary and weatherwise, with a strong experienced paddler, have plenty of time, and are well insured. Obviously, the entire lake was coasted routinely in days of the fur trade. But this was with large canoes by tough hominy-fed voyageurs whose way of life it was, and not all of them came back. The many great ships that lie on the bottom attest that there are times when this cold lake rises in fury and will not be denied its toll.

Probably the most self-reliant woodsmen who ever lived were the "mountain men" of the western fur-trade era, the early 1800s. Their supply base was St. Louis, and they went up the Missouri in well-equipped parties, each under the direction of a bourgeois (corrupted Yankee-style into "bugeway") who commonly represented one of the fur companies.

Normally the trappers acquired horses and scattered out to every part of the West where there were beaver pelts to be taken, allowing for jurisdictional understandings or misunderstandings between companies. Often a man did his spring and fall trapping alone or with a partner or Indian wife. Without map or compass he found his way through strange country. Far upstream he camped, took his peltries, and lived off the land and his own devices.

The mountain man required a horse to carry his traps and other working gear, his catch of fur and meat, and his kit of "possibles," or personal belongings. Without the horse he was in dire straits; "afoot on the prairie" described more than just being out of business. Life depended on the tools of that wilderness culture.

The minimal things a trapper needed were the heavy Green River butcher knife in a belt sheath, his gun with powder and ball, flint and steel, an Indian pipe, and some black twist tobacco. The latter luxury could be pieced out with various kinds of bark or leaves, categorized generally as "kinnikinic." This also diluted its toxic properties. Inner bark of the red-osier dogwood that grows on Isle Royale was one kind of kinnikinic.

These details are mentioned because I often ponder such things while solving my minor problems of getting over the terrain and doing the job that brought me there. How would Jed Smith have kept a beeline through this tangle on a cloudy day with no compass? How would old Bill Williams have "cached" himself and his mules from the pesky redskins? Maybe, like me, they would just get lost sometimes.

Today we take for granted such conveniences as a map, compass, canteen, well-fitting pack, and waterproofed matches. My possibles consist mainly of a folding belt knife (a Buck "Hunter"), a large cup, a small aluminum bowl, and a heavy duty spoon. We modern softies also must have a toothbrush, towel, washcloth, and soap.[3] For work purposes we need no traps or gun, but a camera, lenses, and binoculars are mandatory. On overnight trips from a base like Windigo, I allow myself a 4-pound pop-up tent. On long trips a 2-pound sleeve tent must serve. It keeps off the rain, but water runs down the inside as one's breath condenses on the cold surface.

Needless to say, a buffalo-hide lodge with a fire in the middle would be much better, but in a national park we are not allowed to cut poles.

From spring to fall, any given day can bring blustery weather to the island; a warm jacket is always at hand. There may be half a week of cold and drizzle, or a rousing lightning storm of the kind that can produce a fire in dry summers.

Carrying a pack through the rain is not the best of wilderness experience. As one squishes along in sodden footgear, there is little he can hear inside the crisp hood of his parka. Through wet glasses he sees a blurred approximation of anything. He thinks only of those great amenities of human culture, a dry floor and a tight roof.

That is what he may be aiming for if there is a campground at the end of his trail. There is particular exhilaration in reaching the waterside camp and finding an unoccupied lean-to after a forced march under the imminent threat of a spring thunderstorm. Having outrun the elements, one deploys his equipment, leans back, and the world's problems seem suddenly solved. After all, how can you beat the luxury of a 9-by-15-foot space, closed on three sides, with the fourth screened against bugs?

I personally savor the crash of lightning and the downpour, when I am safely under cover. If it lasts all night, I sleep with a sense of comfort and security. Often, however, Maytime showers are of short duration. They wash down the rocks, muddy the trails, and freshen the woods with wholesome odors. You sit in the shelter, pleasantly enduring a totality of idleness. There are notes to write, but you don't do it. Eventually the winds are spent, the dripping subsides, the clouds thin and part. The forest brightens in a glow of tender green. Sunrays sparkle through the half-leaves in kaleidoscopic hues, and all nature is a picture of loveliness in tears.

Then you are glad you came, and there is no urgency other than holding, for once, the advantage of keeping dry. Perhaps one knows for a while the mental repose of living entirely in the present, which must characterize the outlook of creatures who inhabit this milieu and know no other.

In spring the birds are in their nuptial plumages, and many species are passing through that will not be seen later on. It is a time of inspiration for the young-at-heart with binoculars and field guides. Needing no prompting, but with some help from their elders, our undergraduate assistants take on the never-ending task of identifying the birds (songs and plumages) and the plants of Isle Royale. I must constantly do the same, for I backslide between trips. Although feathered residents and migrants may be at their best in spring, this cannot be said of the mammals, which are moulting their winter coats. At this season a rat-tailed fox or a half-naked fly-blown nag of a moose is definitely at its annual worst.

Both spring and fall, there is considerable variation in the bird migrants. I have never personally seen a whistling swan on Isle Royale, but Woolington and Scheidler saw five on McCargoe Cove on 9 May 1975. Jim said that on the 17th there were 10 on Moskey Basin, and in the evening they "took off north." The following year he reported two on a beaver pond west of Todd Harbor as late as 22 June. Birds that loiter could be subadults,

A shedding cow in spring is not at her seasonal best.

or an injury might be involved. One wonders about the snowy owl that Rolf Peterson saw on 22 June.

Some of the deep-water ducks that winter around the icy edges of the island are still there through most of May. Goldeneyes are common, and we have occasional flocks of old squaws. On Siskiwit Bay I have seen white-winged scoters on 8 June. By that time most of the winter ducks have gone on to their northern breeding grounds. However, goldeneyes and both the American and red-breasted mergansers stay to nest. Other ducks that nest regularly, usually around lakes and beaver ponds, are the mallard, black, green-winged teal, ringneck, and wood duck. Breeding pairs and young of the hooded merganser are seen occasionally.

Among marsh birds, the woodcock nests on the island, and probably also the Wilson snipe, for "winnowing" males are heard at times in summer. Year after year, a few American bitterns frequent the same locations on the same ponds. A small rookery of great blue herons is located on Hawk Island, and another is in a beaver deadening on upper Washington Creek. Herring gulls have always nested on offshore islands and rocks. Historically the eggs of the herring gull were an important food resource for Indians, voyageurs, and others. Krefting et al. quoted records of the early lighthouse keepers in this respect (148).

As would be expected, Isle Royale has an abundant migration of warblers, sparrows, and other small birds, which need not be detailed. Resident nesting species are the usual ones for the northern lake states. Shelton did a particularly good job on the birds, persistently running down and identifying those whose calls he did not know. He pointed out some that were unfamiliar to me (e.g., olive-sided and yellow-bellied flycatchers) and in general gave me a good refresher course. In recent years Peterson has been updating the bird list, which can go on indefinitely.[4]

From open ridges in early spring it is not unusual to hear faintly the shrill drawn-out cry of an eagle. Then the bird will be seen, circling on the thermals high above the lake and island, working its way north. At least a dozen old eagle nests have been commonly known and regularly checked by our field crews, but the last reported successful nest produced one young near Lookout Louise (a mile north of Rock Harbor Lodge) in

1961 (148). Bald eagles can be seen occasionally on Isle Royale at any season.[5]

Our other fish-eating bird of prey, the osprey, has done somewhat better. Stauber watched productive nests in 1963 and 1964 and saw adults in various parts of the island in each of his three

A midsummer aging (by tooth wear) of the jaw collection. A more accurate job will be done in the laboratory by grinding the roots of molars and reading annulations (photo by Rolf O. Peterson).

summers. Observations since the mid-60s have been less frequent. We have had records of nestings or young birds of the redtail, goshawk, sharpshin, and Cooper's hawk. However, the most common nesting hawk is the broadwing.

Of the falcons, there were occasional sightings of the nearly extinct peregrine (duck hawk) in the early 60s.[6] In the past decade they have been few indeed. From a canoe on 30 August 1971, Rolf and Carolyn saw one on a dead snag along the shore east of Malone Bay. The next day they had a distant view of what appeared to be two of these birds, and other observers reported seeing a pair of peregrines at two locations on Greenstone Ridge. A year later John Wiehe, who manned the Feldtmann tower, saw "a migrating peregrine" on 21 August. These may turn out to be the last records of this magnificent bird on Isle Royale. Our two small falcons, the pigeon hawk (merlin) and sparrow hawk (kestrel), are seen with relative frequency, and from one to several nests are found each year.

I have no thought of neglecting several birds that have special interest in our work or which help lend a particular character to this wilderness island. Most notably, these would be the raven and whiskeyjack (Canada jay). They will have appropriate recognition later on.

Summer comes with a rush after mid-June. It is a time of drying, toughening, massive plant growth, mosquitos, black flies, and big squashy toads in the trail dust. Among old-growth conifers there are resinous smells and a sultry midday quiet of which the hesitant semi-warble of the red-eyed vireo becomes a monotonous part. With constant exposure, one's ear also takes for granted another bird, doggedly at practice on a squeaky violin. High and out of view, the black-throated green warbler speaks eternally of "sweet, sweet, Susie!"

These are times that try the souls of our summer help. On windless days the trips can be long and hot, packs heavy, and water scarce. The sticky flies are an abomination. One must concentrate and remember with gratitude that Isle Royale has no poison ivy. No doubt it will get there. Migrating birds will eat the fruit on the mainland, fly 20 miles, and deposit the seeds.

Sometimes there is great effort and little accomplished. In 1963 Larry Roop mentioned briefly that he took *Voyageur* from Rock Harbor to Windigo. A primary mission was to re-examine the bones of a moose that he and Mike Long had seen the summer before. He hiked to Grace Creek and found the kill, a single carpel bone. Larry needed to return to Bangsund Cabin the following day, but no transportation was available. With a light pack, he left Windigo at 7:30 A.M. and arrived at Daisy Farm at 5:30 P.M., nearly 35 miles in 10 hours. Dutifully in his notes he recorded the purpose of that day's hike: "to get back!"

Often on ridge trails cooling breezes from the lake are a beneficence to the hiker. There is satisfaction too in crossing rock exposures where the spectacular view can actually be seen. It may be possible to take compass bearings on such landmarks as the end of Houghton Peninsula, Menagerie Light, or northward across the water to the Sleeping Giant, a ridge on Sibley Peninsula. Then you triangulate on the map. It is reassuring to know, at times, exactly where you are. Frequently you have not come as far as you thought. Somewhere an experienced person said, "The strange trail is the long one," and that is true. On a familiar route you are under no delusions and know how far there is yet to go. The first time over any trail you are, too soon, looking for the end.

The mosquito festival begins in May and is in its full glory by late June. Sometimes only headnets and gloves make it possible to keep up the work. Rolf mentioned how he and Carolyn sat on a rocky knoll listening for the wolves near a rendezvous site. They could hear adults and the yapping of pups, but it was difficult because of the loud drone of mosquitos. Late spring and early summer nights in a tent can be a trial. Between lines written by dedicated field assistants there is sometimes a cry of anguish.

After sunset a swelling tide of mosquitos spreads from the lowlands and seeks out every warm-blooded living thing. Inevitably, as you crawl into the tent you take some of them along. It is too warm to stay in the sleeping bag, so you are prey to the invaders, thankful only that the swarms outside cannot reach you. But then a hatch of no-see-ums (midges) rises out of the muddy edges. They come through the bobinette "without even slowing up." The invisible hordes penetrate clothing, accompany you into the sack, and keep on biting after the cool of night has slowed down the mosquitos.

I knew this situation long before my Isle Royale work, and in these later years I have resorted to a craven expedient — a bug bomb, one not dispensing DDT. Before the critical hour I fog out the shelter or tent and spray the screen or netting. If done properly, and if you enter carefully, the little varmints are foiled. Dorothy and I have slept well through many a plague of voracious bloodsuckers. I should say that I have not been near a wolf denning or rearing site early in the year. Rolf and his help deprived themselves of any insecticide protection on the chance that the odor might affect wolf behavior. Greater love hath no man.

There is a tendency on the part of many to idealize wildlife research; you get paid for *that*? They must be reminded that field work is not all sunshine and birdsong. To get good information — even to enjoy the wilderness — one must be at the right place at the right time. As the price of this, he must inevitably be at the wrong place at the wrong time much of the time.

The blossoms of June become a set of fruit by July. Then we know whether the crop will be large or small. Some Juneberries are already ripe; many of them are borne near the ground because moose took all the twigs above last winter's snow line. This genetically variable shrub is the shadbush of the East; it flowers early, when shad are running in coastal streams. Midwesterners call it serviceberry (there a tree), and it is just "sarvis" in southern mountains. The fruit is sweet and nourishing, but sparingly produced on this island, some being spoiled by insects. As I sit by a moose-ravaged bush and take what there is, I think of the Indians who gathered it in more bountiful areas to dry for their pemmican.[7]

Juneberries and wild strawberries are among the earliest fruits of the season, but from mid-July on there are many others. In a good year, the blueberry crop means mass production for a great variety of wildlife and people. Red raspberries and the similar-appearing thimbleberry are widely distributed over the island. Thimbleberries are seedy and bland when eaten fresh, but they are famous for jam. In the winter camp we enjoyed glasses of this jam sent out by Donna, Bill Dohrn's wife, and we had thick pies containing a quart of Carolyn Peterson's blueberries. How she got it done I do not know.

The fruits of July and August include a new crop of bearberries, wintergreen, bunchberry, currants, red elderberry, red-osier dogwood, squashberry, sarsaparilla, beadlily, baneberry,[8] twisted-stalk, northern commandra, and others. No doubt all of these are taken by birds and some by foxes, squirrels, and mice. In part, at least, fruit crops of the year depend on the prevalence of summer rains.

At this season fireweed touches the openings with spots of pink; there would be fields of it in a new burn. With it are pearly everlasting, several asters, and the first goldenrods. The floor of the forest is bedight with white heads of wild lily-of-the-valley, and the elegant pink of other floral miniatures, twin flower and pipsissewa. From unlikely places — cracks in the barren pavements along wind-whipped shores — rise orderly rows or clumps of the fine-leaved and delicate harebell.

The nesting season is largely over, and one hears fewer bird songs in the woods. The large broods of ducklings we saw at the end of May have been drastically thinned out; survivors are now pin-feathery fledglings skulking about in the shallows. Many adults have sought seclusion in brushy deadenings and reedy lake margins. In drab eclipse plumage they are moulting their flight feathers.

To the relief of the long-suffering trail crew and their clients the hikers, the flies of high noon and the mosquitos of evening have abated. Myriads of tiny toads are hopping about on the upland. Noisy families of young crows move through stands of thick forest, in constant complaint about poor service and short rations. They must have few enemies; any insecure prey species could not afford this publicity. However, after nightfall the great horned owl may eat crow. A few blue-winged teal are appearing on lakes and ponds. Most of them have just arrived from the North as heralds of the advancing migration.

By the end of the month a new crop of beaked hazelnuts is ripening, if there is to be one. Several years may go by without really good production. When the small sweet nuts are plentiful, they are a food bonanza for many creatures. An early frost brings them down, and they disappear within a few days. Now too we know how well the mountain ash has done. Where this tree has grown beyond the reach of moose, its twigs may be weighted with clusters of brilliant red "berries" — not really a berry, but technically a pome, like an apple. The fruit is still hard and not fully ripe, about which we will have more to say.

After the middle of September fall is proclaimed, zone by zone, in the changing forest. High color comes first where red maples, long stubbled by the moose, set flame to open ridge tops: Feldtmann, Greenstone, and Minong. Soon the hard maples follow with every shade from maroon to lemon, and the yellow of birch and aspen enriches the island scene from end to end. Then the green conifers stand out clearly — even more so, I believe, than they do against winter snow.

Our summer help has gone back to school, and with Labor Day behind us, the season of public visitation is largely over. In 15 weeks some 17,000 citizen-owners of the park have come to inspect and enjoy their property. Now *Ranger III* makes but one trip a week with a short crew. Soon a trip will be made to Windigo, and the cargo will include supplies for the winter study: canned goods for the root cellar and barreled aviation gas to be cached under a canvas on the dock.

Important work is to be done. Usually we have a float plane and pilot for a week, and we hope for decent weather. Covering areas of reasonably good visibility, we count a sample of the moose herd: antlered bulls and cows and calves. This is an index of calf productivity for the year. As frequently as possible a fall survey has been made of active beaver workings, which furnishes a rough computation of the population.

The fall bird migration reaches a crescendo in the rutting season of the moose, late September and early October. Out of sight above low clouds that come off the lake on cool mornings, flocks of geese can be heard, the resonant call of Canadas and sometimes the high bark of blues and snows. Small numbers of Canadas may stop to rest and feed. Other waterfowl, including ducks, loons, grebes (mostly horned), and a few cormorants, are staging on Isle Royale bays and lakes.

One sees many small migrants, which need a hard look, particularly the sparrows and warblers in immature and winter plumages. Most of them will not be here for the winter; only siskins, redpolls, goldfinches, and a few grosbeaks are likely to stay if food conditions are favorable (p. 331). Now we see species that seldom occur on the island at other times of year:

tree, fox, and white-crowned sparrows, pipit, horned lark, and Lapland longspur. Several of the latter were on the beach near Senter Point when I began a field trip at Siskiwit Camp on 23 September 1974. As I stood motionless the birds walked around picking up something within a couple of feet of my boots.

When I started out another time, from Daisy Farm dock on the morning of 22 September 1975, a dozen-odd black-bellied plovers in winter attire were running about on the grass and shore. In late afternoon, on arriving at the McCargoe campground, I saw several more of the same species. In each case, a smaller, browner bird was with them, a golden plover.

By the second week of October flocks of snow buntings have arrived on the island, conspicuous in their striking plumage. Like the juncos and sparrows, they are on the way south, and these will spend the cold season in Michigan and Wisconsin where there are fields of weeds to feed on. I have not seen them in spring and suspect that they pass through early before we get there.

It is usual for the leaves to come down soon after mid-October, although it may happen earlier if there are strong winds. Where the leaf fall is heavy, it covers moose pellets and wolf and fox scats, and this provides an approximate dating for those deposited thereafter. In the interest of visibility, our aerial moose count is taken in the last week of the month, when deciduous trees are nearly bare.

One has the feeling, as October wears away, that he is witnessing the payoff of the season. By now the yield of chokecherries and pin cherries has been depleted by migrant birds. Spruce, fir, and cedar cones have ripened, and winged seeds are carried off by the wind. Ninebark, nearly immune to browsing, dangles clusters of hard seeds that will persist to spring; I have yet to learn what good they are.

The plant that, literally, opened our growing season is a late producer in the fall. In swamps and wet edges over the entire island, the large green spadix of the skunk cabbage stands among frost-shriveled leaves. Its soft yellow flesh is eagerly sought by foxes and whiskeyjacks, and no doubt by other species we have yet to learn about. This is a perishable fruit and does not last for long. Also at this season there are cranberries in acid bogs, but we have no information on their fate. They are known to be eaten by a wide assortment of birds and mammals.

Certain fleshy fruits are particularly durable and appear to ripen slowly. They seem intended for use by something in late fall, winter, or spring; possibly freezing helps make some of them more palatable, as is true of the wild crabapples farther south. Lasting well through the fall are the red drupes of high-bush cranberry and hips of the prickly rose. The scarlet clusters of mountain ash endure through the winter, barring an early ice storm. Another red berry is seen commonly in the woods, that ground-level dogwood, the bunchberry. These, as well as bearberry, wintergreen, and ground juniper may be available when the protective snow melts and on through the spring. Any of them may have a short crop or even fail to bear in a given year.

Many fall fruits are abundant or of special interest to us in some way, but the above list is incomplete. One thinks of some localized to the high ridges, such as snowberry and buffaloberry, or the crowberry that vines thickly over the rocks on Scoville Point. The species that stands out with particular significance is mountain ash. From midsummer on we make frequent notes on it, for it has importance in winter to the birds and foxes, as will be described later.

During the growing season, at least part of the time, the sun has beamed its energizing rays over land and water. A small portion of this light power has been tied up through the chemistry of photosynthesis in the chlorophyll of all green plants, from cells of algae suspended in the surface layers of lake water to the hard maples growing on Mount Desor.

In about 5 months energy has been stored that, for the most part, must operate the entire ecosystem during the rest of the year. One should not discount the metabolism that takes place in conifer needles and leaves of other evergreens during periods of favorable temperature through the cold season. However, this probably is a minor part of the total annual production. From October to May, most plants are largely or totally dormant, and animal life is living on seeds, fruits, and the vegetative parts of plants that grew and developed when they could. Some animals eat the plants directly: birds, woodmice, red squirrels, hares, muskrats, beavers, and moose. Higher in the pyramid of life, and fewer in number, are the secondary consumers, the animals that eat the animals that eat the plants: hawk, raven, weasel, mink, fox, and wolf.

This is the gross picture, which we will have occasion to examine in more detail.

CHAPTER EIGHT

New Generations

THE MOOSE IS CIRCUMPOLAR in its modern distribution. The species *Alces alces* originated somewhere in north Eurasia, where today it is represented by three subspecies. During one or more glacial advances in the Pleistocene, it spread across the Bering plain to North America, and we now have four recognized subspecies on this continent.

According to Randolph Peterson, at the time of the last glacial maximum,[1] the subspecies *Alces alces americana* occupied a refugium centering in Pennsylvania and the mid-Atlantic states. Another race, *Alces a. andersoni,* was found west of the ice mass, mainly in Illinois and lower Wisconsin. With melting of the glacier, about 10,000 years ago, the eastern animals accompanied their coniferous forest habitat northward into New England and Quebec. Midwestern moose made a corresponding shift into Minnesota and Ontario. Then, in a great pincers movement, *americana* spread to the west and *andersoni* extended its range to the east. They joined and interbred in the area north of Lake Superior.

On this basis, it is likely that the moose we have studied is an intergrade between the two subspecies. In time Isle Royale will furnish many skulls for a taxonomic review, but it has not yet been done. Our own interests have been elsewhere.

Moose being the principal prey of the wolf, it has been a concern of this work to learn how many there are on the island at different seasons. Reasonably accurate population figures are fundamental to the story of annual recruitment, mortality, and habits. Between predator and prey, and with an eye on vegetation, we wanted to know who was gaining on whom and for how long.

It is generally true that counting animals is one of the most difficult jobs the field biologist has to do. On Isle Royale the moose is no exception. We look forward wistfully to a time when it will be possible to fly over this island and, with some kind of mechanical scanner, tally the moose, all of them, and without duplication. Of that we can only dream, for the time is not now. Our counting has been done the hard way, and it includes errors for which we cannot accurately compensate.

Our most intensive inventory of the big and presumably conspicuous beasts was the one in 1960. From 13 February to 2 March, Dave and Don counted moose on the entire island. In 10 flying days they devoted 45 hours to the "census" and recorded 529 moose. On a map Dave had laid the island off into blocks with natural boundaries, so they could fly overlapping strips 3 to 6 miles long and 200 to 300 yards wide. He knew they had missed some animals, but how many? There was no real way to tell, so we agreed that rounding off the figure at 600 should be reasonable as a "conservative" estimate. The count was a great pioneering effort and a good basis for what was to come.

Unfortunately, many wildlife inventories come out like Dave's moose census; you have factors that cannot be meas-

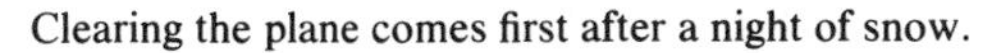
Clearing the plane comes first after a night of snow.

The white vulval patch shows conspicuously in wintering cows, helping to distinguish them from males even when viewed from the air.

ured, so the result must be called an estimate. When we allow for animals not tallied, there may be no basis for stretching the guessed-at error very far, so our inclination is to be conservative. This may be a reason why the final calculation often turns out to be low, and we have suspected that the first island moose count was indeed low.

As a principal reason, accumulating experience indicates that we often fail to see moose even when tracks prove they are there. I had an impressive demonstration of this on 16 February 1962.

It snowed in the morning and cleared up at noon. Shelton and Murray went out after lunch but could not locate the big pack; they did find a duo on the east end of Amygdaloid Island. After

4:30 Don and I flew up the shore and checked Amygdaloid again. The wolves were not to be seen, but there were new wolf and moose tracks. Since no tracks in the fresh snow led off the island, it was evident that two wolves and at least one moose were still there.

Don and I flew the length of the narrow island twice, carefully circling and looking into every covert — no moose, no wolves. Thoroughly annoyed, we did it again. Half a mile east of Amygdaloid Lake, we saw wolf tracks in the moose tracks. Then, there were the moose, two cows with their calves, about 100 yards apart. And down over a bordering ridge on the open snow came two wolves!

Such an incident is especially bothersome in that this narrow island with much open growth should be easy to count. We know well that we miss individual wolves if they are not moving, but the moose is a big animal and seeing it against the snow should be easy. Often it is not.

Other investigators have had similar troubles. To test the efficiency of aerial observations in Alaska, LeResche and Rausch used four pens, each enclosing a square mile of open forest, mainly birch, aspen, spruce, and cutover. They knew the number of moose present, varying from 7 to 23. The fenced plots were censused from the air, using 49 different observers.

Results of this experiment were illuminating. Observers with recent experience, on excellent snow cover, counted 68 percent of the moose present, whereas inexperienced ones saw only 43 percent. Counting accuracy did not depend on the density of moose. It was affected by weather, snow conditions, and time of day. The best lighting was in midmorning and midafternoon, under a high overcast. The favorable effects of a high, thin cloud cover and little wind have been known to us since Mech's work, and we confirm it every winter.

It soon became evident that Dave and Don took their 1960 census in a rare period of sustained good weather. We would not often be able to count moose on the entire island. So, over a 6-year period, a main object of Jordan and Wolfe was to develop a moose-counting method based on an appropriate sampling of habitats. The orderly way to go about this is to divide the island into major areas (strata) based on the expected density of moose. Then you lay out sampling plots in the strata; for statistical purposes more plots are needed in the habitats of high density than in those of low density. The counts of moose are extrapolated to the acreage in different density strata, and these are added up to give the calculated population on the entire island.

After experimenting for two years, that is the kind of system Jordan laid out in 1966, employing four density levels of the moose. With modifications, it has been used since then (Figure 8). Most of the counting has been done in the last two weeks of February and early March, and it has never been possible to cover all of the 70 plots because of weather limitations.

All of us have studied the census figures,[2] endeavoring to account for the inconsistencies and variability. There is leeway for individual judgment of what they mean, but we probably agree that they give the general proportions of the island moose population in midwinter. It is a logical expectation that with each year of experience the count should have been refined

Figure 8. The four density strata, based on preliminary plot counts, used for surveying the moose population in winter 1970 (from original by Michael L. Wolfe and John C. Keeler).

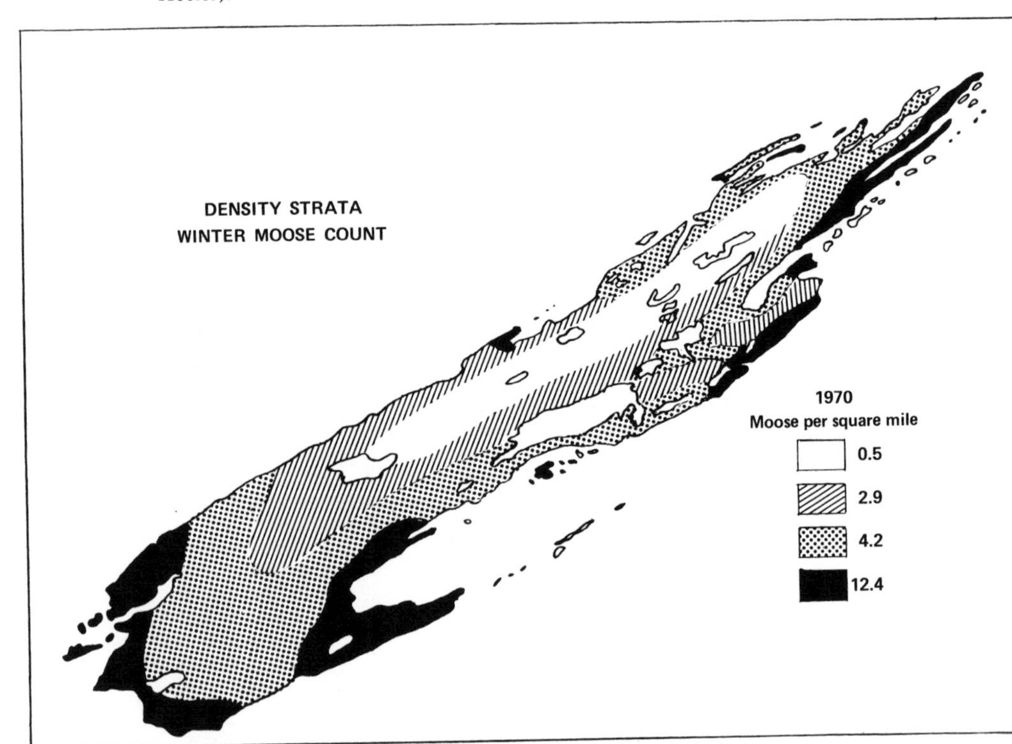

somewhat. We have been steadily more selective of those times and conditions that favor moose activity and visibility. Another reason is that Don Murray made all the winter flights, and he seems capable of continuous improvement in whatever he does. His practice, and that of our student investigators, has been gained not only in census counting but in the day-to-day observation of moose, as weather permits, during the seven weeks of winter study. In any rating of observers, ours are "experienced."

My own conclusions on moose numbers and trends over nearly 18 years have had some presumed benefit from hindsight. Starting with the basic counts, I have postulated that 20 percent of the moose were not seen and recorded. Each count was boosted by this amount, although it could be high for some years and low for others. Based on variance alone in the plot counts, a possible statistical error can be calculated; this is the plus-or-minus (±) figure given with the preliminary population calculation in Table 4, p. 453. It means that, according to mathematical probabilities, we have 95 chances in a hundred of being right if we say that the 1970 moose population is 945, plus or minus 242. If we demand that degree of certainty, there were between 703 and 1187 moose on Isle Royale.

Now we can ask: Within the wide range of mathematical expectation, was the actual population higher or lower than the median figure? Even though they cannot be quantified, there may be clues in the year-to-year trend or in certain biological variables we know about. For example, a high or low productivity and survival of calves, as well as our information on mortality, could indicate in which direction the population is headed. Population changes in the moose were mediated largely by wolf numbers and winter weather. More on this in Chapter 15, but for now we can note that after the onset of a series of deep-snow winters in 1969, the toll by wolves grew year by year. My own field work of the 70s left a distinct impression that I was progressively seeing fewer moose.

My "adjusted estimates" of the island moose population in certain years of the 16-year period follow:

1960	*1966*	*1968*	*1969*	*1970*	*1972*	*1974*	*1976*
700	881	1000	1100	1000	1000	900	800

Moose appear to have increased during the 60s, which in some degree correlates with our lowest wolf numbers in 1968–70. After 1969, both the number of wolves and moose vulnerability built up rapidly and undoubtedly reduced the moose herd by 1976. I have used Rolf's estimate of about 800 for that year. Bad weather prevented an actual count.

The above figures indicate why we commonly think of the late winter moose population of Isle Royale as about a thousand animals, or approximately 5 per square mile. This is a high density of moose on anybody's range. After examining all available North American reports on the numbers of moose (as of the early 50s), Randolph Peterson concluded that a moose per square mile probably is a relatively high density under most conditions, and 2 or more per square mile approaches maximum carrying capacity for major regions. It will be recalled that he was concerned with hunted ranges and that Isle Royale, as discussed elsewhere (p. 44), has the high-density characteristics of an island.

In light of all that has been done on Isle Royale, we may now reconsider Adolph Murie's statement:

> From general observations I should estimate that in 1930 there were at least a thousand moose on Isle Royale, and I think that a count would give a figure far above the estimated minimum. As a rule, wild populations are greatly underestimated, so it would not be surprising if the actual number of moose in 1930 proved to be two or three thousand.

Judged by the fact that vegetation was obviously holding up much better in 1970 than in 1930, and that in 1970 moose were not being decimated by malnutrition, we can assume that the population at the time of the early irruption was well above a thousand.[3] Murie's preferred estimate of 2 to 3 thousand seems realistic.

At the beginning of this project it was evident that we would need all possible information on the annual production of moose calves and their survival. The turnover of individuals in the herd from year to year was the main support of the wolves, and any major trends might indicate how long such an adjustment could hold up. If we could get reasonably large and well distributed counts of the recognizable sex and age groups of moose, then comparisons among different years should be valid.

There were three fairly obvious approaches to the problem. In summer our field people could record what they saw, and this would be an index of how many cows had calves. A main difficulty was that young calves are hidden by their mothers, and a further bias could be introduced by duplications in areas under frequent observation, Ojibway Lake above Daisy Farm being an example. As might be expected, some years the usable information has been far greater than others.

It appeared that a more reliable kind of count could be taken (as previously mentioned, p. 159) after the fall of leaves in October — an aerial inventory at the time when it is easiest to recognize antlered bulls and also cows with their calves. Dave and I took the first count of this kind after mid-October in 1959, with Jack Burgess flying a Piper Cub. It worked out well, and such a job has been done in all but five fall seasons during the

Figure 9. Locations where remains of 1068 moose were found during the period 1958-75. While the distribution could hardly be realistic at any given time, the map conveys a feel for the late-winter density of Isle Royale's moose population about 1970 (compiled by Carolyn C. Peterson).

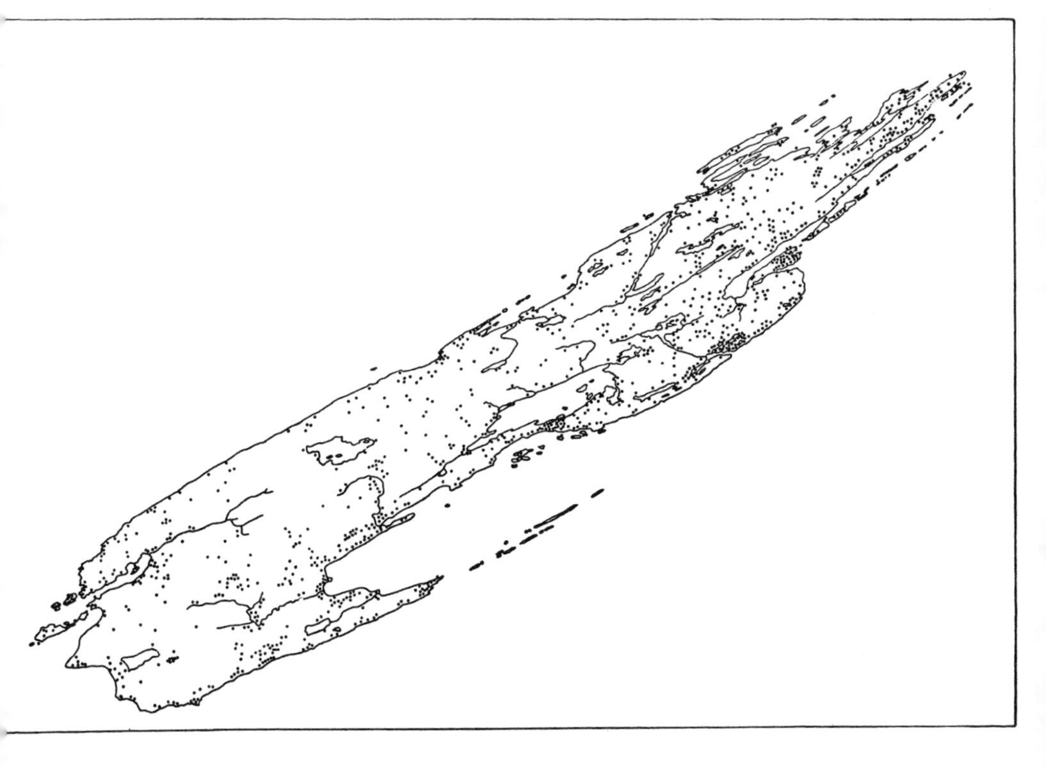

18 years. Heavy winds and other bad weather frequently hampered the flying. Accurate records require low-level circling, and it gets to be a problem for anyone subject to airsickness, as Mech, Shelton, and I well know.

The bulls could be spotted without much difficulty, and yearling bulls could be separated on the basis of their cervine (deerlike, nonpalmate) antlers.[4] Adult cows with one or two calves also were evident. Adult cows that had lost their calves, barren cows, and yearling cows could not be distinguished and had to be lumped initially with the female component of the herd.

However, there were further deductions to be made: Information from all sources indicates an even sex ratio among Isle Royale moose. So it is fair to assume that adult bulls (leaving out the yearlings) are about equal in number to adult cows. From this it is possible to deduce the proportion of mature cows without calves. Yearling bulls should approximately equal the yearling cows. Twinning is a reliable indication of high or low reproductivity, so we were interested in this figure also.

The third kind of herd composition count has been made during the winter study. By then more calves have disappeared, and the bulls are without antlers. However, visibility is good, and since 1969 the records have been particularly reliable. It was then we learned something important from Mitchell's work in British Columbia. Females, including calves, have a white vulval patch that shows up clearly under the short tail, a marking that is absent in males. This had been before our eyes all the time, and the oversight was mainly my own. When I reviewed my photographic files, I saw that in fall and winter this marking was just as consistent for Isle Royale moose as it was for those of British Columbia. We proceeded to use it, as Mitchell suggested, to confirm the sex of moose when the males are antlerless. Antler pedicels and the dark noses of bulls also were helpful.

When Rolf wrote his thesis in 1974, he summarized many of our back records. He found that calves averaged about 17 percent of the moose herd in fall and about 14 percent in midwinter. An average date of birth for calves on the island is about 1 June, and on its first birthday every surviving calf becomes, in our terminology, a yearling. On the basis of their high survival rate, yearlings can be considered a part of the adult herd, but for some purposes we speak of them as yearlings until they are two years old.

If the midwinter herd contains 14 percent calves, and a few more losses occur in spring (the weakest ones are already gone), a "graduating class" of calves at the end of May must compose nearly 13 percent of the population. If the total population is 1000, then 130 calves have survived from the generation born 12 months previously.

How many were born? We have no direct way of calculating this. The best information is that of Simkin, who examined the ovaries of 210 cow moose killed by hunters in northwest Ontario.[5] In the population he studied, slightly over 87 percent of adult cows had borne calves, and 25 percent produced twins (113 calves per 100 cows). In addition, 17 percent of the yearling cows had bred (at 16 months of age) and produced single calves.

Jordan and Wolfe brought together 10 years of our age-ratio counts (124), and I have rounded these figures to yield the following theoretical composition of a stable Isle Royale moose population about the first of May, before the birth of any new calves:

Calves (nearly a year old)	130
Yearlings (nearly 2 years old)	110
Adults (more than 2 years old)	760
	1000

Half of each of these groups would be females, and applying Simkin's reproductive rates we would get, as of early June, a new generation of 439 calves.

If we accept this and the figures are near average, they indicate that 7 out of 10 calves born will be dead within the year. Put another way, about 30 percent of the calves live to become yearlings. If the moose herd were increasing, survival would be higher; and lower calf survival would result in a decline of moose numbers. In a stable herd the number of yearlings recruited in June must equal a year of losses in the adult population.

We are having a look at the dynamics of moose numbers. It is useful to have averages against which departures up or down can be evaluated. It helps to measure what the wolves are using. Actually, the gains and losses of moose would seldom be exactly equal from year to year.

Edwards and Ritcey studied the uteri of cow moose secured from hunters in Wells Gray Park, British Columbia, and identified four well-defined periods of estrus: early September, late September and early October, late October, and late November. The second period encompassed 10 days and accounted for about 85 percent of the pregnancies. The breeding schedule on Isle Royale probably is similar. The rut is at its height during the last week of September and the first week of October. Aside from observing the activity and association of cows and bulls, we have noted another evidence of breeding: The cows show "mounting marks." When copulation occurs, the front hooves of the bull gouge divots of hair out of the flanks of the cow. A cow showing such marks has been mounted, successfully or otherwise.

A few calves are born early and a few are late, but the bulk of the crop evidently is produced in late May and the first few days of June. Our field records check out well against information obtained at the Cusino Wildlife Experiment Station, as reported by Louis Verme.

When Paul Hickie shipped some Isle Royale moose to the upper peninsula of Michigan in the 30s, six of them were taken to Cusino and placed in large pens. They were fed primarily on natural foods and their habits studied for many years.

Records on the three cows indicated that they bred in the period 25–28 September. Six calves born in captivity were dropped between 12 May and 17 June. Three accurately checked gestations averaged 242.3 days. Our generalization that 1 October is the middle of the rut and 1 June the midpoint of borning must be close to actuality, since it indicates a gestation of 242 days.

We appear to have only two acceptable birth dates for calves on Isle Royale. The first was given to Dave Mech by Leslie Robinette, Fish and Wildlife Service biologist, who was on the island helping Krefting with browse surveys. On 18 May 1961, Robinette encountered a cow with a newly born calf at the east end of the Siskiwit Lake outlet. The umbilicus was still evident on the calf. Les climbed a tree and watched while the cow lay and nursed the calf. When he finally descended and became a bit incautious, the cow charged him. He took refuge behind a large aspen, and the cow hit the tree, her hoofs on each side narrowly missing him. Then the moose went back to her calf.[6]

From his description of an encounter with a cow, it appears that Larry Roop came near witnessing the birth of a calf, or possibly two, south of Sumner Lake on 6 June 1963. He was climbing up a rocky slope when he saw a cow look over the edge from a flat area about 15 feet above. The cow withdrew, and Larry moved along the slope a short distance before ascending to the top. There the cow saw him and approached in a threatening manner, so he jettisoned his pack and climbed a spruce tree.

On looking around, he saw that the cow was "down in her hindquarters" about 20 yards from him. However, she moved back to her original location 50 yards away and nudged with her nose what appeared to be a newborn calf; the young animal could hardly stand. Within 15 minutes she had moved it to a position where Larry was able to descend and retreat from the scene. Several times the cow squatted as though another birth were imminent, but this was not verified.

At Cusino, Verme found that the mean weight at birth of six normal-appearing calves was 24.7 pounds. These young grew rapidly, weighing about 82 pounds at a month old, 250 pounds at 3 months, and 350 pounds at 6 months. Most calves probably do not gain much on their winter diet of browse, and in hard winters they certainly lose. On 27 January 1973 (a relatively favorable winter), I collected an 8-year-old cow and also her calf (aut. 117, 118). The cow was in good condition, weighing 800 pounds, and was not pregnant. Her large female calf weighed 410 pounds. Neither animal was particularly fat. The only other calf we weighed was 300 pounds.

Most cow moose on Isle Royale bear one calf at a birth, and reports in the literature indicate that this is true of nearly all yearlings that breed. Circumstantially, the twinning rate is a reflection of the nutritional level of the herd over several years and can be associated with range quality and the severity of winters. These factors appear to affect the rate of multiple births in all members of the deer family. In the first six summers of this study — calving seasons preceded by generally favorable winters — the twinning rate among cows that were seen with young averaged 34 percent. From 1965 through 1973, a period including five severe winters, the average rate was 14 percent.

It is particularly true on a range where wolves are present that field observations of cows with young do not produce reliable statistics on the twinning rate at birth. Two calves are more difficult to defend than one, and both of a pair of twins are unlikely to survive for an entire year. We see a few twin yearlings, but if the cow is not with them, one cannot be sure that they are not simply friends. In Newfoundland, Pimlott studied the moose from 1950 to 1956. He collected pregnancy records from hunter-killed cows, which is about the only source for such information. He found variations of from 5 to 30 percent in twinning rates for different districts — not clearly correlated with range conditions. The average was 12 percent.

In a total collection of 561 uteri, Pimlott found no triplets, although two sets were reported on record cards from hunters. Literature from all sources indicates that triplets are rare in the moose. We have seen none, although one record of a cow with triplets has been published for Isle Royale.[7]

Diminutive tracks in the mud of trails: These usually are the first tangible evidence that a new generation of moose is appearing. The signs increase rapidly after the last week in May. In 1969 on 25 May, Mike Wolfe's first record of the season was in the form of a wolf scat containing the fuzzy red-brown hair of a calf. Although calves are sometimes seen in May, a field worker often does not get his first observation of one until well into June.

As is characteristic of solitary forest ungulates, the moose calf is born in a comparatively weak condition. It must be hidden and nursed for some days before it is capable of moving about and following the mother. Shelton evidently saw twin calves at a fairly early stage on 22 May 1963. They were within a few feet of Tobin Creek; he said the calves were in the open between the water and a stand of alders. They showed no alarm when a beaver splashed only 10 feet away. However, the cow came through the alders, grunting occasionally, and the young animals arose, one of them "a little shakey." They nursed for about five minutes, and the cow escorted them up the hillside. Phil moved on, and when 150 yards away he looked back. The cow was watching him intently.

Altmann, who studied the Shiras moose in Yellowstone, said that for about 3 days the cow stays close to her calf and squats or lies down to nurse it. In the period from 4 to 20 days the cow is still localized, with the calf improving in strength and ability to follow.

During the early period of close contact, a strong social bond is established between the cow and calf, one that will last through the year until just before the next calf is born. Then the cow drives off her yearling, or at least attempts to, and gives attention to the new arrival.[8]

According to Denniston, calves depend mainly on the mother's milk for the first two months and then are gradually weaned. This agrees with our observations. Records of calves suckling are principally from late June and early July, and they are few thereafter. However, calves have been seen to attempt suckling in fall and even winter. Mike and Don were circling a cow and calf on upper Washington Creek on 5 February 1968, when the calf spooked and ran to its mother, attempting to suckle. I recall Don's remark that it actually looked successful, although our autopsies show that cows are dry by that time.

Our summer investigators have always kept Ojibway Lake under frequent observation; it has been a famous gathering place for moose after the spring greening. There, on 24 June 1963, Larry Roop was watching 7 bulls, 2 cows with twins, and an additional cow. He said he nearly stepped on a pair of twins lying in the thick (last year's) Canada bluejoint grass. The calves squealed and ran into the water, where their mother was cropping aquatics. A month later, on making the rounds at Ojibway, Erik Stauber also saw two cows with twins. In each case, as he approached to watch the calves, which were frisking about, the cow led them out to a small island or heavy sedge mat and left them while she re-entered the water to feed. Another day that July, I watched both a cow and her calf swim about in a deep hole of the lake and eat water shield off the surface.

On Isle Royale we have been impressed that cows frequently seek the seclusion (and greater safety from wolves?) of islands for dropping and tending their calves. Lake Richie has been especially favored as a calving area.

On 2 July 1970 Rolf was flying from Windigo to Mott, and they passed over Richie. On Hastings Island they saw five moose bedded down. One cow had a single calf and the other twins. On 9 July 1975 Rolf and Carolyn were in a canoe on Richie. A cow came into the shallows on the north side and fed for five minutes. Then it swam to Hastings Island. In his notes Rolf said, "Within 30 seconds of the time the cow entered the woods . . . a calf came scampering to her and began to nurse, butting with its head."

That year he received other island records from Joan Edwards, of the University of Michigan, who was studying moose browsing effects at the east end of Isle Royale. On Tallman Island, near the head of Tobin Harbor, 19 May, she watched a cow with twins that appeared not over a day old. One of the twins was smaller than the other and seemed to get little attention from the cow. The larger twin was seen to push it away and it fell down. The next day it was dead. This dead calf was one of two that Edwards found on islands in 1975.

A calf that is distressed or wanting may utter a plaintive bleat, and when they are moving through heavy cover the cow sometimes grunts softly to maintain contact. She may call loudly when a calf is lost, or she may simply walk ahead and let it catch up if it can. In October feeding, the calf often wanders 50 to 100 yards from its dam, and by the following May that distance will sometimes be 300–400 yards.

When young calves are at heel and still dependent on milk, they sample a wide assortment of green stuff, and by late August and September they are feeding heavily on vegetation. Much of the early nibbling is in or near water areas where their mothers commonly feed. When barely able to follow, a calf may be led into water and even forced to swim. On 21 May 1974, Scheidler saw a cow with a very young calf in the bog at the west end of Chickenbone Lake. At first the calf was nursing, but later it followed the cow out into the lake. Joe said it hesitated before entering the water, and he surmised it was the first time.

Later the cows and calves swim extensively. The juveniles are given little quarter and their strength is obviously taxed. Some of them drown when waves build up during the crossing of larger bays.

The first pelage of a calf is a coat of fine, faded-brick-red hair, almost woolly in appearance. This may have a special function: The young animal must survive the hoards of flies that ravage adults during the April–July moulting of their winter hair. In contrast, the calves are little bothered; both Stauber and Peterson remarked on this in their notes.

Shedding of the juvenile coat takes place principally in late July and the first half of August. Rolf saw a calf that was getting its dark mane on 21 July, and he recorded another that had the dark coat of the adult on 10 August. By the third week of August nearly all calves are dark. Perhaps significantly, the fly season is largely over by then.

Both young and old moose sometimes engage in what appears to be *play*. I make no difficulties with this word, nor any technical definition. Here it describes what most people call play: activity that to anthropocentric observers appears to be a pleasurable and exuberant working off of energy, and possibly the practice of arts useful to an adult.

One of the calves watched by Roop at Ojibway climbed out of the water onto a solid sedge mat and began to caper about. It would lay its ears back, kick out with both front and hind feet, and make short charges. Between times, it fed on something among the sedges that could not be identified at a distance.

Near Lake Richie, Stauber saw a cow with a calf he described as very playful. It ran through the thimbleberries, making oc-

In October a calf may weigh 300 pounds. It still shows the short head, in contrast with adults.

casional high jumps, until it was out of breath. Erik said that several times the cow got into the spirit of things, taking quick steps forward and to the side.

The most unusual antics of adult moose that I have witnessed were seen in Yellowstone National Park on the evening of 30 June 1972. Dorothy and I were in our canoe after sundown at the mouth of Grouse Creek, which enters Yellowstone Lake at the extreme southwest corner. A well-antlered (in velvet) bull was lying with head up, evidently ruminating, beyond the south bank of the creek. When we approached within a hundred yards, it got up and crossed the creek to join two other large bulls that appeared to be feeding in the marsh. As it approached, one of the two began to dash about in the shallow water, grunting and splashing. Then all three engaged in an obvious frolic, running through the shallows, roaring, leaping, and throwing water in every direction. Momentarily they would stand, and then go through the performance again. We watched for a few minutes and, since it was getting dark, paddled off to the east. The evening was quiet, and when we were a mile away we could still hear the gamboling bulls, grunting, roaring, and thrashing about in the marsh.

The calves of June are stub-headed and gnomish; as summer wears on they gain in size, strength, and agility. After the August change in color they look much more like moose. By October most of them are rounded and thrifty, probably near to 300 pounds in weight. However, at this season the developing animal still has a conspicuously short muzzle. This may well represent the ancestral condition of the species. It can be seen in midwinter and helps to identify a calf that is standing alone, where size is difficult to judge.

Among the winter-camp folk wisdom we invent for our intellectual improvement is Don Murray's Indian legend about how the moose got its long nose:

In the earliness of time, when the Great Spirit first designed the moose, all of them had short heads, and the calves have remained so to this day. However, along about the second summer, there is a time when these young animals learn that, for their kind, sex comes but once a year. And that is why the adult moose has such a long face.

Each spring and summer since the beginning of this study, we have accumulated a few field records of dead calves. They bear witness that the thinning out of the new generation begins as soon as the animals are born. Some of the losses are kills by one or more wolves. Undoubtedly where several wolves are involved there is little left for anyone to find and report to us, a common way such records originate. A single wolf that eats a small calf may leave the legs, as Shelton discovered on Tobin Creek, 3 June 1963.

John Coble found the intact skeleton of a young calf on 14 August 1966. There was no evidence of predation. Such remains would often be found by scavengers — ravens, gulls, or foxes — and disappear without a trace. Drowning victims turn up occasionally; one of the first we knew about was a month-old calf washed up on Scoville Point and examined by Mech on 25 July 1959. Since the chances of finding dead calves are small, the records we have probably indicate fairly high losses, at least in some years.

In the summer of 1963 Phil accounted for four calf kills, two of which were actually observed, or nearly so. One of these was reported by Erik Stauber, who was near the southeast shore of Lake Richie on 25 June. As he said, he "heard the cry of an animal in distress." The clamor of some crows led him to the spot, and he heard a deep growl as a large animal — no doubt a wolf — ran off into the brush. A spruce tree impeded his view of what was happening. Erik found a calf, estimated to weigh 60 pounds, still alive but with extensive muscle damage on both sides of the neck.

He left the scene immediately but returned the next day with Phil and Larry Roop. The calf was alive but helpless, so they killed and autopsied it. All evidence indicated wolf attack. Five days later Erik and Larry visited the area again, and the calf had been totally consumed, evidently by several wolves.

Another calf killing of that year was on 10 July. According to Roop's notes, seasonal ranger Dave Kanannin was on his way past Baker Point (west of Bangsund Cabin on Rock Harbor) about 9 A.M. On the shore he saw a wolf attacking a calf, biting at its neck. He turned his small boat toward shore, and the wolf was frightened away; then the ranger went on to check the Moskey Basin campground.

On returning, he found that the wolf had come back, killed the calf, and was dragging its carcass about 80 yards farther along the shore. Again Kanannin spooked off the wolf, after which he went on to Bangsund Cabin and brought Stauber back to the site. They looked at the calf, without touching it, then Erik was put ashore on the opposite side of the harbor to watch through his field glasses. Some time later a cow came by and swam across to Baker Point. There she called repeatedly, obviously looking for the calf, which she did not find. She then swam back across the harbor.

After dinner that evening Erik and Larry went to Baker Point and found two men from Mott Island moving the calf about for pictures. This being the case (in terms of human scent), they examined the calf. It weighed 80–100 pounds, had a fractured foreleg, and there were wounds on the head and neck. A 6-inch hole was torn in its side, and the liver had been eaten. The attack had begun in a bed of tall grass at the end of the point.

Erik and Larry placed the calf so they could see it from across the harbor, from which vantage point they watched until dark. The cow came back near them and stood grunting on the north shore for nearly an hour. Then she again swam across the harbor — taking about eight minutes to cover the quarter-mile — and when the men gave up she was still searching for the calf on Baker Point.

A short time later, Larry went back in a small boat and anchored 60 feet offshore from the calf remains. He said the night was clear, and the moon came up about midnight.

Between 10 and 11 P.M., Larry could hear the cow coming from the woods grunting. She went to the calf and was quiet for about half an hour. Then he heard something else approaching, which at first he took to be another moose. However, the cow rushed forward and there was a great commotion. Then she retreated into the shallow water, and Larry became concerned about having a moose in the boat with him. Regaining shore, the cow stood still then charged again, as a smaller animal ran through the leaves, emitting a doglike growl. After that, things appeared to be at a stand-off, and Larry went to sleep, to be awakened at 3:30 A.M. when he heard more of the chasing and rustling.

At 5:45 in the morning, the cow stood up from a bed in the grass about 40 yards from the calf, then walked directly to it. From there she went to Baker Point and swam north across the harbor. As she passed along the shore, Larry saw a bloody gash on her rump. After that Erik took up the vigil until noon and saw nothing. Phil said the calf remains rotted away with no further attention by a wolf. This probably is a case where human scent, in an area much disturbed by passing boats, discouraged the wolf from returning.

Although young moose take a drastic weeding out during their calfhood, the situation becomes much different once they are through the natal year. When Rolf was working out demographic statistics on the moose herd for his thesis, he found that a calf at one year of age (now a yearling) has a life expectancy of 7.3 years. Between ages 1 and 7, there is an average loss of about 10 percent in each year class. After that the rate of mortality progressively increases.

As mentioned before, cows drive away their yearling dependents when the new calf is produced. The rejected adolescents appear to seek company, which sometimes is an older bull.[9] They are "curious," and may wander into an occupied camp site. The strong attachment of the yearling for its mother is not easily broken, and some probably stay in the vicinity of the cow and new calf. In the course of the summer, the cow's irritability may subside as her calf develops, for cow-calf-yearling combinations are seen at times. In the breeding season it is not unusual to find a cow and calf with a bull, and nearby a yearling hovering about.

The yearlings do not appear to outgrow their calfish behavior any faster than is strictly necessary. On 13 June 1976, Scheidler was watching four moose at the mouth of Washington Creek. Two were yearlings. Joe saw one of them approach the other head-on and walk by it. Then it turned, placed its head between the other's hind legs, and nudged "as if it wanted milk!"

If a cow loses her spring calf or does not produce one, it probably is usual for the association with her yearling to continue. Stauber mentioned what was evidently such a case on 16 June 1964. Near Lake Ojibway he saw a cow "with last year's bull calf, which still seems to depend greatly on the cow and is protected like any new calf that I have seen." All of our summer observers have seen these privileged yearlings accompanying their calfless mothers, and we have recorded some examples during the winter study.

On 15 February 1969, Don and I were heading out on a flight, when we saw a large moose and a smaller one lying in the sun

The light-colored nose of this cow is typical of adult females.

on the north side of Washington Harbor. As we flew over they got up, and the large moose moved off, followed closely by the other. We circled to check them and found that the lead animal was a cow and the other a bull; I noted that it might be her calf of a year ago. Several times we saw the cow run at the male, chasing it a short distance. Each time she turned away, the young bull — evidently a yearling — followed her. Five days later, on a snowshoe trip, I watched the two on the north side of Beaver Island, and the cow was still trying to drive away the insistent youngster.

That same winter I saw another cow and a much tick-bitten young bull standing closely side by side near Lake Desor. In 1971 on 9 March Don and I saw two moose standing together at the east end of Siskiwit Lake. The larger was a cow and the other a bull yearling with an unshed antler on the right side. My notes mentioned that this was another case of a cow having no calf of the year and still attended by her yearling. It may be a matter of chance, but most of our field records indicate that the associated yearling was a male. This was not so in an observation made by Ron Bell and John Vanada at Wallace Lake on

An antlered bull, showing the characteristic dark nose of males, is ready for the rutting season.

11 June 1970. They saw an exceptionally small cow and with it a female yearling. They deduced that this was a young cow that had bred as a yearling (a year ago last October) but had not borne a viable calf in the current season.

It is evident that cows are strongly attached to their calves of the year, and they defend them vigorously. However, unlike an adult wolf, which seems to like *any* pup, the cow is discriminating. She is partial to her own calf and no other.

This was illustrated to me in mid-October 1976, when I was moose watching in the Windigo area. On a game trail in the woods, I had slowly approached a cow and calf to determine what they were eating. I was somewhat surprised when the cow turned into the trail about 30 feet away and walked directly toward me. I saw that her head and ears were up and her mane was not erect. There was no threat in her attitude, so I stood still and let her approach. When 8 feet away, she turned aside and went around me.

Then I saw that she had been looking, not at me but beyond

me, toward a second calf that had been coming toward us in the trail. My immediate conclusion was that this cow had twins. But not for long; she rushed the approaching calf with every evidence of antagonism. The calf turned and ran into the brush, at which point I realized that it had become separated from its own mother; I had seen them a short time before.

Our studies have made it evident that, on a range inhabited by wolves, an open season on cow moose will mean the loss of every orphaned calf. In the absence of wolves, where moose have become overpopulated and cows are hunted, it has commonly been assumed that the calves survive.[10] Actually, there is no evident reason why they should not. However, social factors could operate against them.

In Sweden, where both sexes of moose are hunted, it was estimated in the 60s that some 2000 calves were orphaned annually. More recently, this number may have been reduced as a result of a campaign to persuade hunters to take calves and protect cows with calves.

During 8 years, Gunnar Markgren studied the survival, movements, and condition of 17 of the motherless calves. The survival of calves is greatest in southern ranges and least in deep-snow areas of the north. He found that their diet is similar to that of adults, but other factors indicated that the life of an orphaned calf is precarious.

> The orphans seemed socially inferior and were generally not accepted by other moose. Sometimes aggressive behavior was recorded. Flight distance from different disturbances was varying but generally the distance covered in running flight was long. Two of the studied 17 specimens were close to death but one of these was injured in an accident. Six of the rest were very lean or scrawny while five appeared normal for the season . . . There is reason to believe that surviving individuals may be permanently retarded in growth and development.

Judging by what we have seen on Isle Royale, the wolf would introduce a critical factor into this situation, and it would be an exceptional and lucky calf that could survive at all.

Markgren's findings seem to mean that, without the authoritative backing of mother, a calf has a difficult time fitting into a home range where it can live at peace with the existing establishment. We have had numerous demonstrations of the "devotion" of a cow to her offspring, some of which catch us up short and remind us that we really know very little of such things.

In the winter of 1969 we were asked to accommodate an MGM film-making party in our camp. They were doing a documentary on wolves and the people who were studying wolves in North America. This was later on television as "The Wolf Men."[11] In addition to our regular field plane Don had brought in his Piper Supercub and employed his friend Donald E. Glaser, a Minnesota Conservation Officer, to fly it. On 15 February Glaser and I were flying east along the Minong Ridge. As we arrived at Lake Harvey, we saw a cow standing over a dead calf on the ice. Surrounding her were 8 of the 9 wolves of the big pack, the black male showing up conspicuously among them. The cow was obviously holding them off, and when she charged the animals in one direction several others would close in behind her to get at the calf. This went on, as we circled, for about 10 minutes; then we continued east to Lake Eva to pick up a kill. Forty minutes later we arrived back at Lake Harvey.

Six wolves were feeding on the calf, and two others were lying 50 yards out on the ice, about 20 yards apart. The two were watching the banquet scene, and it was evident that they were subordinates who would not be allowed to eat until the others had finished. Then we saw the cow emerge from the thick growth of the shoreline and advance on the wolves, scattering them. She stood over the calf for a few more minutes, then walked away and gave up.

On other occasions we have seen a cow defend the remains of a calf, but the young one might have been alive for a time after it went down. In at least one case we knew this to be true. On 24 February 1973, Don and I circled a kill west of Francis Point on the Houghton Peninsula. Here a wounded moose had stood for 8 or 9 days and then either died or was killed.[12] The west pack had fed and moved on; a wolf still on the carcass probably was a trailing subordinate. We followed tracks north onto Siskiwit Bay. Then we saw five wolves coming toward us, scattered for a quarter-mile over the ice. They had rounded the tip of Senter Point. As we came over the point, we could see a moose down on the snow in a bloody tracked area at the south end of Carnelian Beach. Its head was up, and the animal was still alive.

Don selected a smooth stretch of ice along the shore to the north, and we landed. Then we walked half a mile to what we soon could see was an injured calf. As we approached, a cow emerged from the trees on shore and came to the calf, staying 10 to 20 feet behind it on the bank. When Don and I got within 50 feet her hackles went up and we stopped.

The calf was bleeding from wounds on its left rump, shoulder, and possibly the neck. As we watched, its head sank and it ceased to move. We took a few pictures and left the cow standing over it. We noticed that she was browsing anything within reach.

The next day Don and Rolf saw the cow still there, holding

On Harvey Lake a cow defends the remains of her calf against the west pack (February 1969); the black male is at left.

off a fox and several ravens. However, on the 26th she was gone, and seven wolves were lying around the kill, which had been dragged farther out on the ice. Our observations indicated that the cow had guarded the remains of her calf for a period somewhere between 30 and 48 hours.

On the day of the Siskiwit Bay killing, Don and Rolf saw an injured calf on the ice in a bay on the east side of North Gap. This was 3.5 miles southwest of our camp, near the mouth of Washington Harbor. They had no time to investigate, but the next morning Don ferried Rolf and me to North Gap. With the collecting gun, we walked around the shore into the small bay and approached the calf. It was lying on its belly with head up

After more than an hour, she gives up and walks away. The single wolf on the ice is a subordinate that must feed later.

and hind legs stretched out. As it tried to move, the hind quarters seemed to be immobilized. We watched it from 50 yards off, and since it appeared helpless I killed it with a shot in the neck.

Rolf unpacked the autopsy kit, and we began to dismantle the calf, especially examining the hind quarters. We could find no damage and suspected that the spine had been injured as the animal tried to climb up the steep shoreline into the woods. Then I heard Don coming back, so I left Rolf to finish. Don picked me up in the harbor and we flew to Senter Point, where I was to do a job on the other calf. However, we found the cow still on duty, so Don took me back to Windigo prior to picking up Rolf. Before landing we swung back out over North Gap to see how things were progressing. There was the dead calf, with autopsy gear strewn about, but no Rolf!

We circled, becoming increasingly concerned; then Don spotted him in a tree 75 yards down the shore. The large birch leaned almost horizontally from the bank over the ice, and at its base was an irate cow keeping Rolf effectively treed. Later Rolf filled in the details of what happened, a part of which we saw from the air.

He was checking the spine of the dead animal when he heard a moose coming through the snow. Then he realized it was the mother of the calf. She trotted toward him and he ran down the shore, where he scrambled up the bank and out onto the tree. The cow followed. Rolf saw her urinate, but she did not wet her hocks as a disturbed moose often does. He shouted at her, but she only stared, with ears cocked. After about ten minutes, she walked back along the bank calling to her calf. Reaching a point above the carcass, she stood for a time looking down at it. Rolf worked his way to solid ground and stood watching. He saw the cow move down the bank to the remains, which she sniffed over carefully, her nose only a few inches away. In his notes he said,

> Finally she began to lap something up, and I realized she had begun eating the rumen contents of the calf. The cow completely filled her mouth, and when she raised her head "steam" issued from her mouth. The feeding went on for about 15 minutes. She tugged at the viscera a couple of times, and once pulled up a 5-foot piece of omentum, chewing as she pulled.

Don put me down at Windigo, then went back and buzzed the moose repeatedly, eventually causing her to move away from the scene of carnage and depravity. Rolf gathered up his tools and made a run for it. When putting knives hurriedly in the pack, he stuck one of them into his hand. He was still bleeding when he and Don landed in front of the dock, and we had to patch him up at the shack. We have had other such incidents.

The behavior of the cow is mystifying. It recalls Schaller's record of a Serengeti lioness who returned to a cub that had been killed by a marauding male. "She sniffed at her remaining cub, licked it briefly — then ate it . . ."

Such doings place a strain on tender human sensibilities. Our intolerant eye sees tragedy in death, then we sorrow at the spectacle of maternal despair, and finally the episode closes on a level that leaves one hanging in a vacuum of his own ignorance. No doubt the cow was responding to primordial thrusts in her concern for a living calf. Almost certainly she had left it to go inland and feed. At the last, was her putative "grief" overridden by a superseding impulse toward survival?

We have far to go in understanding what goes on inside the skull of a fellow creature: processes of cerebration that could have governed our own behavior in eons past. Within widely varying limits the higher animals are capable of learning and responding to new patterns of stimuli. It would be folly to oversimplify and regard any of them as a genetically controlled automaton. But then . . .

As we reflect on the hungry cow and the lioness, one can only conclude that in our terms nature is often callously frugal.

CHAPTER NINE

Moose Ways

ON THE TRAIL FROM MOSKEY BASIN to Lake Richie an alert hiker in spring is likely to find a scattering of small, cone-shaped morel mushrooms. It is a spare crop, which I have gathered on occasion, always looking for more than I found.

No doubt they occur elsewhere on the island, but I have not seen them, and a particular circumstance has impressed me. These fungi spring from the bare hard-packed trail itself and never from the looser soil to the side. Why?

The most plausible reason is that the trail has been fertilized by moose over many decades. In time someone may account for the mushrooms in some other way. But for now, that is where I come out.

Like everyone else, moose travel where the going is easy, and in fall and winter we are accustomed to seeing their pelleted droppings in every trail. The animals defecate while walking, and I have counted 173 pellets in a single fresh passage scattered out over 20 feet. Often they leave piles of droppings in their beds or where they have stood. The piles are an indication of how long the moose has lingered in a particular spot, perhaps to browse on a blown-down tree or because of disability. Droppings in the pelleted form still are the rule in early May, and those of calves are easy to distinguish by their smaller size. Some pellets are nearly round and others the shape of an egg — a snake egg, that is.

After the ground is exposed in spring, the first greenery eaten will be a few plants kept reasonably fresh under the snow since last fall. This is especially true of the plentiful barren strawberry, of which many leaves are still green, although they turn dark and deteriorate as new ones appear in mid-May. By that time moose begin to take aquatics and sprouting herbage of many kinds. With the increasing use of succulent foods, there is a corresponding change in scatology. Pelleted droppings become fewer, and fecal deposits take on the cow-flop module that is seen through summer months, often referred to in notes as "pies." There is an intermediate stage in which the pies contain a few pellets, but by the middle of June the change is complete. In September, when moose start browsing again, there is a conversion back to pelleted droppings, much in evidence by mid-October. I have noticed this a week or two earlier on the ridges than in the lowlands, possibly correlating with the earlier leaf fall and drying of vegetation on higher ground.

A complete catalog of summer foods probably would include a majority of the island vegetation. Some species are taken only in early stages as tender greens — for example, bracken and thimbleberry. I saw a cow nipping 4-inch shoots of Canada bluejoint at the edge of a beaver flowage on 8 May 1976. This and other grasses, as well as sedges, rushes, and a wide assortment of broad-leaved herbaceous plants, get most of their use

Left: Winter food of the moose: buds and twigs of woody plants.
Right: Droppings of the moose in late winter, calf left, adult right. The pelleted form is a result of feeding on browse.

during the first half of summer. They seem to have little attraction for moose when they become coarse and dry. This certainly applies to the widely abundant bigleaf aster.[1]

Moose feed on succulent aquatics from June to October, but in this case also, the heaviest utilization is in June. By the end of July the aggregations of 10 to 20-odd animals at Ojibway Lake are seen no more. Moose are well scattered over the island in midsummer, and it is possible to be in the field for a day or more without seeing one.

The food mainstay of the moose herd from late May to late September is the annual production of leaves on certain favored woody plants. These are all broad-leaved species, now that the ground hemlock stands of 60 years ago are gone. The bulk of summer browse is furnished by aspen, birch, beaked hazel, mountain alder, juneberry, mountain ash, red maple, fire cherry, and a few lesser species. Except for the first two, these are in no particular order.

Although we constantly make observations of moose at their feeding and interpret the effects on vegetation, we have undertaken no quantitative studies. We have been glad to leave the intensive work in this field to Laurits Krefting, who has made periodic visits to the island since he first joined Aldous in the browse survey of 1945. This long-continued project resulted from an invitation to the Fish and Wildlife Service (by the National Park Service) to make moose and vegetation studies. As part of it, in the late 40s, Krefting established four 50-by-50-foot fenced exclosures in different parts of the island. These kept out the moose, and the differences in plant growth inside and outside the fences have been evident. They have helped us all to appraise the extensive long-term influence of browsing.

Krefting and his coworkers made detailed plot surveys of herbaceous and woody vegetation and their use by moose, as well as plot counts of pellets as an index of moose numbers. They mapped the forest types from aerial photos and summarized historic and recent information on fires and vegetation changes. Results of this work became available in the early 70s[2] and have helped confirm or modify our own conclusions based on more general field observations and comparisons with published findings on other areas.

LeResche and Davis made intensive observational investigations (counting the bites taken) of foods used by three tame

The bulk of summer food is leaves of many kinds of shrubs and small trees. Beaked hazel, browsed by this yearling bull in June, is a favorite species on the island.

moose in fenced ranges on the Kenai Peninsula of Alaska. They found that over half the summer diet was leaves of birch; woody plants in general accounted for 65 percent. The remainder comprised many herbaceous species, including grasses, sedges, aquatics, lichens, and mushrooms. While that appears to represent our situation on Isle Royale, it is likely that more species are taken. The vegetation surveys of Krefting and others showed the wide array of ground vegetation available in different habitats, and Murie's notes on moose food habits demonstrate, in particular, the variety of foliose browse taken in summer. Although the bulk of this food from woody plants is leaves, it often includes the growing tips of stems. Murie, too, noted the use of herbs and grasses, some of particular importance, such as horsetail and shallow-water rushes. It must be remembered that his observations were made after the island vegetation had been devastated by the moose irruption of the 20s, and it is possible that second-rate foods (which I make no attempt to identify in this case) might have seen greater use in 1929 than they do today.

Moose make heavy use of aquatic plants in early summer.

Murie called particular attention to the near extirpation of succulent aquatics — as all agree, highly preferred moose foods — from Isle Royale waters. He cited the abundant aquatic flora found earlier in the century by the Adams party (1905) and Cooper (1913). "Today," said Murie in 1929, "water lilies are practically gone; only an occasional plant is seen."

At that time Walter Koelz was making fish surveys of Isle Royale lakes, and he too commented on the relative scarcity of water vegetation as a result of moose activity.[3] He said it probably had reduced the food of fish.

With this as background, Stanley A. Cain, of the University of Michigan, decided to recheck the lakes described by Koelz and determine what had happened in a 33-year interval. In August 1962, using a low-flying plane, he made a reconnaissance of 30 named lakes and a number of ponds. Among the more important aquatic plants, all sought by moose, were yellow water lily, water shield, pondweed (*Potamogeton*) of several species, eelgrass, and bur reed.

Cain's appraisal, compared with that of Koelz, indicated that,

In late September moose feed on the last of the green foliage. With the coming of frosts, they shift over to a diet of browse.

while there had been some improvement in certain lakes, the general status of aquatics was much as it had been in 1929. In short, moose had continued to feed in the lakes and ponds of Isle Royale and kept the stock of water plants well thinned out. We have found that this applies also to subsequent years.

A lake that has held up well is Richie; it has fairly deep beds of pondweeds and myriophyllum, which are washed up along the shore after being uprooted by feeding moose. In the main beds the animals must swim, and they go completely under water for half a minute or longer in seeking this food. As one watches through glasses from shore, he sees a cloud of flies above the moose. When the moose disappears under water, the flies hover over the spot and settle down again when the head and back break the surface.

It is generally conceded that moose seek water mainly for food, but relief from flies and the heat of summer probably is a by-product or primary motive in some cases. Kellum[4] said that an old mill pond was partially in the pens at Cusino, and the moose turned this into a wallow. In hot weather they would lie in it for hours with only a part of their heads above the mud and water. On 27 July 1967, John Keeler saw a bull at Feldtmann Lake deliberately bed down in shallow water and lie there ruminating.

The longer one observes the moose, the more he comes to regard it as a semiaquatic species. The round-barreled, splay-toed beast (with flexible dew-claws) wades and wallows through soft marshes, the bottomless mud of beaver floodings, and even the edges of quaking bogs, which it turns into an unsightly morass. A powerful animal, seemingly with no fear of unstable ground, it rests betimes, and plows ahead where other large creatures (a buffalo, for example) would be hopelessly mired.

In this connection, I have never seen a place where a moose used one of the plank causeways built by the trail crew across swampy lowlands for the benefit of hikers. Wolves and foxes happily trot across this dry footing and leave their sign along the way. The moose slogs through the mud at one side, invariably. Which brings to mind a statement by that master woodsman Thoreau in recounting his experiences in the Maine woods:

> . . . I noticed where a moose, which possibly I had scared by my shouting, had apparently just run along a large rotten trunk of a pine, which made a bridge, thirty or forty feet long, over a hollow, as convenient for him as for me. The tracks were as large as those of an ox, but an ox could not have crossed there.

My only possible comment is that no Isle Royale moose would ever do such a thing. But then, there were large trees in the Maine woods of the 1840s.

There is mention in various literature of a feeding habit of moose assumed to be widespread and common. The animal is reputed to push over tall saplings with its chest to get at the top twigs and leaves, called straddling, or riding down.[5] Over the years, to my best knowledge, no one has seen this on Isle Royale. In some places there are, literally, acres of broken-off aspens, birches, maples, cherries, and mountain ash. These commonly measure an inch, or slightly more, at the point of breakage.

Many times we have watched a moose feeding in this way: Turning its head sidewise, it grasps the offending stem in the "diastema" of the jaw, the gap between the lower incisors (the animal has no upper incisors) and cheek teeth. With a twist, it breaks off the young tree, then proceeds to nip the tips of twigs with an upward jerk of the head. Or in summer feeding it may slide long, wandlike suckers through the diastema stripping off mouthfuls of leaves. In a note of 1970, Rolf said that John Vanada had seen a bull grab a 15-foot aspen sapling and bend it over to browse on it. In the McCargoe Cove area I once saw 6 slender aspens in a group that had been bent over and fed upon. In a way it looked like riding down.

When food becomes a limiting factor for moose, it is almost invariably the winter browse supply that is deficient. Thus, winter forage and its degree of utilization get our major attention in judging the carrying capacity of range.

In Appendix II (p. 431) I have made a rough appraisal of woody plants furnishing the bulk of the diet of Isle Royale moose. There are two classes of preference, and the first of these is divided into three levels of estimated availability. In addition to a single list of secondary, or "acceptable," species, I have included a third class of woody plants significant for the fact that they are hardly used at all.

The most important conclusion from this appraisal is that by far the greatest dependence of wintering moose is on balsam fir, aspen, and birch. My rating of preferred and important species is in general agreement with previous work on Isle Royale and investigations in other areas of eastern North America, most notably those of Pimlott (1953) in Newfoundland, Des Meules in Quebec, Peterson in Ontario, and Peek in Minnesota. As would be expected, individual browse species are not equally available and perhaps not equally palatable in all areas. Nor do all investigators judge and measure things alike. However, the moose shows evident characteristics of diet that are consistently different from those of other members of the deer family. Its regional adaptations are associated with vegetation types. Moose of western ranges appear to have tastes much like those of the east, although they feed on various willows in proportion as these are more plentiful than other preferred browse (cf. Hatter in British Columbia, Houston in Yellowstone).

On Isle Royale it has been something of a mystery why the eastern third of the island, a pattern of ridges and narrow beaver-flooded valleys, supports stands of young balsams that largely escape heavy browsing for many years. Such a stand may then be riddled in a single winter, but enough trees outgrow the moose so that this species will be an important component of the forest that is taking over the openings. This is in contrast to portions of the island farther west, where it is usual for balsam to be suppressed, so that the unpalatable spruce is taking over.

By no means do we fully understand why moose eat certain things in certain areas. When I first did field work around Windigo in the early 60s, I was impressed that the scattered undergrowth of mountain maple was being passed up in favor of almost anything else. I had expected this species to be highly palatable. By the early 70s I could see a much heavier use of mountain maple. This might be related to the years of maximum moose numbers and more pressure on the vegetation in general, although I could not be sure of any associated depletion of preferred species. Most of the vegetation at Windigo is hit hard by moose every winter.

Almost any generalization on food habits will be found invalid sooner or later. With each period in the field I had become more convinced that ninebark rivals spruce in being almost totally unacceptable as moose food. Then, on the morning of 23 September 1974, I left Siskiwit camp headed for Lake Desor. Where the trail bordered Carnelian Beach north of Senter Point, a moose had thoroughly stripped the leaves and tips of a shrub of ninebark; and beside it was a red-osier dogwood that showed only minor browsing. In 16 years I had seen nothing like it.

It should be added that what grows along trails is subject to heavy usage, especially evident on such sites as Feldtmann Ridge, where the route is through portions of the '36 burn. Here both highly preferred and secondary foods (e.g., red elder, hard maple) have been browsed back by generations of moose. Old fields dating from the CCC camp occupation in the 30s are still in timothy and other grasses, although spruces have gained ground on the edges, and spruce seedlings are showing even in heavy sod. Balsam has been nearly wiped out.

Barking is sometimes conspicuous on aspen, speckled alder, mountain alder, mountain ash, red maple, and fire cherry. It is obvious that moose eat bark by preference and not because they are starving. This was recognized at Cusino, where Kellum said the crew cut poplar and red maple poles for the penned animals to bark. The most intensive bark feeding is in early spring in years when there have been violent fall or winter storms. Then there are aspen windfalls across the trails, and limbs from 4 inches in diameter down to twig-size are peeled clean by the moose. Some of this is seen each year, but in May 1976 it was more prevalent than ever before. Near Todd Harbor I watched a cow for half an hour as she gnawed away at a fallen aspen.

The liking of moose for bark of the "popples" brought to mind something I had read in several accounts of early western travels. When forage was scarce or snow covered, it was a habit of the resourceful plainsman to cut down cottonwoods that grew in stream bottoms and allow his horses to feed on the bark. It was well known that the animals relished this provender and it sustained them adequately. As an example, Osborne Russell was with a trapping party headed by Jim Bridger on the Yellowstone River at Christmas time in 1836. He said, "The bottoms along these rivers are heavily timbered with sweet cottonwood and our horses and mules are very fond of the bark which we strip from the limbs and give them every night as the Buffaloe have entirely destroyed the grass throughout this part of the country." It is not surprising that to the moose the fallen aspens are a source of nutritious food, evidently in many cases preferred over browse growing along the trails.

It has been said that moose rear up on their hind legs to feed on the limbs of trees. We have not seen this. They stretch high with the neck, standing on all fours, and neatly trim the balsams to a height that usually is 8 or 9 feet, often depending on the depth of snow and ice, since this is winter feeding. They also leave a pronounced browse line on waterfront cedars. White cedar is not a preferred food, as it is for deer, but often it is the only thing available along the heavily browsed lake edges where ice makes easy traveling.

Winter ice storms sometimes break the slender tops out of many balsam trees, and moose are quick to take advantage. In old-growth forest, where there is little browse, one or more animals will localize around a downed balsam for days at a time. In the third week of February 1972, our park man was Arnold J. Long, of Grand Portage National Monument. He said that in the previous fall there had been a storm and prolonged blow that thoroughly weeded out the forest along the north shore. We found similar conditions on the island, and many old balsam trees were down. Half a mile east of the shack, Arnie found a cow and calf at such a tree. He said the cow was bedded down, keeping contact with the calf by occasional grunts. He made trips for three days to check on them. After Arnie left the island, I investigated the site and found the snow packed down, many piles of droppings, and a fallen balsam tree well trimmed of its outer twigs.

That winter Don mentioned that he habitually looked for a moose when he saw one of these windthrows, which often were easily visible from the air. He and I noted something else that seemed correlated with heavy feeding on balsam: the color of urine. In winters of heavy blowdown, or at times when there was so little snow that moose could feed on the bottom rosettes of small balsam trees,[6] we frequently saw moose urine in the snow that was so distinctly orange-red it looked like blood. This was especially true from the air, and we well remembered an episode in February 1966 when all of us were betrayed by circumstances.

It was the winter of the great sleet storm. Alders and other slender trees had their tops bent to the ground under a load of ice; for many years the effects would be seen. Trails were blocked by bent-over saplings, and tops were broken out of large trees, including balsams and poplars that provided a "windfall" of food for moose. The stems of small shrubs were hooped wicketlike under the snow, a deadly snare for anyone incautious enough to slide down a slope on skis or snowshoes.

A mile-plus north of Washington Harbor we had found a kill, blood sprinkled plentifully over packed snow under a clump of balsams. On 21 February I went out to pick up the specimens. The site was a saddle above the large rift that splits the bedrocks from the harbor to the north shore. I thought that by staying on high ground I would cross a wolf trail that would take me to the remains.

I ended up high on the rocks looking down on the icy channel west of Hugginin Cove. So I decided the wolves had come from the east and I could intersect their track by following the deep rift itself back to the harbor.

No wolf track. I got back to the shack empty handed, with Murray and Jordan obviously wondering what was wrong with me.

The next day Pete and I went out together, but we started from the northernmost point of the harbor across from the ranger station, and that proved to be a mistake. We carelessly got confused in the broken terrain and saw practically no wolf sign. At the social hour that evening Don opined that we needed some professional help.

In the morning we left Pete to saw bones, and Don and I flew over the kill to spot it with great precision. After doing so, he and I snowshoed north from the harbor following pre-selected landmarks. We still found only one wolf trail, but we located the saddle and moved through the balsams eagerly looking for the remains. Don was going to show me the kill at last.

All we found were many moose tracks, piles of pellets, and an unusual amount of reddish urine showing conspicuously on the packed icy snow. Years later, Don and I would be spinning yarns about the little kill that wasn't.

Collecting its daily food is a time-consuming job for a moose. In the pens at Cusino, Kellum found that a wintering animal needed 40 to 50 pounds of browse. Des Meules calculated the food intake of a cow at 47 pounds and that of her calf at 31 pounds on a winter range in Quebec. As a rough index, if we assume that 40 pounds is a year-round average for all ages, 1000 moose would consume 7300 tons of vegetable matter per year. It is evident why the effects of browsing are so conspicuous on Isle Royale.

In the early 60s I noticed that many of the moose we saw from the air were bedded down during the midportion of the day, obviously chewing the cud. In 1964 I began to record whether an observed animal was standing or in its bed. After 1970, Rolf and I did this consistently. No statistical analysis has been made, but the trend is evident. In the morning most moose move about and feed until about 10 o'clock. After that increasing numbers are bedded. In midafternoon most of them become active again, feeding until nightfall. We know little about rhythms in activity during hours of darkness, except that at times moose do at night almost anything they do in daylight. No doubt there is every kind of variation with individuals, with weather conditions, and possibly with the phase of the moon (degree of darkness). Several times I have seen moose bedded at daybreak around Windigo, but we were never in the air at that hour.

In winter moose get water, along with their food, by eating mouthfuls of snow. We all do this, and we know it takes much snow to make a little water. Moose probably prefer water to drink when it is available. During periods of thaw, we have seen holes pawed in the slush of the harbor where moose have drunk.

Mineral licks are used in many parts of the island in summer, and some are kept open in winter. A lick on Washington Creek a mile above the harbor has been used year-round since our studies began, and I am sure it was known to park people and Krefting before 1958. Murie was impressed with the frequent use of licks, where moose drink the water, lick up mud, and eat soil. He had analyses made, and we have had several, indicating the presence of salts (sulfates and chlorides) of sodium, potassium, calcium, magnesium, and iron. Rolf and his assistants made many visits to licks, camping and watching from trees and other vantage points. They found these locations ideal for observing behavioral interactions. Moose sometimes took the waters from a kneeling position, or even lying down. Often the long-legged animals are seen to spread their forefeet wide apart when they must nip or drink something at ground level. Less often they kneel with one or both front legs.[7]

Like other ruminants,[8] the moose has a stomach of four fairly discrete chambers. The largest of these, the rumen, is a compartment where partially chewed twigs, leaves, and other vegetable material are stored for the first stages of processing. After several hours of gathering, the animal settles into a bed where the substrate (loose snow in winter) and exposure to sun and wind provide favorable temperature conditions. Then, for several more hours, the rough food material is brought back up the esophagus a cud at a time — this can be seen at close range — for mastication and a second swallowing.

The moose does not initially digest its food, as we do, by exposure to enzymes that break it down into proteins, carbohydrates, and fats in such forms as can be absorbed through the intestinal wall into the bloodstream. Nor can a ruminant digest the abundant cellulose and other fibrous plant tissues in its diet. These must be converted to usable compounds through fermentation by a specialized microbial flora and fauna (mostly bacteria) living in the stomach.

Chewed food passes from one chamber to another and on into the duodenum. The microbes multiply enormously in the nutrient slurry and are themselves digested in the process. They produce fatty acids that fuel major energy requirements of the animal. They utilize nitrogenous components of the vegetable material to synthesize proteins, which eventually become all the amino acids needed by the moose. Final steps in this chemical recombination take place in the intestine, where essential vitamins also appear. Through tiny villi in the lining of the gut (an acceptable technical term), the refurbished nutrients pass into the blood and on to assimilation in every living cell of the massive creature.

By such a complex processing the crude fibers of the forest are converted into animal form and become a proper diet for the wolf and a host of scavengers.

Most of the published weights of wild moose have been calculated from dressed carcasses in the fall season. Accordingly, we made a special effort to get total weights of animals we collected in winter on Washington Harbor.

In 1968 Mike and I assembled some portable gear that could be attached to a large tree and make possible the hoisting of a whole moose clear of the snow. We took weights, before bleeding, with a dynamometer reading to the nearest 25 pounds.

Weighing a moose collected for autopsy, a 950-pound bull. This work was done across from the Windigo dock, where remains served as a bait station to intercept traveling wolves and hold them for observation. Jim Dietz, Rolf Peterson, and Don Murray.

Incidentally, the purpose of bleeding was to get rid of gallons of blood that otherwise complicated the job when veins and arteries were cut in the body cavity. Even so, bleeding did not entirely solve the problem. We always wore "old clothes" to an autopsy.

From 1969 to 1976, we got the following weights on moose:

Aut. no.	*Date*	*Sex*	*Age*	*Weight (lbs.)*
447	21 Feb. 69	M	9 mo.	300
504	8 Feb. 70	M	7.5 yrs.	936
537	4 Feb. 71	M	7.5 yrs.	950
643	15 Feb. 72	F	6 yrs.	880
644	21 Feb. 72	M	6.5 yrs.	990
817	27 Jan. 73	F	7.5 yrs.	800
818	27 Jan. 73	F	8 mo.	410
1068	14 Feb. 75	M	8.5 yrs.	850
1220	26 Feb. 76	M	10.5 yrs.	850

In this list ages are in accordance with our practice of adding half a year when adults die in winter. Actually, in February they are 2 to 3 months older.

Kellum said the heaviest bull kept at Cusino weighed 1200 pounds in the breeding period. According to season, adult males varied from 900 to 1200 and females from 600 to 800 pounds. No doubt these seasonal changes also occur in the wild.

One of our motives in collecting a moose or two each winter[9] was to examine animals of the young age class that wolves were not killing. This helped provide a standard by which to judge the condition of remains we saw in the field.

Two of the specimens served the purpose particularly well. The first was a 3.5-year-old cow that Pete, Bill Dunmire, Don, and I autopsied (aut. 283) on 20 February 1965. Pete shot the animal out of a group of three on Beaver Island. The cow had a thick, glossy coat with heavy underfur. There was fat under the skin — few moose retain much fat in midwinter, although they probably have it in the fall. We found heavy fat deposits around the kidneys and the large intestine; mesentaries and the great omentum were laced with it. Mammary tissue was undeveloped, and the animal was in her first pregnancy with one embryo. There was "normal" fat in the bone marrow. Pete went through the lungs and found only one small hydatid cyst. On the surface of the liver were two bladderlike cysts of some other species.

Probably the finest specimen we examined (aut. 643) was the cow I collected near the powerhouse on 15 February 1972. Six-plus years of age, she had borne and suckled a calf that evidently had been lost. She was pregnant with a fetus weighing 4.5 pounds. Her weight was 880 pounds. There was an inch of fat over her rump and generous masses of it in the body cavity. Marrow fat was first class.[10]

The heavily browsed Windigo area is hardly a top-grade range, and 1972 was a winter of deep snow (as was 1965). This cow in the prime of life had been largely relieved of maternal drain. She had only four discernible cysts in the lungs and carried a light infestation of winter ticks. It was not the kind of moose the wolves were likely to kill.

The paucity of fat on most of the moose we examined — this applies to old and young, and especially to bulls[11] — suggests that the winter diet is usually suitable for maintenance, or something less, and functions to hold the animal over from fall to spring. Since the stress of rutting tells most on the males, these probably go into the winter with less nutritional reserves than do the cows.

It occurred to me that if any wintering moose were fit to eat, it would be the fat cow of 1972. So I cooked a chuck roast, but we found it as tough and dry as others had been. I have concluded that any moose nourished on browse for four months is strictly survival food regardless of age or sex. This does not apply to the heart and tongue, which we sometimes salvaged from an autopsy. I had read that moose muffle (nose) was a favorite dish with the Indians but never got around to trying it.

A further comment on bone marrow may be in order. A healthy animal, though spare and lean, may have a last reserve of fat in the marrow of the large leg bones. This would be the case in many deer, elk, bison, or moose in late winter and spring, or on poor range. In my early readings on frontier life I did not understand statements, encountered in different sources, that might have read like this: "It had been a hard winter and the game was poor. We killed an old buck but he was unfit to eat; we took only the marrow bones." In my ignorance I thought that meat, regardless of quality, would be better than a few bones.[12]

I did not know about the marrow-fat situation and the fact that people subsisting on lean meat got "grease hungry." The fat in those bones was the most satisfying food they had. This accounts also for a widespread practice among primitive peoples of cracking up bones to boil out the fat.

Moose shed much of their winter coats in April, and possibly earlier. At this time we are not on the island, but the process is well along by the first of May, when we see them. The neck, shoulders, and much of the sides are bare, and the moult spreads rapidly to other parts of the body. In spring the animal is ragged and unkempt, with little protection during much of the fly season.

In the main shedding period, and also later in summer, some cows appear to be without the white hairs of the vulval patch. The patch may show clearly on individuals, but it is not consistently present until late fall. We have no respectable hypothesis as to the function of this female marking. The same must be said of the bell, an outgrowth of the skin, which hangs from the throat of both sexes.

Moulting of the coat may be hastened by rubbing against trees, although this is not seen frequently. On my field trip in May 1976 I glimpsed the top of a bent-over balsam tree thrashing about on a hillside. On getting a better look, I saw that a cow was under the tree vigorously scrubbing her mane and back.

A new growth of short, dark hair is evident on some old bulls by the end of May, and many have much of the dark coat a month later. Our field records indicate that cows commonly are growing new hair in June and July; the moult seems to be delayed in females with calves. Nearly all adults are dark brown and glossy by the first week in August, when young of the year are completing their juvenile moult.

Antlers of the bulls also get an early start and are well along in the stubby velvet stage by the beginning of May. A note I made at Ojibway Lake on 6 June 1962 said that 14 moose were in view at one time and I estimated at least 30 using the area. Yearling bulls had antlers several inches long, and the old males already showed heavy palmations.

Development of the antlers is rapid through June and July. It is evident that in their growing period of about four months the bull must take in great amounts of food. On a range such as Isle Royale, where the igneous bedrocks have little calcium, the source of this essential element must be the deposits of glacial

A yearling is no thing of beauty during the spring moult, but its survival prospects are excellent. At this age life expectancy is about 7 years.

till. In their studies of nutrients in browse, Peek et al. found that beaked hazel concentrates calcium. This might help account for the many bulls one sees in spring and early summer industriously browsing the thick stands of hazel that grow on south exposures in burns of the past century. This understory is extensive around Ojibway Lake.

In the thick, vascular, velvet stage, antlers are never seen to make contact with anything solid, although they may be rubbed against a leg. They are obviously fragile, and damage sometimes results in a misshapen antler.

In late August the blood supply cuts off; the skinlike coating dries up and is scraped off in shreds. The bone antler, white-tipped at first, is honed, polished, and stained brown by rubbing small tree trunks, breaking up signs, and mauling the shrubbery. Cringen says the largest antlers of Ontario moose are to be expected on bulls from 6.5 to 10.5 years old.

We are not on the island when antlers are shed, but they will be much in evidence along trails before visitors arrive in May. In 1976 I saw a rubbing tree at the junction of the Minong and Hatchet Lake trails: a young balsam with bark and branches skinned off in an area 3 to 4 feet above the ground. A bull had polished and stained his tines here the previous September. But at the foot of the tree was an antler; how did that happen?

I could visualize what probably occurred: Weeks after the rut, this bull was making habitual rounds, and he stopped to eye his old rubbing tree. Impelled by lingering impulse, he gave the tree another clip and this time knocked off an antler.

Three days later I was on Greenstone Ridge above Angleworm Lake. Fifty feet apart in the trail I found two antlers, the same size and from opposite sides. It is not often both are shed at the same time.

The literature on antlers indicates that most old bulls drop their heavy racks in December, and younger males are progressively later. Each winter in late January and February we see a few antlered bulls, nearly always young animals with cervine tines. On an excellent day for observing, 13 February 1960, Dave counted 87 moose from the air, and nine of them were antlered, none palmate. Pete and Don saw four moose on Thompson Island on 6 February 1966, of which three were antlered, one young and two with large racks. The following day, at the opposite end of Isle Royale, there were eight moose on Minong Island, of which two were young antlered bulls. In my notes I asked the question, "Why should moose with horns stay together?"

A certain young bull that had not yet dropped his head gear will be long remembered by Dave and me. In 1961 we flew from Eveleth to Windigo for my first winter landing on the island. From separate planes, Dave and I saw a moose under a balsam tree 30 feet from the back door of the bunkhouse. After unloading we went up the hill on snowshoes to check the situation. What we saw from a distance was hardly believable.

A young bull was standing beside the heating oil tank-trailer that had been left for our use at the corner of the shack. With his antlers he had worried a canvas bag that hung on the trailer and extracted a heavy-duty rubberized extension cord used to plug in the motor that pumped the oil. This cord was hopelessly wound and tangled around the bull's antlers, head, and neck; he was tethered for keeps.

Dave and I approached out of sight behind the shack, then stepped into the open 50 feet from the moose. The animal faced us, lowered his head, and came off the blocks in full career. In an instant he used up 20 feet of slack and did a ground loop on the snow in front of us. Dave and I were thankful the cable had a steel core and those antlers were on tight.

We could not liberate the moose or use our back door. So this 3.5-year-old bull became our first "collected" autopsy, number 74.

CHAPTER TEN

More about Moose

MOOSE ARE AT THEIR BEST in both condition and appearance during the fall. However, by midwinter the pelage of some may be thinned and scruffy looking, or there are patches of bare scabby skin. This results from infestation by the winter tick.

Larvae of this tick spend the summer on the ground and then become active at the onset of cool weather in the fall. They climb the vegetation and wait for a passing moose to get hooked on their waving forelegs. On attachment they moult into a subadult stage and feed on the host through winter. They cause scabs and open lesions evidenced by blood on the snow of moose beds, as well as hair and the ticks themselves. In irritated spots the moose probably rubs off the hair against tree trunks and windthrows; Phil saw a tick-bitten moose scratching its back under a limb on 27 February 1963. However, I was mystified by the fact that some of the bare areas were located symmetrically on both sides of the animal, and occasionally a moose would be naked posterior to a straight line across the belly, which was true of autopsy 172 (p. 457).

These problems finally cleared up. I found places where moose had jumped over blowdowns of the right height to drag the belly, leaving much hair to show for it. The cross-log was hitting the belly at about the same place each time, producing that straight margin of the bare area. The other question resolved itself when Don and I saw an abandoned calf (p. 359) pursued and injured by a single wolf on 22 February 1971. We finished the calf off (aut. 550) and did a field examination, bringing the head in for further attention. Working on the kitchen drainboard, I found a cataract in one eye, and something else unexpected: The incisor teeth had a thick collection of hair around their bases. On each side of this calf I had noticed a large bare spot on the flank, precisely the area it could reach with its teeth. Now I knew.[1]

Two years later, on 28 May, I was on Benson Creek above Daisy Farm and found a place where a moose had stood beneath a balsam and left several piles of droppings. It also left a pile of hair 2 feet across and several inches deep. The hair was in gobs, just as it had been sheared off by the teeth of the moose. I suspect that this was part of the regular shedding process, although tick irritation could have been involved.

In spring the ticks are actively feeding, and they mature in June, according to Anderson and Lankester. When a moose dies in April or thereabouts, it is common to find the ticks on vegetation surrounding the carcass, evidently waiting for another host on which the maturation process can be completed. In June the blood-engorged females drop off, lay masses of eggs, and die. The eggs hatch into larvae that stay on the ground for the summer.

The winter tick does not ordinarily bother humans. However, when Shelton, Roop, and Stauber examined the carcass of a bull (aut. 192) near Lake LeSage on 9 June 1963, they found

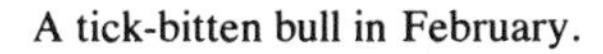

A tick-bitten bull in February.

mature ticks festooning the surrounding bushes and small trees. The moose was a 9-year-old, with a broken leg and no marrow fat. It had been dead about a month, was loaded with maggots, and had apparently been fed upon by a single wolf. That evening Stauber found one of the ticks attached to his leg. Shelton said the ticks were "thin" (usually the females are engorged and fat), and that could have accounted for an interest in substandard food.

For reasons not known, we recorded many more tick-bitten moose in the early 60s than were being seen 10 years later. I found that Mike mentioned this in his notes of 1968, but it did not impress me until I discussed it with Rolf in the 70s. He had seen only a few such animals; then I realized that in recent years I had not seen many either. This contradicts all the rules, because logically there should have been more tick infestation in a period of abundant moose when environmental troubles were catching up with them.[2]

In June moose begin to show open lesions on the back of the hind leg above the hock. These may be several in number and as much as 1 to 2 inches across. They appear to be opened and maintained by the attacks of biting flies, and they are referred to often in the literature on moose studies in other areas. Many adults (but not calves) have these fly-infested hock sores in early and midsummer. The lesions heal in August and September, and bare spots of scar tissue can be seen in the fall. Our observers made dozens of such records.

I have suspected that the site of the sores is an area of short hair and thin skin that the moose cannot reach with its teeth. The animal has no long tail for switching. Possibly on the hocks that special pest the moose fly,[3] as well as horse and stable flies, can feed with little interference. On 19 July 1970 John Vanada saw a bull at Feldtmann Lake whirling and biting at himself, evidently because of insects.

If it occurred only in summer, this would not seem unusual, but we have a number of winter records of such behavior. In my first February on the island I saw a bull standing quietly in a patch of aspens. Suddenly it whirled and appeared to be biting at the hock area. In warm weather I would have ascribed this to a fly attack, but there were no flies abroad in zero weather.

On 30 January 1963 Phil saw a cow come by in the woods behind the shack. Forty yards off it stopped, looked around, then "whirled around two or three times, stomping with front feet and kicking with hind, blowing and making a vibrating horse-like snort." Trotting on, it came out 50 yards up the Greenstone Trail, "still occasionally stopping, snorting, and kicking with the hind feet."

On a marginal flying day in 1965, Pete and Don made a circuit of the harbor to look for moose that we might observe further on the ground. On a ridge to the north they saw two, one a small cow (yearling?) that was bedded; it got up as the plane went over. Pete recorded that on each of five or so subsequent passes this animal flailed out "with all four feet in all directions, as though surrounded by wolves. Head was down and ears back." Even when the plane went over higher up they witnessed several similar displays. "Most noticeable was the flinging of hind feet sideways more than . . . backward . . . Also body arched to one side, then the other."

And yet again: In the winter of '74, Rolf and Don flew over a moose on Lake Ahmik that struck out with its front hoofs and shook its head "as if attacking wolves." Another on Washington Harbor "kicked out with back and front legs as we passed over at about 500 feet."

Explanation? It is guesswork, but I have thought of some possibilities. The winter of 1963 was one when we saw many tick-bitten moose. Do tick irritations remind moose of fly-biting? Is the scar tissue of hock sores subject to stinging pains? Does the buzzing of a plane remind a moose of a horsefly? Do moose get excited and give way to "automatic" reactions, as when repelling wolves? On and on . . .

In its everyday behavior some of the things a moose does are reasonably consistent, but after being wrong on many assumptions, we have come to regard the creature as "unpredictable." Various authorities have mentioned that its eyesight is not the best, but it has keen senses of smell and hearing.

Of course, one can judge what the animal sees only by how it reacts. Often it does not react at all until something is verified by sound or scent. This impressed me on a late February day in 1962 when I snowshoed up the Greenstone Trail northeast from the bunkhouse. It was sunny and quiet, and I was looking for a moose to photograph. Don had taken Superintendent Hank Schmidt out to show him his park.

I went out a mile or more, looking and waiting, without seeing or hearing a moose. On the return trip I did the same until about 250 yards from the shack. Then, looking ahead, I saw a moose bedded down within 20 feet of our back door. I decided to make a cautious approach, moving at times when the animal's head was turned away. But the instant my snowshoes stirred on the somewhat crusty trail, the head would turn back my way with ears cocked. Then I would freeze, and the moose would look long and carefully, evidently without recognizing me for what I was.

The bull was on the alert, and I had worked my way about a hundred yards closer, when I heard the plane coming. As I was certain he would, Don spotted the moose. Then he came in low to give Hank a better view. Finally, they saw me on the trail and realized they had fouled up my stalk. Later, at the social hour, Don said he and Hank thought they probably "should just keep on going." Of course, the moose spooked and went off over the hill at that lumbering trot that is so characteristic.

Except on rare occasions, this trot is as close as a moose gets to running. Only momentarily have I seen one break into anything resembling a bound or gallop; in his book Randolph Peterson describes the gaits of a moose similarly. In summer it is not unusual on Isle Royale to have a moose stand in the brush without moving, as someone passes nearby on the trail. If the moose is in the open and aware of being discovered, it may walk slowly toward cover; then, when out of sight, it is likely to change direction and break into that rapid trot. I have adopted a strategy of my own in such cases. When the moose disappears I run quickly ahead on the trail. Often I catch only a glimpse of the hurrying animal.

Some moose behave in an unaccountable manner, like the cow that walked deliberately toward three of us who were standing in the trail above Daisy Farm on a July afternoon. She had been 50 yards in the brush, and went by within 10 feet of Phil. She gave no indication that she even saw us, which she certainly did, for we had been moving about and talking. In a somewhat similar episode, Simpson and Bell had a cow follow them for 200 yards on the trail south of Lake Richie in June 1972. The next year on 20 June, Rolf and Carolyn literally got "caught" between two nervous-appearing female yearlings north of Tobin

Harbor. The young animals approached them repeatedly within 50 feet. Rolf thought they probably had been rejected recently by their mothers.

Several times in winter I have stood at a strategic point, with my camera on a tripod, waiting while someone drove several moose toward me over a knoll or out of the brush. On occasion I have wondered how one climbs a tree in snowshoes. However, it was always a comfort to know that moose are never aggressive in winter. To be more accurate, I knew this previous to 13 March 1970.

Zeb McKinney, Don, and I expected to close camp and fly out that day, but at 2 P.M. we heard that Martila was weathered-in and could not make it. After working in the shack all day, I took a late-evening walk down to the ranger station. I was wearing my insulated leather (going home) boots and no snow-shoes, since we had a trail well beaten by the snow machine. The weasel had given up for the winter.

At the edge of our opening, and headed down the hill 60 yards in front of me, was a cow moose. I hurried a bit to push her along. Twice she broke off the edge of the trail and seemed to panic as she floundered in the crusted drifts. Then she struggled back onto the hard surface and went on.

The second time she did not go far. As she stopped and looked back at me, I stopped about 50 feet behind her. Then she turned in the trail and walked toward me. I shouted, but she kept coming. At 30 feet she put her head down, cupped her ears to the side, raised her hackles, and broke into a trot. I had no place to go. When she was within 10 feet I leaped up, waved my arms, and clapped my gloved hands, yelling my loudest.

The cow swung to the side, broke off into deep snow, climbed back on the trail, and started away from me. Forty yards down the hill she left the trail and lunged through the snow upslope into the brush.

We have little systematic information on moose behavior toward their own kind (see Altmann, De Vos, Denniston, Geist), but random observations suggest some common relationships. Cow-calf habits have been referred to. The fact that different recog-nizable moose or groups of moose reappear in the same area at intervals of a few days indicates that home ranges overlap, but we do not know how large they are or how many moose is a tolerable density at various seasons.[4] Winter aggregations will be discussed later (pp. 349ff).

Licks probably attract moose from several nearby home ranges, and dominance relationships are evident. Interactions may be intensified under such conditions. Rolf and Carolyn frequently changed off with Jim and Joe to keep a particularly active lick under observation. They described and named the moose so that both parties could recognize them and make consistent records.

In early July 1975 Woolington took notes on the comings and goings of both cows and bulls, some of which appeared especially cautious and watchful. As a moose drank, its ears would rotate, evidently to pick up sounds. I have seen this when the animals were feeding in shallow water in the fall.

Two aggressive cows had difficulty deciding who was to occupy the lick. As one drank, the other approached, head high and ears back. Then both might lower their heads with ears cupped down and to the side. When the dominant moose approached more closely, the other turned her head away, which seemed to suppress aggression. Later another cow came in and rushed at one of them, striking her with both front hoofs.

On another day Joe told how a lop-eared bull that Rolf had described came to the lick or approached it and left 10 times. It was spooky and sometimes would retreat when it saw Joe and Jim. Twice it urinated on its hocks, rubbing them together. This may be a way of distributing scent from the tarsal glands, which are evidenced by a tuft of hair on the inside of the hock.[5]

Both cows and bulls have been seen to urinate on the hocks at times of stress or alarm. Jim Dietz saw a cow do this in June 1971, and Rolf has recorded it a number of times, usually involving females.

However, on 5 March 1973 he immobilized a bull at Windigo with a dart gun to mark it with a colored collar.[6] When the bull recovered and had regained its feet, Rolf approached it for another close look. The animal urinated on its hocks, rubbing them together, then moved off through the brush.

Before noon on 28 May 1973, Rolf and Carolyn saw a bull a quarter-mile off on a ridge by Newt Lake. It stood and watched while Rolf howled in an effort to get a wolf response. Then the bull urinated on his hocks, rubbed them together, and trotted up the hill.

Rolf recorded another example, perhaps qualifying as stressful, on 10 August 1975. A cow was lying in a foot of water at the head of Pickerel Cove, evidently cooling off in the 80-degree weather. The cow got up as Rolf approached. He saw that she was beset by flies and had anal lesions, swollen hock sores, and a calf to tend. She moved away with her offspring, stopped momentarily, and gave her hocks the treatment. It seemed to express the way she felt about the whole rotten business.

There are indications that a lesser degree of stress or excitement simply stimulates urination. On 10 May 1976, on Greenstone Ridge, two successive cows that I approached paused to hump and urinate before retreating. One was attended by a yearling that kept about 100 feet away; evidently it had achieved the unwelcome status of the parturition season.

Moose are generally regarded as stolid and silent, but they have a wide repertoire of vocalizations. There are soft maternal grunts and moaning moos, calf bleatings, what seem to be alarm and threat calls, and the more obvious signals of the rutting season. Like much else, our knowledge of these behaviorisms is in the descriptive phase, and we record suggested meanings that need review by the more intensive kind of work that Altmann and Geist have begun.

On 12 February 1963, when I alerted a badly tick-bitten bull, it uttered a strident high-pitched sound that "probably is the counterpart of a deer snort." I heard this again two years later when attempting to approach a moose that was breaking saplings in its browsing on the hill west of the shack. Pete came up the trail from the ranger station and the animal detected him; then it headed south toward the crest of the hill. Twice it made a noise like a heavy exhale of air. I described it as "sshow!" Then it stood and emitted a barklike "arrgh!" before going on. "The call was loud and sharp."

The next winter on 8 February I spooked three moose out of this same browsing area at Windigo. They went back over the hill into a swamp, with one going west and the other two east. I followed the latter and soon heard the other animal uttering a call that sounded like "mrrahh!" with a bovine quality. I guessed it was a location call resulting from the separation.

As is well known, moose are heard from most during the breeding season, and Stauber may have witnessed something anticipatory of the rut on 11 August 1965. He was camped south of Moskey Basin, and he said a cow stood around near his tent all night emitting "a loud moaning call." This might have been related to what happened later on:

At 7:00 A.M. a cow and calf walked into my tent lines and pulled down half of the tent. When I crawled out . . . to check what happened, the cow seemed to be quite irritated. Fortunately cow and calf moved on. A few minutes after this episode a young and an older bull appeared on the opposite side of the beaver pond — shortly thereafter a tremendous bull came. This last bull made loud roaring sounds when he discovered me and took off into the thicket.

The calling of the cow in breeding season is justly famous as a result of its widespread imitation by hunting guides, who attract the concupiscent bull to be shot. In late September and early October I have heard cows calling at various times throughout the day and night; rutting is a 24-hour preoccupation. The cow call is a sonorous appeal, sometimes with a tin-horn quality, that usually comes two or three in succession, then is repeated at intervals of seconds or minutes. It varies in sound and sequence with different cows in different situations.

In the rutting season of late September and early October, stylized dominance contests take place among the bulls. Animal on the right shows neck wound.

On 22 September 1975 I was on my way from Daisy Farm to McCargoe Cove. The day was cloudy with occasional showers. I heard a cow call intermittently for eight minutes in the Lake Benson area. As I noted it, this one had a whine in her voice, with now and then an "ooo–wah!" I heard two other cow calls on that 8-mile trip.

It was obvious that breeding activity was building up. I saw a young bull with long tines belaboring the bushes. He paid no attention as I hurried by 30 yards off. Two other bulls were milling about close to the trail, and again I got by without recognition. All of these bulls grunted occasionally. I saw two cows, neither of which showed the vulval patch — unusual.

Where Greenstone Trail descends into the lowland near Chickenbone Lake, it takes a sharp bend around a mass of tumbled rocks at the broken end of the ridge. In approaching this I saw a cow 20 feet below the trail, and beyond her a calf. I intended to pass her by, but 50 yards ahead a massive bull with a great rack rose ponderously out of the brush and stood in my way. The cow came up to the trail, cocked her ears toward me, and emitted four sharp bellows. Her attitude bespoke excitement but probably not threat. Then the bull took several steps in my direction.

Leaving the trail, and keeping my pack well balanced, I climbed up over the jumble of rocks on my right, quickly gaining 20 feet of height on the bull. I knew that no moose could negotiate those rocks, slick with wet moss, without breaking a leg. My judgment was verified farther on as I climbed back down into the trail, for I was supporting myself by grasping large stems of American yew! Moose obviously could not get to it summer or winter.

Two nights later I lay in total comfort by a window on the top bunk of the Hatchet Lake trail cabin. It was 11:50 when something awakened me. The lake lay misty calm in the quiet darkness. Then, farther east along the shore, I heard the cadence grunting of a bull. He was still at it as I went back to sleep.

"Cadence" grunting is my own term, and it describes something I have heard at what seemed to be the height of the rut. The most impressive occasion was on 25 September 1974 on the Minong Ridge fire manway(unimproved trail) several miles west of Lake Desor. A little after 5 P.M. it began to rain. On the crest of a low hill, I hurriedly put up my sleeve tent among the trailside thimbleberries. I hung the pack, in its rain cover, on a spruce sapling, unrolled my sleeping bag, and slid in.

A few minutes later, as I munched a handful of pecans, I heard a familiar sound: a cow calling somewhere in the lowland south of the trail. Nearby, to the northeast, a bull grunted; at intervals I could hear him flailing brush with his antlers. After about 40 minutes of this, another bull was heard from, evidently a hundred yards north of me on the downslope.

This was the cadence grunt, a steadily repeated "ungh - ungh - ungh - ungh . . ." It went on for minutes at a time, a grunt about every three seconds, as I timed it, not very loud.

At 6:24 the cow began to call rapidly, an insistent drawn-out whine in her voice. This, I thought, will do it. It appeared my tent was on one of the lines of communication, which was not a good prospect. Then the northeast bull became quiet, and so did the cow. My "big bull" was still grunting at intervals as I went to sleep.

In early morning there was grunting possibly 300 yards to the northwest. The rain had stopped, and I went back to sleep. Shortly after 6:00 I was startled awake by four hair-raising roars, evidently by the "big bull" of the night before. A little later there was a clash of antlers, and nothing further.

The sun came out and dried things somewhat before I packed up at 8:30 and headed for Windigo. Three hundred yards down the trail a young bull trotted across in front of me from the north. I suspected he was retreating from alien territory.

At about noon I was picking my way across the 60-foot dam of a new brimful beaver pond. I carried a pole, planting it every few feet ahead for stability. Balanced on muddy sticks at the center of the dam, I stopped as a great thrashing of water broke out behind some flooded alders to my left. A cow rushed through the flowage to the end of the dam ahead, and following close at her rump was a large shovel-horned bull. As they dashed up the bank and into the woods, the cow uttered two loud calls and the bull grunted. I was glad to be left behind.

An experience of Rolf and Carolyn ties in well with these observations. They were canoeing across Lake Richie on the evening of 25 September 1975. At the west shore they saw a bull with medium antlers. "He grunted every few seconds and stared directly at us . . ." At sunset they camped on top of a steep hill. They had just got situated when a nearby bull "uttered many loud roars — first time we've heard it so well. Two other bulls and a cow rounded out the chorus. Fortunately none ventured up on our small knoll."

It was shortly after noon on 9 October 1972. I was in a small opening somewhere south of Feldtmann Ridge. Eastward in the woods a cow called. This one had a groanlike quality. Leaving my pack, I walked to the edge of the opening and heard it again, closer.

A cow came striding rapidly through the tall aspens, grunting and calling; behind her were two bulls. Eighty yards away the group stopped, and both males sniffed the anal area of the cow. Then she whirled and bounded in a tight 20-foot circle before trotting briskly uphill toward the ridge. The bulls followed, one probably a 2-year-old, the other with a modest rack. As they disappeared, the young bull was ahead of the cow, the larger one close behind.

I heard another sound, and a big bull with spreading antlers came along, head down, obviously trailing them. As he disappeared, I had a feeling that destiny was catching up with the cow.

At the time, I was puzzled by the cow's circling but discovered later that Murie mentioned it, and it is a fairly stereotyped kind of behavior. Denniston said both cows and bulls do it. As Geist described the performance,

> Frequently associated with extreme defense in moose, is circling. The moose jumps away from the opponent, whirls and bounces around in a tight circle. The head is kept low and is frequently shaken. This was observed in 10 instances of aggression. It was also seen in the play of calves and yearlings.

Another behaviorism that I have witnessed only once is *flehmen*. It is seen in nearly all hoofed animals (pigs excepted), and is most commonly provoked by vulval scent or that of urine in the breeding season.

At Windigo, on 28 September 1975, I heard at least one bull grunt at intervals in an area of brushy poplars. On investigation I saw a cow standing quietly, and 20 yards away were two bulls. The males did some maneuvering, and when one came into an opening it was followed by another cow. The cow made low wailing noises; the bull grunted and beat the brush with his antlers.

This bull approached and smelled the genitals of the cow. He then raised his head high, with an expression that can only be described as a "dazed stare." His lip curled down so that I could see the full row of incisor teeth. He held the position for a few seconds and then followed the cow into the brush.

Later I encountered in Jordan's notes a description of something similar he had seen on the part of a yearling bull at Ojibway Lake on 14 June 1964. Pete said the teeth showed clearly, the animal flattened its ears backward, and its "eyes appeared to be rolled aside so more white . . . was evident. This exaggerated position was held for about 10 seconds."

Although most behaviorists know this posture by its original German name, *flehmen,* Geist has also called it lipcurl. On the basis of two observations, I think it is much less common in the moose than in some other species, for example the buffalo. However, one needs caution in such interpretations, because moose are commonly well hidden in the brush and difficult to watch. We see signs of another habit that various investigators have described: the use of pawing, or rutting, pits.

I have found them most often on game trails in open woods — shallow oval diggings 2-plus feet long where a bull scraped away the topsoil with his hoofs. These probably were experimental, test holes, you might say. However, Rolf recorded that some pits smelled of urine, and these could well have been the real thing. In the fall of 1972 he saw the first pits, four in number, on 23 September and three more the next day. South of Conglomerate Bay, most of these were in cedar groves, and one had a strong odor. On 2 October, near Long Point, Rolf found a pit 3 feet across and a foot deep. He thought this one could have been used for many years.

The pits are urine wallows, and one of the best descriptions of how they are made and used is by W. K. Thompson. The event was witnessed by J. E. Gaab and Robert Neal of the Montana Wildlife Restoration Division.

They saw a large bull pawing in a depression at the base of a cliff, after which he straddled and urinated in it. He did this seven times. As a cow approached, the bull padded the pit with his front feet and then wallowed in the depression. The cow struck the bull with her front feet until he moved away, and she too wallowed, covering herself with mud. Then the bull struck the cow with a front foot and made her stand, following which he smelled her genitals and rubbed his neck and bell along the cow's back and sides. She stamped her rear legs and tossed her head. When the cow tried to walk away, a grunt of the bull stopped her. Finally, both moose walked off into the forest. The pawing pit was 36 inches in diameter and 8 inches deep. It had a musky-urine smell.

It appears that the association of bull and cow is a one-at-a-time affair of short duration. In two or three weeks of rutting a bull has ample time to mate with several cows. Observing animals previously tagged, in Jackson Hole Houston saw one bull with three different females in 11 days. Another was with three different cows in a 14-day period. Although yearling males are capable of breeding, it is unlikely that the competition permits it.

Bulls sometimes engage in sparring matches wherein two animals put their heads together and shove each other around. We have seen this activity from mid-September to the third week of October.[7] In such challenges one function of the large spreading antlers of the moose is sometimes evident: They are for display and the intimidation of rivals. On 18 September 1973, from the roof of the Siskiwit Camp trail cabin (an old CCC building), Rolf and Carolyn watched the best-ever routine of this kind. It involved three bulls of about the same size, but with small, medium, and large antlers, respectively.

At 8:25 A.M. they saw the medium-antlered bull standing in the trail 100 yards west of the cabin. In the woods on both sides, other bulls were grunting. Then the "large" bull emerged from the birches on the north and approached the other. When 50 yards away,

> he began to display his antlers — walking slowly and stiffly, with head and antlers swaying from side to side with each step. The big bull walked in front of and around the medium-sized bull, who merely watched while remaining motionless. The bull with the larger rack continued to walk back and forth in front of the medium-sized bull . . . Soon the third bull appeared from the south. He had the smallest antlers of the three . . .[8]

Rolf said that as soon as the "small" bull appeared, the other two walked toward him side by side, *both* displaying their antlers. The smaller one retreated to the north, as the others followed. A couple of times the young bull "pranced ahead a few steps and turned to face the two briefly — " then he would walk on. When they reached some small birches, the big animal paused to thrash his antlers, then lost interest and walked into the woods.

The medium bull continued to follow the small one, but with the greatest threat gone (?), the latter turned and for 25 minutes engaged the larger-antlered animal in a shoving contest:

The bulls would slowly make antler contact and push slightly, ensuring a firm 'hold', then each would vigorously push with the head low toward the other. A tremendous amount of energy was expended in these contests, with the bulls arching their bodies and all muscles tensed . . . After a few seconds they would disengage and stand still, their breaths fogging the air . . . once when the small bull was walking away in retreat the medium-sized bull poked him in the rear with an antler tine, which caused the smaller bull to run a couple of steps then turn and resume the sparring.

At intervals the larger animal would stop the action to thrash his antlers or paw the ground with a front hoof. Rolf noted that several times the smaller bull emitted a high-pitched grunt as he broke away and jumped back defensively. The affair ended when the young animal walked away. The other followed briefly, then went by the cabin, stopping for several minutes to stare at the two figures on the roof, obviously trying to get their scent. He went on south over the ridge.

Like this one, most dominance contests end with no one hurt. However, the dueling is not all in fun, and any deviation from the rules could have serious consequences. In watching two large bulls go at it near Windigo on 13 October 1972, I saw that one of them had a 5-inch gash in its neck — wide open, with red muscles showing. And there are other hazards:

On 1 March 1969 Mike and Don saw part of an antler protruding from the ice of a beaver pond on the north drainage of Mud Lake. They landed and looked at it — two legs had been dug up on shore by the pack of eight. Mike recorded the location, and in the following August Ron Bell and Steve Ruckle went after the remains. They found two bulls with locked antlers. They carried out the skulls (aut. 452, 452a), which are now on display at the Rock Harbor Visitor Center. The animals were found to be 7 and 8 years old. It must have taken great power

Our only example of locked antlers: two bulls, age 7 and 8, died in a beaver pond.

to lock the racks of these well-matched bulls. There is no possibility of pulling them apart.

Some evident risks and ills of moose existence have appeared in the remains we examined from season to season over many years. At times of freeze-up and break-up heedless animals come to grief on thin ice. They break through and cannot make it to shore. Each spring, floaters must be salvaged for autopsy on harbors and lakes. It does not pay to put this off. Some are not found until warm weather, but at least the bones can be examined.

Many mortalities are dated approximately, like the 5-year-old cow found frozen in the ice of Rock Harbor on 2 May 1972 (aut. 744). From the size of the fetus, compared with those found in winter autopsies, Rolf thought the cow had died in late January or February. That part of the harbor froze about the end of January. Ice casualties include fall calves, spring short yearlings, and young adults in good condition.

These are "accidental" deaths, like moose that fall into mine pits, or a bull electrocuted by the new powerline to Ojibway Tower (aut. 881). An old rock-lined well near the Island Mine Trail is something of a calf trap. When Scheidler and Woolington hoisted out a defunct month-old calf (aut. 1178) on 19 July 1975, they found beneath it what was left of a 9-month-old (aut. 1179), a winter casualty. These young of the year can be aged fairly well by size, the degree of tooth eruption, and sometimes the stage of the moult.

Because of wave action and possibly temperature, Siskiwit and other outer bays pose problems for young calves that must follow their swimming mothers. Two have washed ashore at the head of Siskiwit Bay, one in Malone Bay, two in Rock Harbor, and one in Belle Harbor — a partial listing, I believe.

Not all floaters are drowning victims; some have fallen off rock shelves or cliffs onto the ice. If we cannot get to them for an examination, wolves are likely to perform the rites, and the bones will disappear with no autopsy record. If the carcass is intact and floats in spring, we may have a chance.

Three cliff falls in 1974 included the oldest moose yet found on Isle Royale. This 19-year-old cow (aut. 982) was on the ice below a 15-foot icy slope northeast of Huginnin Cove. When Rolf got there on 9 March he found that wolves had pulled out the smelly (i.e., dead for a while) viscera. The head and a hind leg were twisted by the fall. Lower teeth were in good condition, but there was food impaction and bone deterioration around the upper premolars. Bone marrow was class 2.

The following day Rolf examined remains of a calf at the base of a 40-foot cliff on the ice west of Lake Desor. A loner had found this one; later the west pack would clean it up. A month earlier a calf had met its end by falling 30 feet off the sheer north side of Blake Point. No wolves present. This one had depleted bone marrow, class 3.

These records are a sampling of casualties occasionally found in winter. The high precipice at the west end of Feldtmann Ridge is a site we check in our flights, and also from the trail in summer. A moose that starts its glide at the edge of the abrupt overhang is likely to lose 200 feet of altitude before it fetches up behind a well-anchored tree on the steep slope footing the ridge.

Sometimes they have help in going off such places, but that will come later (p. 353). Usually, they pay the penalty for brash carelessness. A moose will walk to the edge of an icy, snow-covered, or wet-slippery cliff edge and casually reach out over the abyss to browse a twig of mountain ash. As a matter of odds, the animal may lose its gamble, and the wolves have another grubstake.

As would be expected, not all accidents are fatal, and if a wolf confrontation is avoided for a while, a moose can show remarkable recuperative powers. Broken bones of the lower leg may heal satisfactorily; we have also seen several that were partially healed, well stabilized by a mass of fibrous tissue. Usually they are misshapen in some degree. Our observers have encountered at least a dozen moose limping badly enough so that there would be little question of the outcome if discovered by the wolves. I found a lame cow (September '74) that had stayed in one place for many days at the tip of Senter Point. That wolves do catch up with these invalids is evident in the gnawed skeletons showing arthritic afflictions of the spine and joints. In several specimens the hip joint had literally disappeared (e.g., aut. 284), the clubbed head of the femur being anchored to the sacrum by a fibrous growth. Deformed joints and vertebral bridging are fairly common in old animals.[9]

The most frequent and specific bone abnormality for which

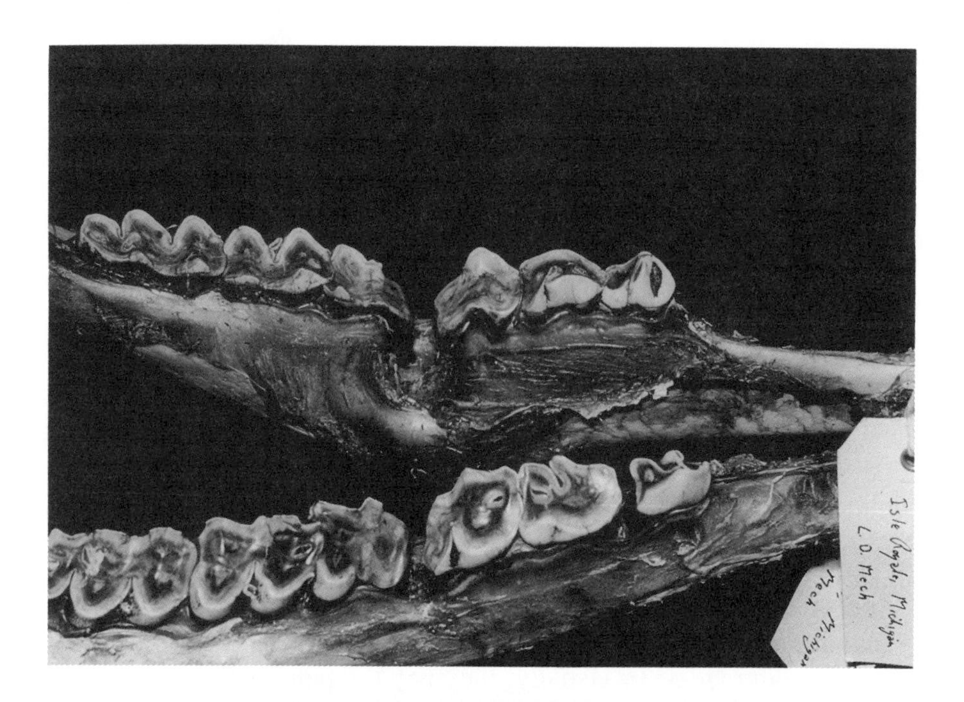

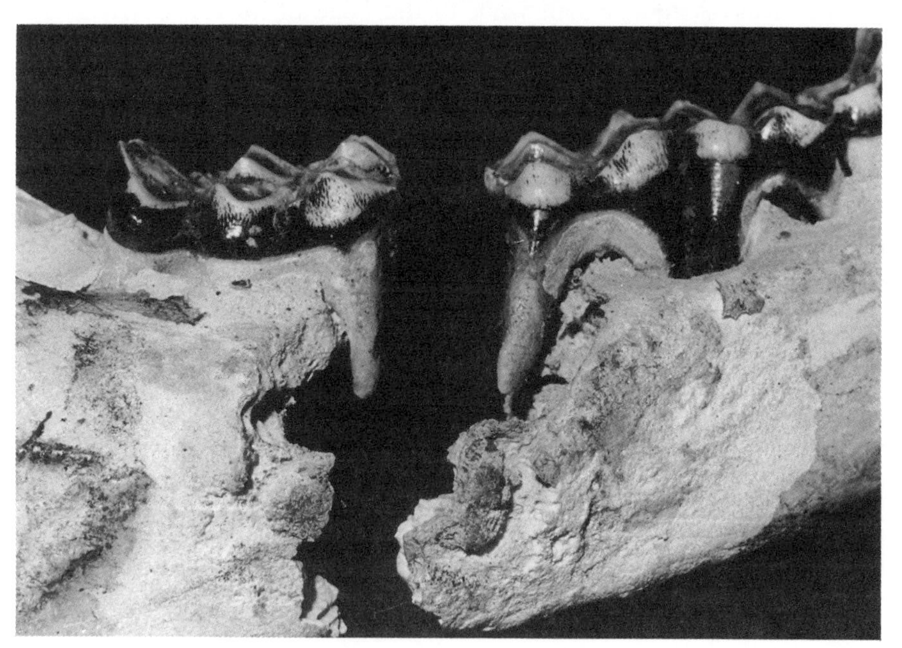

Top: The most common bone abnormality is "lumpy-jaw," which primarily affects moose over 10 years old.
Bottom: This one fell apart in our hands. The moose was killed by two wolves.

we look is lumpy-jaw, probably originating as a food impaction. The usual site is between the third premolar and the first molar. A mycotic infection (i.e., actinomycosis) probably is involved, but we have not verified this.[10] The condition is often reported in big-game animals. This kind of jaw necrosis causes malformations of the mandible or maxillary bone, erosion of teeth, and the probable impairment of feeding and digestion. It is associated with age, as indicated by a statement Mike Wolfe prepared for our 1969 annual report: "Of 186 specimens examined, no animals younger than 9 years were found with periodontal necrosis, and of those with an incipient condition, 95 percent of the males and 100 percent of the females were 10 years or older." A total of 51 specimens showed some stage of this pathology.

Moose have a complement of internal parasites usual to hoofed creatures,[11] although only one has been evident as a possible cause of mortality on Isle Royale. It has an apparent ecological tie-in with old age and the vulnerability of moose to the wolves.

The hydatid tapeworm (*Echinococcus granulosus*) lives as a minute adult attached to the wall of the wolf's intestine. Millions of eggs are shed with the feces, and these are picked up by the moose in water or on its upland foods; this phase of the parasite's life history is little known. On being swallowed by a moose, the eggs hatch and the resulting larval stage passes into the portal venous system. The larvae usually lodge in the lungs (sometimes in the liver) and form the hydatid cysts that are so conspicuous in old animals. Based on autopsies, we are confident that every young moose has begun to accumulate its inevitable load of hydatids by the time it is 2 years old (p. 123). By the 5th year the lungs may have half a dozen, some of marble size, although the pathological effect probably is minimal. The cysts continue to accumulate and grow, the larvae multiply asexually during the life of the moose, and healthy lung tissue is progressively displaced. In old animals some of the cysts are the size of Ping-Pong or golf balls — reportedly they get much larger — and dozens of various sizes may be present. In a moose injured by the wolves and finished off for autopsy (aut. 286) by Jordan and Murray in February 1965 Pete estimated the total load at 74. A moose killed by two wolves in February 1967 and field autopsied (aut. 386) by Wolfe and Johnson had so many cysts they brought in the lungs for further examination. We estimated 85 cysts, many of them large. This was the most we ever found.

The hydatid tapeworm represents a subtle life-history adjustment to the predator-prey relationship. It probably has little effect on the wolf, or the moose in early years of infection. However, the accumulation of cysts in the lungs of an old animal — one that must labor through deep snow in escaping its enemies — can hardly be without effect. A heavy burden of cysts must inevitably impair the critical functioning of the lungs and increase vulnerability to the wolves. In feeding, the wolves remove the lungs, swallow the living scolices of the tapeworm, and renew their own infection of adults.

There is much in the ecology of this parasite that is not known, and it is not an attractive species for research, since the eggs can be infective in man. This is one reason for the park admonition to boil one's drinking water. However, no human case of hydatid disease is known for the island, although staff members and our crew have been tested.[12]

The wolves of Isle Royale probably are being supported in the manner to which most wolves were accustomed in eons past. Presumably they are protected from the threat of manmade contrivances, and they harvest in the ancient manner an excess of production by the prospering moose. In the use of habitat, in reproduction, and in their congenital disabilities, the moose have adapted to their ecological chore. These kine of the wild have been at it a long time, and husbandry by the wolves is finely attuned.

The same can be said of other prey of the wolf. The predator tends to stabilize relationships between the herbivore and the vegetation on which it lives. However, stability is never reached, and fluctuations with changing climatic trends can be expected.

On a basis of wolf-free and wolf-inhabited areas of deer range in Alaska's southeastern coastal region, Klein and Olson concluded that, while local factors complicated the situation, certain generalizations could be made: In areas without wolf predation the deer stabilized or slowly increased at levels that were beyond the carrying capacity of the winter food supply. This produced heavy losses and the exhaustion of winter ranges. In

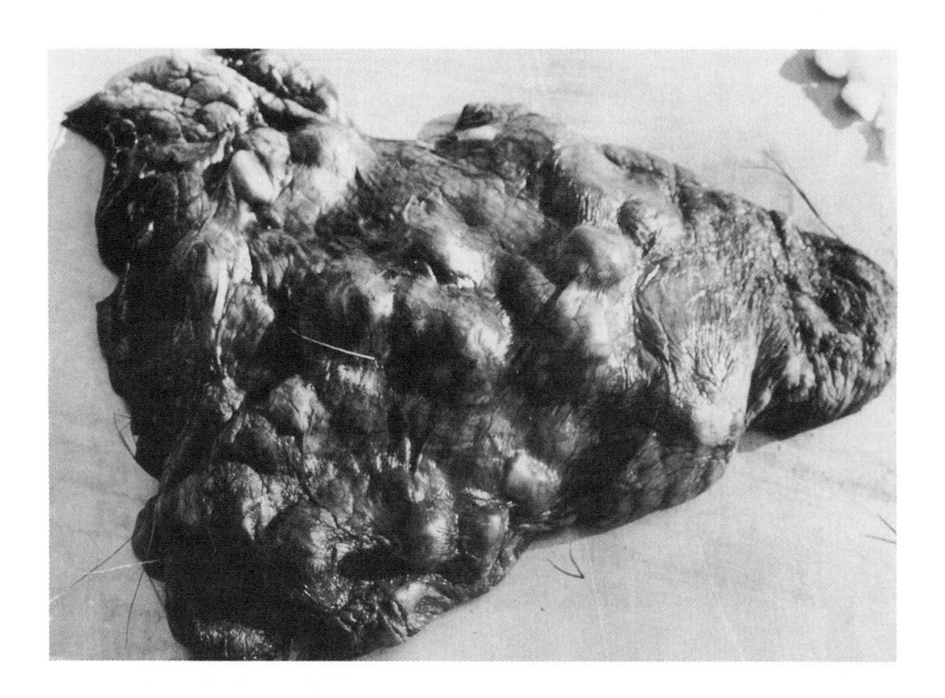

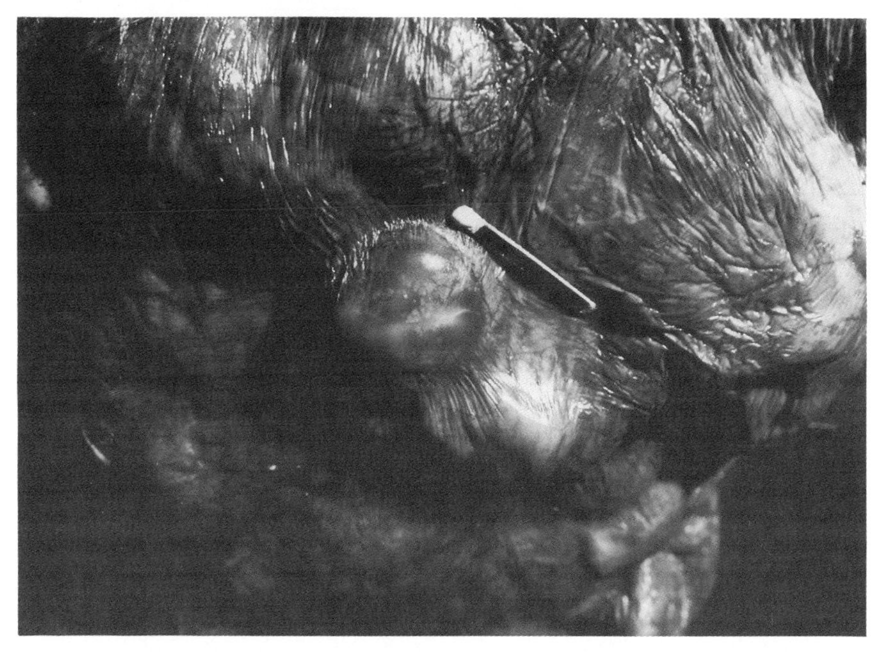

Top: The lung of an old moose (14.5 years) heavily infected with cysts of the hydatid tapeworm. The wolf is the primary host, harboring the tiny adults in its intestine. Bottom: Close-up of a hydatid cyst within the wall of the lung.

contrast, where ranges were supporting both deer and wolves, the deer had a higher rate of annual increase and few losses from starvation. Winter ranges were in fair-to-good condition. "The wolf-populated ranges, as a general rule, during the period of the study supported a greater annual hunter harvest of deer per unit of area under comparable hunting pressure."

If wolves were absent from Isle Royale, the moose probably would increase, as they did early in the century. They are not totally using the available browse (which would mean exhaustion) or, presumably, their living space. The trend toward expansion is countered by wolf pressure, of a special kind.

It begins for the moose in calfhood, where the least vigorous, a majority, become surplus. For some that pass this trial the productive years are short, and for the most durable they stretch out to age 12, 14, 16 . . . Probably for most the wolves do not allow the processes of aging and decrepitude to run their full course.

There is a point where usefulness in the population (efficient calf production) is on the decline. The keen sensitivities of the wolf are brought to bear. A particular moose performs its last function in the phenomenon of death and disintegration. There is one less moose, displaced in favor of another to come. Another that will be, for its time, productive.

To endure in vigor the prey animal must have this eternal culling out and cutting down. Individual loss is the corporate gain. Like all else in nature it is a system based on averages. In the set of evolution, essentially unlimited time and numbers, the odds are starkly realistic.

CHAPTER ELEVEN

Heavy Industry

MID-SEPTEMBER IS A TIME of great activity for beavers. This was plainly evident among the narrow ridges in the northeast section of the island in 1973. I was on one of my 5-day excursions. It had rained earlier, and I was doing some late afternoon exploring. About 6 P.M. I retraced the route to my sleeve tent beside an old flowage.

On the way out I had passed a new dam and pond. There were stumps and logs of freshly cut aspen. Much recent work had been done on the dam and a large lodge that rose from the water about 30 feet off the north shore. There on a steep slope I stopped and sat quietly to see what might be going on. It was the hour when beavers should be at work.

They were: Alders below the bank — flooded but still alive — obscured the view to some extent, but I could see two beavers swimming along the dam. A large one and a smaller one, the latter probably a yearling. Seventy-five yards to my left, in the upper shallows, brush was thrashing where another large beaver pulled a limb of aspen through a narrow waterway; it appeared to be a freshly dug canal that had not yet been cleared for easy navigation.

The beaver got his burden out of the tangle and towed it, butt first, to the lodge, which rose about 4 feet above the water. There he took the branch down, and I saw no more of it. The large beaver by the dam now came along with a stick that appeared 2 inches thick and 3 feet long. This one also was submerged just west of the lodge. That would be the location of the winter food storage pile, I thought. There was a fourth beaver on the far side that I never did see clearly.

One of the big animals swam to the head of the pond and

began hauling on a large spreading bough, probably more aspen. It hung up in the thick stuff, and I could not see what happened. But later I could hear gnawing from that direction.

The other large beaver barely stirred the water in a smooth swirl as it dived beside the lodge. Presently it reappeared on the gently sloping east side. The button-eyed creature was tar-black with mud, and under its chin, clutched with both front legs, it carried a huge gob of mire. On its haunches and propped by the flat tail, it shuffled up the incline on the spreading black webs of its hind feet. Near the top of the lodge it dumped the burden. Then with its forepaws, it pressed the mud into the jumble of sticks that formed the solid mass of the structure. I watched the laboring creature carry seven loads of mud onto the lodge and pack it into the surface layers. The lodge was about 12 feet across at the waterline, and it was not difficult to see why its frozen mass would be impregnable to enemies, meaning wolves, in winter.

Once the beaver went under and had its hod of mud in 14 seconds. Another time it took 20 seconds. The seven loads were delivered over a period of about 20 minutes. Toward the end of my watch the beaver I could not see well slapped its tail on the water with a resounding whack, and went under. A bit later the one nearby (now out of sight behind the alders) slapped its tail

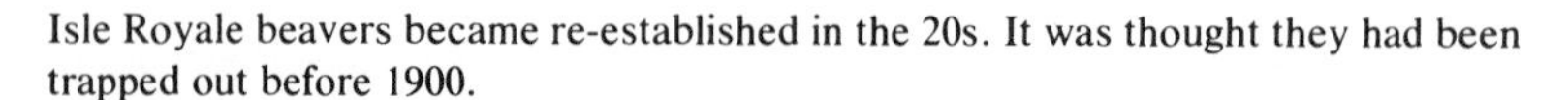
Isle Royale beavers became re-established in the 20s. It was thought they had been trapped out before 1900.

directly below me. Nothing moved for several minutes, and I realized it was getting dark. In passing around the end of the dam I saw the small beaver lying in the water a few feet from shore watching me. Twice I heard one of the animals make little whining calls; no kits of the year were in evidence.

Improvidently, I had brought no flashlight, and in the deepening shadows I picked my way carefully along the faint wolf trail that led back toward camp.

Dave's first two years on the project had made it evident that beavers were the secondary prey of the wolf. Various scat collections ran 10 to 15 percent beaver, and in a period of thaws in late February and March 1961 we had convincing evidence that winter, too, could be a time of jeopardy for venturesome beavers. Weather was clear, and daily temperatures were in the mid-20s.

On 3 March wolves of the big pack were scattered in the vicinity of Washington Island. Near a dock of the Sivertson fishery Dave and Don saw a hole in the ice with beaver tracks leading toward shore. The tracks ended in a blood-spattered area packed by wolf tracks, and nearby a wolf was working on a piece of hide.

That same day similar signs indicated a kill near a small island in Tobin Harbor. There were fresh beaver tracks at the outlet of Hatchet Lake, at the southeast corner of Todd Harbor, at the Chickenbone Lake outlet, and at Fishermans Home on Houghton Peninsula. Two days later another beaver was killed at the Tobin Harbor site. Dave and Don thought this one had used a tunnel through the snow from its lodge against the bank.

These were our first observations of winter beaver kills, of which there would be more in late-winter warm periods of subsequent years. The thaws of February and early March were the kind of weather we liked the least. Ice went out of the bays, and slush lay over inland lakes, so that landing was a problem anywhere. Rain packed the snow, and great pools of water gathered on Washington Harbor.

Such times set the beavers astir; they seemed hungry for anything green. The wolves were sensitive to the situation. They prospected the ponds and lake-edge lodges and patrolled shores where signs were promising. Sometimes we could neither fly nor snowshoe and were reduced to wading slush on the harbor to find something of interest. In 1964 a late-February thaw continued on to mid-March. On the 13th Don and Bill Bromberg were skirting the north shore of Beaver Island, and they found tracks of a large and a small beaver that had been working in a thicket of young birches. The animals had dragged cuttings down through a crack in the ice.

A few days later Don visited the place again and spooked a large beaver into one of three new holes. Under the open crack he could see evidence of a food pile, and he said he could hear faint mewings under the ice. There were wolf tracks all about, and a fresh scat contained beaver fur.

Phil Shelton enrolled at Purdue and appeared in my spring class in 1960. He had a wildlife degree from Montana State and had taken courses during the fall at the University of Chicago. As it happened, 1960 was the last year I would teach in the second

Shelton and Mech setting a beaver livetrap. The animals were examined, weighed, and ear-tagged.

semester.[1] I offered Phil an assistantship and the opportunity to study beaver-moose-wolf relationships for his Ph.D. Overlapping with Mech for the summer, he went to the island from June to September and then returned to the campus for two semesters of course work.

In the spring of 1961, Dave was in the field for an additional month (it beats thesis writing), and he helped Phil experiment with the Hancock live traps we purchased for trapping and ear-tagging beavers. It is testimony to Phil's care in setting and tending traps (he checked them between 10 P.M. and midnight) that in three years he caught beavers 357 times and killed only seven. In 1963 he was employed by the park as a summer naturalist — continuing the beaver work in off hours — and in 1964–65 we engaged William K. Seitz to go on with the trapping. Bill got additional records on marked beavers and made valuable observations of other kinds.

Traps were set beside lodges and dams and where drag paths or canals entered ponds. Phil followed the historic practice of trappers and used castoreum (from beaver scent glands), in addition to bait, for the first two years. Then he decided it was not necessary and did just as well with bouquets of plants the beavers were feeding on at particular times and locations. He found the abundant thimbleberry to be a preferred food until the third week in August, at which time it seemed to lose palatability. Other useful baits were aspen, birch, red-osier, willow, mountain ash, bigleaf aster, and green alder.

In piecing together the history of beavers on Isle Royale, Phil made use of a few old records and the more recent recollections of long-term residents. The earliest reference was that of John Tanner, previously mentioned (pp. 48–49), who found beavers plentiful on the island about 1793. In the park files were maps by William Ives, who made the original land surveys in 1847–48. In perusing these, Phil found records of beaver works at four locations in the southwest end.

It appeared that beavers had been trapped out, as they were nearly countrywide, by the time of the Adams expedition in 1904–5. There were no recent reports of the species. However by the early 20s beavers were becoming re-established, and they were much in evidence at the time of Murie's moose work in 1929–30. Fires of the nineteenth century had left extensive stands of successional forest, and beavers found the aspen and birch they needed adjacent to many inland lakes, beside the slow waters in narrow valleys favorable for dams, and around protected bay heads on the island margins.

Shelton also made good use of a set of aerial photographs that were in the park files, taken in 1930.[2] On the 600 prints he found 27 sites identifiable as beaver works. These were the more conspicuous ponds, and they did not include the less-evident lake, harbor, and bank locations. However, Murie had described all the beaver signs he found on a small portion of the island. So, extrapolating from these sources, Phil made a rough estimate that there were about 103 active sites in 1930.

Using the Mott Island bunkhouse as his base, Phil did most of his beaver work in the eastern half of the island, for reasons explained by a statement in his third annual report:

> Beaver distribution is influenced by vegetation types and water resources. The growth of young trees, especially aspen, birch, and willow, in the central quarter of the island burned in 1936 is producing abundant beaver food for several colonies. But the most extensively

Liberating a marked beaver.

used habitat is the forest created by regrowth of aspen-birch-conifer forests following fires that burned over the northeastern half of the island about a century ago. This area also has the most favorable water resources for beavers — many small streams, lakes, and narrow swamps — and by far most of the beavers are found in this half of the island.

Mature spruce-fir-birch forests, which make up most of the forests at lower elevations on the southwestern half of the island, have little food for beavers due to lack of aspen and young birch. Usually birches present are either large, over-mature trees, which beavers seem not to favor, or heavily moose-browsed sprouts. This part of the island also has fewer streams and lakes, and the extensive swamps are commonly too wide for easy damming and surrounded by little if any good beaver foods.

G. W. Bradt had studied beaver populations for the Michigan Department of Conservation during nearly a decade that began in 1929. By trapping out colonies he had found that a fairly reliable average number of animals per colony was 5.1.[3] Using this figure, Phil calculated a 1930 population of between 500 and 600 beavers for Isle Royale.

Ten years after that, the park was dedicated, and henceforth the record was filled out more satisfactorily. The 40s brought normal rainfall, and in a great central area the vegetation was recovering from the '36 burn. Park manuscripts by G. H. Gensch and K. T. Gilbert in 1946 and Laurits Krefting and F. B. Lee in 1948 described field surveys in which beavers were found in practically all suitable locations on the island. Food supplies were being heavily taxed. Krefting used this information, plus observations made in the course of his moose work, to estimate 150–200 colonies and a population of 600–800 beavers (144). Shelton regarded this calculation of numbers as low, based on his later finding of 6.4 beavers per colony.[4] He preferred an estimate of 1000–1300 for the late 40s.

In the years following 1948, beavers underwent a drastic decline — as much as 75 percent, Krefting thought. Food depletion seemed a logical explanation. It also became evident that this correlated with the coming of wolves to the island. But Phil was more impressed with another influence that had appeared about that time.

In 1949–50 beavers were dying in the Sioux Lookout area of

Ontario. A year later hundreds of muskrats died in Orr County, Minnesota. In the following spring Minnesota trappers reported many dead beavers in streams and under the ice of ponds. Some colonies escaped the blight; others were wiped out. Autopsies showed internal lesions indicative of tularemia, and in a few the organism, *Pasteurella tularense,* was isolated. It became evident that an epizoötic disease[5] was sweeping the aquatic rodents of the north country (263).

By 1953 the contagion had spread to northern Michigan, and trappers were finding many dead beavers. It was difficult to get specimens fresh enough for reliable laboratory work, and the bacterium itself was not found in specimens from the field. However, three trappers who had skinned beavers found dead or ailing contracted tularemia. Later, the disease (which also affected muskrats) was transmitted from dead beavers to laboratory mice, and its identity was confirmed.

Lawrence, Fay, and Graham reported that the epizoötic spread eastward in the upper peninsula of Michigan and largely ran its course by 1955. They reviewed circumstantial evidence that the invasion of a tick, *Ixodes banksi,* affecting both beavers

During the years of high water in the 60s and 70s, beavers increased their works and numbers on inland lakes and streams.

and muskrats, was associated with the spread of tularemia. Ticks are common carriers of this disease, with which they inoculate such host animals as rabbits and rodents. Phil mentioned in his thesis that it could also be spread by several kinds of biting flies that are plentiful on Isle Royale.

He noted that Isle Royale lay in the path of beaver die-off from Canada to Michigan and concluded that disease was the most likely cause of the decline on the island. The reduction was rapid from 1950 to 1955, but by the early 60s the population probably had recovered its level of a decade before. Lingering evidence of what happened was the prevalence of abandoned and washed-out ponds having still-adequate food supplies. Often these sites were reoccupied during the 60s.

Pelts of the widely distributed (found in every state) and plentiful beaver were a main incentive in early exploration. Markets of Europe were eager for the seemingly unending supply. Beavers were a staple food of Indians and also the trappers and voyageurs of New France, especially since the Pope had classed the animal as fish and it could be eaten on fasting days. Peter Kalm wrote in 1749 that vast numbers of them had been killed (27):

> All the people in Canada told me that when they were young all the rivers in the neighborhood of Montreal, the St. Lawrence River not excepted, were full of beavers and their dams; but at present they are so far destroyed that one is obliged to go several miles up the country before one can meet one . . . skins from the north are better than those from the south.

Kalm himself had little enthusiasm for the eating qualities of beaver flesh, but a tradition was engendered in the backwoods that beaver tail was a special delicacy. This probably is another example (p. 456, note 10) of people cherishing a food of necessity. The black scaly tail of the beaver has a high fat content, and during late winter and spring trapping it could be a main source of this essential nutrient, for which a craving developed when nearly all game was poor.[6]

The beaver of North America is much more of a dam- and lodge-building expert than its near relative in Europe, which lives habitually in bank dens; all beavers do this on large rivers. No other creature of the continent was so influential on upstream hydrology (flood control, ground-water recharge, streamflow maintenance) and on the habitats of other aquatic life. The animal stirred the interest of early naturalists and became the subject of the first authoritative monograph on a North American mammal.

Lewis H. Morgan was a director of the Marquette and Ontonagon Railway, which was completed in 1858 westward from Marquette to upper Michigan's iron mines. He became intrigued with the extensive beaver workings of the district and made careful studies of the species and its habits for more than 20 years, culminating in his landmark publication, *The American beaver and his works,* in 1868. Morgan's epic record is still outstandingly useful, and it stood practically alone in its field until the present century when Bradt and others brought beaver investigations into the realm of modern scientific research. Thus, there was much background information to draw upon in our island studies, but Shelton had the specific objective of learning how beaver numbers and habits related to the environment and the support of moose and wolves.

Trapping on occupied sites — some records covering several years — largely confirmed the findings of Bradt farther south in Michigan. A "complete" summer colony consisted of a pair of old beavers, their kits of the year, and their yearlings born 12 months earlier. The number of young usually was 2 to 6, averaging 4. It is an unusual circumstance that in this species the mortality among kits is extremely low; often a litter will survive intact to the second summer. Thus the number in a family group may be as high as 10, or even more.[7] Any number down to a single animal could be found, although kits were never without parental support.

Over a 5-year period, Shelton and Seitz handled 298 individual beavers. They found the sex ratio to be even, as were the weights of adult males and females; both sexes averaged slightly less than 40 pounds.[8] In the total sample, the following age groups were represented:

Kits (117)	39 percent
Yearlings (82)	28 percent
2-year-olds (28)	9 percent
Adults (71)	24 percent

Since all beavers are about equally easy to catch, the above listing probably indicates the composition of the total population. Single individuals and wandering animals are least likely to be trapped and could be under-represented. The low percentage of 2-year-olds may be realistic because various studies indicate that usually they are driven from the home pond in spring before birth of the next litter. However, on Isle Royale some of these young were known to stay through a third summer. Bradt found that, on occasion, beavers made long cross-country trips away from water. Probably many of these wanderers were the expelled 2-year-olds, and outside the security of the natal pond they probably are subject to high mortality. It is a good guess that on our wolf-inhabited island beavers follow water courses as much as possible in their explorations for new homes.

In following the movements of individuals, Phil found that three couples of yearlings (at least two pairs were male and female) dispersed from the parent pond and established their own winter quarters: a dam, lodge, and food pile. One pair crossed nearly 5 miles of open water to Passage Island, northeast of Isle Royale. In all cases they were gone the second summer.

Based on one record obtained by Bradt on captive animals, the gestation period of beavers is approximately four months (Morgan said three months). In this case breeding took place in January, and the most common borning time probably is late May. Knowledge of beaver breeding and early care of young is sketchy at best.

By late June kits may be swimming about close to the lodge and feeding on plant materials brought by an adult. On 1 July 1964 Bill Seitz saw four kits on a lodge in a pond northwest of Chippewa Harbor. The adults were working on the dam, and one of them took a piece of aspen bark to the kits, which they ate. When swimming the young animals stayed within about 4 feet of the lodge. Seitz remarked that the adults paid slight attention to them.

There is circumstantial evidence that when parturition is imminent the adult male finds temporary quarters, possibly a bank den, if available. Before the onset of winter he probably rejoins the family.

At various times a kit has been heard to cry like a young child. On 25 July 1964 Seitz saw two of them transported across the shallow waters of Hidden Lake by riding tandem on the lower back and tail of an adult. The young were deposited on a sedge mat, where they sat and called loudly while the adult dived nearby for a time and then brought aspen twigs for them to eat. Bill had watched other kits feed on the bark of aspen twigs at Chippewa Harbor on 16 July.

At the Hidden Lake lodge he saw what evidently was some kind of social communication between the adult and a yearling that swam to meet it. This consisted of a nibbling behind the ear and a touching of noses. "Usually the nose touching was followed by a sort of push off." Evidently this is what Seton was describing when he said, "When two meet in the pond I have several times seen them nibble each other's cheeks, at the same time uttering a chattering noise . . ."

A more elaborate behaviorism, strongly resembling copulation, was seen by Seitz and Stauber on 20 June 1964. The animals were about 200 yards from the lodge, which was built against the shore behind one of the islands in Malone Bay. One of the beavers was smaller than the other, and they had been feeding on green alder along the shore. As Seitz recorded it:

After the feeding, the two beavers started calling and swimming about each other. The larger of the two then mounted the smaller in shallow

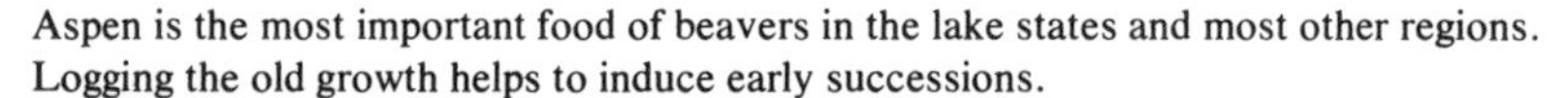

Aspen is the most important food of beavers in the lake states and most other regions. Logging the old growth helps to induce early successions.

water (2-3 feet) about 10 feet from shore. The larger animal bit or seemed to bite the side of the neck region. The beavers were almost out of water at one time.

There were three copulation [?] periods lasting from 30 to 50 seconds in length. The series of copulations was followed by calling from both the beavers. They lost all sensitivity to sound and sight.

On visiting Wright Island, Seitz and Stauber learned that the Holtes had seen approximately the same kind of interaction between two beavers near their fish house on the 20th and 21st.

In June it is unlikely that copulation was involved in this behavior, although it might have pair-bonding significance. Schramm described a wide variety of "play" engaged in by a pair of adults he observed through summer months. Some of it resembled the mating described by Bradt as witnessed in January. Beavers are difficult to observe, and only a start has been made on the study of their behavior. Seitz remarked that when the above activity was witnessed the clarity of the water made it easy to see how the beaver propelled itself under water by using hind feet and the tail. The front feet are pressed close to the body.

The beaver habit of leaving small piles of mud strongly scented with castoreum[9] about their shores is also a subject for behaviorists. On 28 June 1965 Seitz found many of these about a new (from the previous fall) lodge built against the bank of upper Washington Creek. He had earlier mentioned finding three scent piles by the old lodge on Baker Point (Rock Harbor). Shelton said a pair of beavers had moved in there in 1964, and they had kits in '65. One suspects that the scent piles say to a strange beaver, "I am here and in business. Please move on."

Except for anxious mothers with young kits in the lodge, most beavers are sanguine about being trapped and will sit quietly and eat all the bait in sight. Phil noted that animals with full stomachs were least likely to hurt themselves, so he habitually loaded the traps with favorite foods. He handled a trapped beaver by inserting a burlap sack between the two halves of the trap (which operate like a clam shell) and working it down over the head of the beaver. The sacked animal could then be weighed, sexed, and tagged in both ears for future identification. If a tag got lost, the second one served as a back-up, although in a few cases both were lost. Strangely enough, neither Shelton nor Seitz ever got bitten.

The ability of a beaver to survive a trying experience was put to the test on 16 November 1962. It was the fall Phil stayed late on the island and went home with the fishermen.

He was trapping one of the ponds southwest of Moskey Basin. The trap was set at the mouth of a canal and had caught a 23-pound female. Wolves had paid her a visit, wading in ankle-deep water for about 15 feet.

> The trap had been pulled back by one side several inches and enough water had been splashed up that the trap was frozen to an alder stub. The beaver looked like she had had a bad night; the hair on her muzzle was matted and there were ice balls hanging in the fur of her rump. There was ice on the tail also. There was plenty of fight left in her — she ran at me while in the trap. In the bag she whined like a kit several times.

He said the released animal swam out a few feet, dived, and was seen no more.

Its almost incredible feats of construction — dams, lodges, canals, and burrows — have given the beaver a proverbial rep-

Beavers that built this lodge (upper Tobin Creek) were living on birch and the thick understory of beaked hazel.

utation for hard work, even intelligence. Avoiding any anthropomorphic comparisons, we certainly can say that this creature deliberately manages its habitat in greater degree than any other wild vertebrate.

Beavers living on the relatively stable waters of bays and inland lakes commonly build their lodges against the shore or an island. Where conditions are favorable they may dig bank burrows. The entrances of these are under water and they are difficult to locate; Phil found only three, one of which was revealed when a pond washed out. Bank burrows extend upward to a nesting cavity above water level. Seitz said that two beavers were living in a "tunnel" den on Raspberry Island in July 1964. There was no lodge.

The lodges are piles of sticks, stones, and mud rising from the floor of the pond well above the surface. Some of these are added to and grow in size for decades, becoming islands overgrown with alders and other shrubs. Pete Edisen said the Baker Point lodge was there when he came to the island 54 years earlier.

As new materials are added to the outside of the lodge, the inside is hollowed out in a low dome-shaped cavity with a platform above water level. Morgan, who dug out several lodges, found "marsh grass" used as nesting material. Usually several tunnels lead down out of the lodge into deep water, providing safe access summer and winter.

At the top of a lodge the tangle of sticks is not plastered tight with mud. Its loose structure forms an air conduit, and in winter warm air can sometimes be seen emerging from it. Such lodges are always carefully investigated by any passing wolves, but to our knowledge none of them ever found a solution to the problem.

Closely associated with the lodge or bank burrow — if it is to be used as winter quarters — an underwater food pile is built in the fall. This consists of branches and small logs of the best foods available, most commonly aspen, birch, hazel, mountain ash, green alder, and mountain maple. The speckled alder that is so abundant in wet ground is much used as building material, but it is not a preferred food.

Evidently the beaver has no difficulty in sinking the green boughs of the food pile and anchoring them to the muddy bottom. They quickly become water-logged. It is a hazard on open bays that ice or currents may take out the storage pile, which is washed away in the fall, no doubt producing a disaster for the resident beavers. Under normal conditions, in winter a beaver can leave the lodge, go to the pile, cut off a short length, and take it into the lodge where the bark is gnawed off and eaten. The bare stick is then turned loose under the ice of the pond, where it will be available as building material. Cutting under water is not a problem because the animal's lips fold in behind the four chisel-like incisors, which are thus exposed for action. The large orange-colored incisors are anchored in a skull of great strength. They are faced with a layer of flint-like enamel backed by softer dentine that erodes more rapidly, thus leaving a tapered cutting edge. The teeth grow constantly to keep up with the steady wearing away that a beaver's way of life brings about.

Wherever there is a sustained trickle of water and a food supply in the form of aspen groves, birch, or other acceptable woody plants, a pair of beavers will inevitably build a dam and create the pond they need. A dam usually is 2 to 5 feet high, and the longest dams Shelton found on Isle Royale were 300 to 400 feet, on lower Tobin Creek. Dams described for other regions have often been longer, and the highest dam recorded by Morgan was 12 feet; it was 35 feet long.

Dams are started by laying innumerable sticks across the site. Most of them parallel the water flow with the tops downstream. They are weighted with more sticks, and on the pond side mud, sods, and stones are worked into the structure forming a solid bank that holds back the water. Streamflow percolates through the sticks here and there, preventing a build-up of water force at one point. Commonly several dams are built below the main one. Morgan thought this was to create a back-pressure on the base of the large dam and help prevent washouts.

The water level in a pond is nicely adjusted to the height of the nest and feeding platform in the lodge. Dams in good repair are full to the topmost level, where they are chinked with sods and many small sticks that gather floating leaves and other detritus. They require constant maintenance, and if a pond is depopulated for any reason, the dam soon goes out. When a dam is broken, this calls for immediate (usually nocturnal) repairs, and the old-time trappers used this situation to good advantage in catching the inhabitants of a pond. Trapping was also done at lodges; both devices are now illegal.

When one hikes cross-country, a dam often is the only way to get across a series of impoundments or wet marshes. On a relatively new dam this calls for something of a balancing act. Old dams tend to be wider and well grown up with bluejoint grass. They often support a stand of alders or other shrubs. Well-marked game trails may cross these old structures, which have come to resemble substantial earth fills, with new construction and maintenance only at the outlets.

Where terrain favors (flat and wet), beavers dig canals for hundreds of yards to stands of timber that can be harvested for food. Canals on Isle Royale tend to be short. Usually they are about 3 feet wide and 18 inches deep. It is unlikely that they offer much safety from wolves.

Ordinarily well-maintained ponds are stable and last as long as adjacent food supplies hold out. However, there appear to be times of quick runoff in early spring that are too much for some — or something has happened to the tenants. The winter of 1971-72 was one of deep snow, followed by heavy late-spring rains. As a result, at least three major dams were washed out. After a more nearly average winter, in May 1976, I found two drained ponds with washed-out dams in the Todd Harbor and Pickett Bay area.[10]

Sometimes such ponds are not restored for many years. In the spring of catastrophe small groups of yellowlegs and sandpipers patter across the dank mud flats, and the borings of woodcock will be found along soft edges. Before the season is over, sedges, bluejoint, bur reed, iris, and other herbaceous vegetation will begin turning the muddy bottom into a meadow. As years go by a fringe of sweet gale and leatherleaf develops. Labrador tea, skunk cabbage, and marsh marigold spread; speckled alders take over the drainage, effectively hiding the old lodge. Then, if that mound of earth and sticks is still large enough, it might become a wolf whelping den for a season or two.

With the improvement of drainage and the annual deposition of plant material, the meadow becomes less moist, permitting the invasion of other trees of low ground: white cedar, black ash, black spruce, and balsam. On surrounding slopes, once heavily logged, aspen suckers may spring again from the underground root system. Birches can seed on any disturbed ground.

However, this new growth is likely to be suppressed, along with seedling balsams, by moose browsing. The cutting of broad-leaved trees by the beavers may accelerate the invasion and growth rate of the conifers, in which spruce and cedar probably will predominate. Eventually, in some dry summer a fire is likely to burn again and a new belt of successional forest will develop around the old bottomland.

Then another generation of beavers can move in, lay low the alders, and build their series of dams. A deadening is created as the waters rise, sun-loving water plants become established — food for both moose and beavers — and the cycle of usefulness begins again.

Shelton found that in summer there was little tree-cutting by beavers, except for aspen, which is a year-round favorite. Leaves, twigs, and bark are eaten.[11] Where heavy growths of thimbleberry are near water, it is a mainstay. Many kinds of herbs and water plants are taken, and Phil had indications that in any particular location the animals probably developed a preference for familiar foods.

This was illustrated when I arrived, one summer day, at Mott Island to find Phil and Larry Roop reclining on the grass, with a 3-pound beaver kit sitting complacently before them. It had been in a trap long enough to be hungry, and they were offering it a succession of foods and recording its preferences. Some it would eat readily; some it rejected. The animal was about 6 weeks old and could have had experience with a wide assortment of foods. Phil's list of accepted species (twigs and leaves) was:

Aspen	Thimbleberry
White birch	Rose
Red-osier dogwood	Raspberry
Bush honeysuckle	Strawberry
Mountain ash	Barren strawberry

This picky young 'un did not particularly like bigleaf aster, and it definitely turned down the following:

White clover	Solomon's seal
Cow parsnip	Pyrola
Viburnum sp.	Fringed polygala
Apple (fruit)	Sarsaparilla

It was especially surprising that the kit did not eat clover, since the animals at Cemetery and Caribou islands were mowing it down regularly. Such observations implied that customary habits had much to do with what beavers ate. I believe this particular young animal was one of a family that had a nest under the old log dock at Mott.

During the years of this work our field men have recorded the use of many species of woody and herbaceous plants by beavers — some only occasionally, but in all a major portion of the island flora. Nearly any woody plants growing around a pond may be cut for construction purposes, but we have had little evidence of feeding on speckled alder, red maple, sweet gale, black ash, white cedar, spruce, and fir. However, balsam and cedar have been barked by beavers that evidently were on a quick sortie for green food in winter. What a beaver eats at such times may be quite atypical.

In listing the plants cut by beavers, Shelton placed 12 species in a special category as "ecologically important." Next to the first-ranked aspen was paper birch. This was often the main dependence of colonies around the shores and islands on Lake Superior waters, where the animals sometimes ranged as much as half a mile from the lodge in gathering food. At both Mott Island and Snug Harbor (the Rock Harbor Lodge area) wire netting had to be placed around the old birch trees to protect them from late-fall cutting by beavers.

Among the shrubs on Shelton's list, the most commonly taken were thimbleberry, mountain maple, beaked hazel, red-osier dogwood, green alder, and speckled alder. The last-named species was included because it is so often cut, though seldom used as food. Of the succulent aquatics fed upon, Phil considered both yellow and white water lilies to be important, as were the pondweeds (*Potamogeton*) and water shield. Rootstocks of some of these probably are available to beavers foraging under winter ice, and the buds and foliage are eaten in summer.

Practically all studies of beavers have given aspen or other species of poplar top ranking as a food. Birch is another staple wherever it occurs. On Isle Royale this means paper birch for the most part, although Seitz found substantial cuttings of yellow birch on Grace Creek. In May 1975 at a new pond west of Lily Lake, Rolf recorded the use of both white and yellow birch as well as hard maple.

In working up his records on trapped animals, Phil compared weights and measurements of animals living primarily on aspen with others that depended mainly on birch. He found that 17 adult beavers in the first category had a mean weight of 42 pounds, and 15 adults from the second (birch) group had a mean weight of 36.4 pounds, the difference being statistically significant. His sample of 5 adults that fed mainly on aquatic plants was too small for significance, but the mean weight of 42.3 pounds compared favorably with the aspen-supported animals.

Another measure of preference was how far from water beavers would venture to cut trees. Here again aspen was first choice. Nearly all birches cut by beavers were within 50 feet of water, while aspens have been taken regularly up to 100 feet from safety. Steep slopes appeared to be favored because of the ease of moving heavy chunks of wood to water. In the Windigo area Phil recorded cuttings up to 210 feet from the shore. Later (1964) district ranger Bill Bromberg gave Seitz a written record of a beaver "run" uphill into a stand of balsam poplar. "It extended 127 paced yards from the water's edge. Many of the poplars had been cut from the termination of the path in all directions for about 25 feet."

Beavers will cut a tree of almost any size; I have measured several that were more than 20 inches in diameter above the cut.[12] Morgan said that two beavers will work on a tree at the same time, but we have no records on that. After studying the matter, he also decided that the chiseling out of chips is done entirely with the lower incisors used one at a time. There is no indication that the animals attempt to fell a tree in any particular direction. When a tree does topple it sometimes hangs up in an adjacent tree, whereupon the beaver patiently undertakes another cut. I found a 19-inch overmature aspen that, by a prodigy of labor, had been hewn down four times before it finally came to the ground. Even then, it was only partially used.

Old rough bark on a large tree is not suitable for food; only the smooth green bark of the aspen and the smaller limbs of birch are taken. Phil found a section of birch 11 inches in diameter that had been peeled, but that is exceptional. He observed that where food trees are plentiful the beavers are most wasteful and utilize less of the trees they cut. Where the supply of good foods is scant they tend to be more efficient.[13]

How readily these animals will take the risks of winter feeding above the ice evidently depends in part on how successfully they have provided for themselves in fall. In the winter of 1962 Shelton saw that the colony on Outer Hill Island (just north of Mott) had opened a hole in the ice and were feeding on fall-cut birches and freshly cut aspens. The food piles this group had built in September and October had washed away in a November storm. Phil deduced that this was why these animals had to feed on land in February.

It may be that on the outer bays and rocky islands a food pile is especially difficult to anchor on the bottom. On 21 June 1964 Bill Seitz found such a pile that had been washed up on the shore of Malone Bay. This one was 60–70 percent mountain maple, the remainder being mountain ash, birch, and green alder. Mountain maple is especially plentiful on Malone Bay islands and shores, and Phil had noted its extensive use by beavers there.

In their relationships with other species, beavers are mainly a positive influence. They create the aquatic habitats sought by moose in their feeding, and the thinning of forest edges around shorelines induces sprout growth that is taken by the moose. As Phil pointed out, however, browsing largely prevents the suckers and sprouts on these sites from growing into trees that could be useful to the beaver.

As another advantage to moose, they are quick to find the trees felled by beavers. They compete for aspen bark in early spring, and later they take leaves and twigs as browse. I have tried to think of something the moose does for the beaver, but it seems largely a one-way street. In immediate contacts between the two species there appears to be a recognition that the other is harmless. At Newt Lake on 8 July 1963 Larry Roop was watching a moose feed on water lilies. ". . . one of the beavers would swim around him within 10–15 yards and slap its tail on the water. At first the bull would jump each time the beaver slapped its tail, but after a few times it paid no attention."

For the marsh-inhabiting muskrat there is little good habitat on Isle Royale, but a scattered population of these smaller rodents share many water areas with the beaver. There is no antipathy between the two species; Seitz mentioned seeing a beaver and a muskrat only a foot apart. Over the years, our

Mount Franklin beaver pond in 1962; the colony was active at that time.

observers have made many records of the use of beaver lodges by muskrats. Muskrats have been seen frequently carrying nesting material in the form of sedges, grass, or other vegetation into lodges, and they obviously use them as both summer and winter quarters. In 1974 Rolf saw six muskrats on an old beaver lodge on 18 October.

It should not be surprising that abandoned lodges are useful to muskrats, but in some cases the two species are joint tenants, entering and leaving a lodge with no evidence of antagonism. After watching such activities at several locations, Seitz raised the point of primary interest: "Whether or not the two species occupy the same chamber is not known. The muskrats might have a separate chamber." Obviously, someone with great curiosity and plenty of help could solve this problem.

Isle Royale's beaver ponds are the best of its duck habitats. In these shallow waters there are no northern pike, and the ponds are little frequented by gulls, both predators on ducklings. It is likely that the stands of bluejoint grass on low ground

The Mount Franklin site in 1974. In 12 years the pond had become a meadow.

around ponds are good nesting cover for the dabbling ducks, mallard, black, and teal. I have several times seen beavers and ducks active within 20 feet of one another without showing alarm. Deadenings created by the flooding of forested bottoms are a retreat for waterfowl in the flightless eclipse period of late summer.

As one observes how the beaver conserves runoff water and diversifies habitat for many creatures, he can only reflect again on the far-reaching influence this animal must have had, continentwide, in primitive times. Today the wetlands are more than half gone, and we work away at what remains.

In 1960 and 1961 some preliminary counts of occupied beaver sites were made in connection with the fall aerial moose "census." As other workers had noted, food piles were the best evidence of preparations for winter — in other words, of established colonies. In 1962 Phil devoted 17 hours of flying, in a float-equipped Aeronca Champion, to an intensive count of active sites.

Both aerial and ground work accounted for approximately 140 colonies with food piles, which converted to a total population of about 900 beavers. It was evident that the animals had substantially recovered from the low level of the 1950s, and Phil suspected that they were still increasing. After his field work was completed in 1963, he was employed elsewhere for more than five years with no further opportunity to observe the Isle Royale situation.

We were busy with moose and had no hard information on the beaver population for several years, but they seemed to be doing well. In winter warm periods we continued to see evidence of activity on the snow, and occasional animals were killed by wolves — despite the fact that wolves hit their low point of numbers in 1968. Comparisons from one winter to the next were difficult because so much depended on the widely varying weather conditions.

Then 1969 introduced a period of four years three of which were characterized by deep snow, abundant runoff, and ample summer precipitation.[14] Water was trapped in narrow basins well up on the ridges, and beavers were logging areas to which they had not had access before. Some of these sites were abandoned when water levels dropped in late summer. By the time

Peterson began his work in the early 70s it was evident that both beavers and wolves were on the upswing.

In 1969 Phil returned to the island for three weeks in October, which included a brief job of flying with Bill Martila (limited by weather). His estimate was that the beaver population was up at least 25 percent over the level of 1962. He did further fall field work in 1973 and 1974, which included more adequate aerial surveys with Dale Chilson in a Piper Cub. Each year they flew 12.5 hours, and conditions were especially good for the flights in '74.

More trapping also was done during the period of fall field work, 10 September to 24 October. Phil found that Isle Royale beavers had been on the move since his early work, when he concluded that there were 140 active colonies and a population of about 900 animals. In 1973 he estimated 315 colonies. "The 1973–74 figures indicate an increase of 125 percent since 1962. In other words, it appears that Isle Royale now supports a fall beaver population of about 2000 animals."

It was of particular interest that occupied sites on the large lakes and on Lake Superior waters had not changed significantly. The big increase was on stream habitats, supporting the view that abundant runoff water was a favorable factor. Actually, beavers appeared to have increased rather steadily since the die-off of the 50s, and how much the changing weather pattern influenced this we are unable to say.

Rolf found that the wolves had adapted to this situation. In the summer of 1973 he gathered and analyzed 540 scats, the first such collection in 10 years. Beaver constituted over half the prey items, and they were used consistently from spring to fall.

It is evident that the beaver population was not noticeably limited by this drain on their numbers. The hard-working animals were harvesting the herbage of waters and woodland edges, converting plant material into a summer food supply for wolves, in the season when pups were being reared.

As we shall see in a later chapter, this may have done something very special for the wolves.

CHAPTER TWELVE

Wolf Society

BY THE TIME we got to the island in late January or early February, it was always evident that the breeding season of the wolves was in progress. This is a period of great social activity, when behavioral relationships show up most clearly. We were originally quite "cold" in the field of ethology, which is the biology or comparative study of behavior, but with the help of our friends at Chicago (p. 446) and with the coming of Klinghammer to Purdue, we began to understand more clearly what we had been seeing.

One advantage most of us have in recognizing the more obvious behaviorisms of wolves is our preconditioning on the dog. As Scott and Fuller conclude, the dog has had about 10 millenniums of domestication, starting with a wolf ancestor somewhere in central Europe or possibly Mesopotamia.

Thus, as we flew over the well-fed pack and found them in a trifling mood, we saw many familiar antics, posturings, and lecherous sallies of the kind one could expect in a gang of neighborhood canines. However, it is clear that the wolf is not just a runaway domestic on a tear; he is a landed proprietor, organized for the serious business of earning his way in the world.

The well-understood (by wolves) relations of one pack member to another — filial, parental, sexual, dominant, submissive — are in evidence the year round, but they are most keenly sensitive and elaborately displayed during the time of courtship and mating. Essentially this is the month of February, as shown by the record of pairs observed in copulatory tie during the years of our study:

1959 —	21	February	1972 —	20	February
	24	″		27	″
	27	″	1973 —	16	″
1960 —	4	″	1974 —	4	″
1965 —	23	″		4	″
1968 —	13	″	1975 —	20	″
1971 —	7	″		26	″
	8	″			

Based on records at the Brookfield Zoo, Woolpy said that gestation, commonly 62 days, may vary by three to four days either way. It appears that on Isle Royale most litters, averaging about six,[1] are born during the second half of April or early May. The above copulation records do not necessarily indicate a precise date of birth, since a breeding pair may couple repeatedly over a period of a week.

The genetic inheritance of wolves decrees that there will be a pack and there will be dominance hierarchies, usually separate, in males and females. What wolf occupies a particular position is decided by individual characteristics and associations. A main function of animal societies and their dominance and territorial relationships is the spacing out of numbers over the range for optimal use of habitat resources. No species can be successful (as men will learn) without effective means of population control. In a top carnivore like the wolf, little subject to natural predation, the control of numbers is largely built into the societal structure.

The dominance order in a pack is a social security system, the mechanism for governance and peace-keeping. The way dominance is expressed and how it operates vary widely in different species of vertebrates, but it is always present. Without dominance regulation or the human counterpart — good manners — we have individual isolation and social nihilism. Health and survival would be impossible.[2]

In the pack of mid-February, we see frequently the manifestations of rank and social category. On crisp bright days when the animals are traveling, they ignore the slowly circling plane and proceed with affairs of the season. Usually one could say they are hunting, but often they are more intent on their doings with one another.

The original big pack of 15–16 animals was a good one to

exemplify social relationships at the beginning of our study. It was markedly stable for five years and clearly kept its identity as a continuing unit for at least eight years, 1959–66. Its record of maximum winter numbers for this period was as follows:

16 — 16 — 15 — 17 — 16 — 22 — 18 — 15

Two qualifications may be in order. After an initial count of 22 on 2 February 1964, the pack numbered 20 for the rest of the winter. In the last year of the series, 1966, after a single observation of 15, the pack declined and one animal was killed, as will be told.

The big pack was larger than most wolf packs, which we attribute to the fact that it was free from man-inflicted losses, to our best knowledge. In his book of 1970, Mech summarized published information on pack size in various ranges. It is evident that breeding packs commonly number 5 to 8 animals, and a few are 10 to 12. Groups larger than this are unusual, especially in regions where the wolf is subject to fur trapping, hunting, and control programs.

Zimen concluded that pack size is affected by a variety of factors including food supply, size of prey, prey density, and wolf density. Conditions on Isle Royale — especially food supply, as will be shown (Chapter 15) — explain wolf population density much better than they explain pack size. Our best conclusion is that the maximum number for a viable economic and social unit is 16 to 18 animals. Higher numbers tend to drop back to this level, and some packs seem to reach their limit at 8 to 10. These figures apply only to breeding packs, not aggregations of nonbreeders. The latter have varied from 2 to 4, the most common nonbreeding animals being loners.

A wolf pack originates with a breeding pair, and it builds in number with the addition of successive generations of pups. Initially, it can also build, to a lesser degree, by the addition of adults acceptable (through previous association?) to the breeding pair. In recent years we have seen this happen on the island. Pups develop a strong social bond to their parents, and they have amicable relationships with other adults in the pack, mostly older siblings or aunts and uncles.[3]

In this way the pack is integrated, and it does not accept strangers on any basis that we know of. It occupies a territory

A portion of the big pack in February 1962. The second animal in line with its tail up, plus courtship activity at lower right, indicates the presence of both alpha and beta males.

(a defended range) that supposedly is exclusive. However, this ideal is not achieved on Isle Royale because of the incursions of loners, some of whom wander widely. They scrupulously avoid the pack, for they will be killed if caught.

A pack does not always stay together, and some split temporarily more than others. Our records on seasonal numbers and splitting during the first eight years were summarized in the paper of 1967 (125). For most packs it is not unusual for one or more animals to drop off and not be seen for a period of days or weeks. This complicated our counts of the Island population, for how do you distinguish such individuals from "true" loners?

In large continuous ranges, this habit of separating probably could bring potential mates together and lead to the establishment of new packs. Also, not all dropouts return to the parental group, as some of our counts have shown. In radio telemetry work in Minnesota, Mech and his cooperators obtained location records on a lone male in a dispersal movement that took him a straight-line distance of 129 miles within a 2-month period (26 February–24 April). The total minimum distance traveled was 226 miles (181).

I have referred to them as dropouts, but sometimes the departing wolves are "driven-outs." Also, our experience with low-status individuals that trail the pack (p. 99) indicates that they get less welcome as time goes on and eventually lose all or most of their social connections. These become what we have regarded as true loners, who live out their days scavenging and making a living as best they can. Some are definitely old and emaciated ("hammerheads," Don called them). In deep snow Shelton spooked one off a kill on the north shore of Rock Harbor that he suspected he could have caught had he been on snowshoes.

Elsewhere (p. 99) I have suggested that loners may aggregate and form the duos, trios, and foursomes that we have seen in some fairly stable associations outside the range of a breeding pack. Loners may be found on anybody's property.

From what has been learned through various studies of wolves in captivity, the social organization of a complete and typical wolf pack is fairly well known. In this field Schenkel is the pioneer who developed much of the terminology now used. His findings have been confirmed and added to in subsequent work by Rabb, Ginsburg, Woolpy, Fox, Klinghammer, Zimen, and others.

This is the background by which we interpret our many years of observations of wild packs. After the first five years, mainly owing to Mech's work, we knew reasonably well what was taking place in the big pack (p. 98).

A table of organization for a "mature" (several years old) wolf pack might look like the diagram in Figure 10.

Obviously the administrative establishment in this representative pack is the group of three dominant individuals at the top, two of which are males. In the field a dominant animal often can be quickly recognized by its elevated tail — especially evident at times of social interaction. Most of the time a dominant carries its tail straight out behind, and since a "lead" wolf is not always in the lead, it takes a favorable period of observation to sort out a "who's who" of the pack. Sometimes from the air we cannot do it at all.

The alpha male exercises dominance over all the others, but most particularly over other males. The beta male shares the privileges of leadership, but he is subordinate to the alpha male. Often this relationship is low-key, and the beta is the more aggressive in throwing his weight around among the lower-ranking wolves. All packs considered (new and old), it is usual for the dominant male to be sexually bonded to the alpha female. However, the beta male will mate with the alpha female if the alpha male does not assume the breeding role.

The female hierarchy is headed by the alpha, who ordinarily is the only one who breeds and bears young. Other females may come into estrus and are physically capable of breeding. However, both males and females have a psychic inhibition of their sexual lives in the presence of alpha animals. Klinghammer found that the estrous periods of subordinate females were much abbreviated under these conditions. Removal of the alpha re-

Figure 10. Schematic representation of the social relationships in a winter wolf pack.

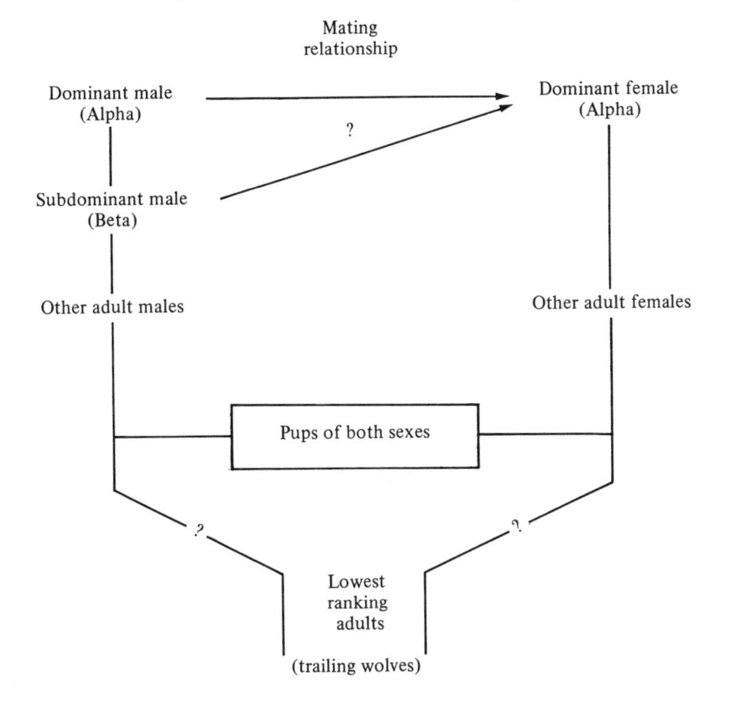

In February 1970, the dominant gray male (center) was bonded to the small alpha female, who had a crook in her left front leg. The large black male associated closely and was obviously the beta animal of the west pack.

moves the physiological blockage, and lower-ranking females can then have a normal long (a month or more) estrous period. This undoubtedly accounts for the high frequency of breeding in the Alaskan wolves, whose packs were often broken up (R. A. Rausch).

Among those Alaska females the first breeding took place at 22 months of age, and they bore young at the age of 2 years. Within a large and well-organized pack, this would be unlikely. As will be noted later, there are times when subordinate females breed, but these probably are not young individuals, as indicated by the records on captive packs.

Breeding among subordinates is repressed by more than psychic means. Among the Brookfield Zoo wolves, a subordinate estrous female receiving attention from a male would be attacked and severely punished by the dominant female. In the 70s on Isle Royale Peterson saw several examples of this in the wild. It appears to be usual procedure, although Klinghammer observed an alpha female who did not forcibly interfere with sexual goings-on among subordinates.

It is evident that social inhibition of breeding activity is an important birth control device in wolves. When inhibitions are removed, the animals can rise to the occasion and produce at a higher-than-normal rate (cf. Alaska). That their ability to compensate goes even further than the breeding of second-year animals was revealed in a report by Medjo and Mech. The records concerned five captive female pups and one male that were removed from their parents at an early age and kept without adults at various locations. At 9 to 10 months old the females showed signs of estrus, and one bred, giving birth to a single young. The latter female had mated with the 10-month-old male.

The "compensation" principle, which means that wild animals react to a thinning of numbers by accelerated reproduction, has been well recognized (7, 78). Relative to the past record of persecution, it could help explain how wolves have managed to survive at all.

In February a bonded pair usually can be recognized by their close association, the male staying beside the female and half a length behind. Under these circumstances the tails usually are "flagged," as we have come to refer to it. Isle Royale's big pack always contained a breeding pair, although we are unable to say what changes in individuals may have taken place. The dominant male was not always involved in the breeding.

On 21 February 1959 Dave and Don saw the pack crossing Siskiwit Bay from Houghton Point to Crow Point. The dominant pair were traveling together. But back in the van another pair were tied for about 15 minutes, lying back-to-back on the ice. Other wolves that were scattered out for 100 yards returned and there was much milling about before they left the pair. After the tie, the two wolves ran to catch up. Dave thought that at least three females were showing signs of estrus.

Several times in the next two years, Dave saw mating activity and copulatory ties in which the lead animal took no interest. In 1961 we saw no coupling, but a breeding pair produced a litter of pups. In the following year Don and I saw and photographed two dominant males in the pack, indicating the presence of a beta, who probably was breeding. Although 1962 and 1963 were extremely poor years for observation (p. 100), breeding certainly took place; in 1963 the big pack contained at least two pups, and in 1964 there were five or six.

In the following two years we did not specifically identify a beta male. However, in 1965 a new breeding pack of five — with two yearlings (?) and a pup — evidently split from the main pack, which had appeared to be at maximum number (20). Pete and Don saw the adults of the new pack coupled at a kill on

Locke Point (northeast end) on 23 February, and Don and I found the five there four days later. We never were certain that we saw these wolves again.

I suspected at the time that they crossed the ice to Canada, plainly visible from there. However, the finding in the next fall of a dead pup somewhere in the east end and the appearance of a nonbreeding group of four the following winter suggested that the pack of five produced a litter and then lost one of the adults.

In our table of organization we have lumped "other" subordinate males or females in their respective categories because there is seldom any basis for doing otherwise. It is probable that relationships in these groups usually are more complex than a linear peck order, although both Zimen and Haber mention such a linear situation (97, 299). On Isle Royale we could not recognize many individuals, and it would not have been evident if a linear arrangement were present. The number of wolves involved could have a bearing on this.

Pack organization is not totally stereotyped: There are a few records of a beta female (Haber in Alaska; Klinghammer at the Wolf Park), and a female may occasionally become the dominant wolf. Females seem to be weaker in their authority than males.

In their first winter with the pack, the surviving pups of a litter must learn about adult social behavior, hunting, and other problems of wolf existence. Among themselves they have already established a dominance order, and certain individuals show signs of leadership that are likely to be confirmed in maturity. In all respects a 3-year-old can be considered fully mature.[4] Stenlund said that male pups could weigh 50 pounds or more by November. Of 156 Minnesota timber wolves on which he had weights, the largest male was 112 pounds and the largest female 80.

During the early years of our work there were serious questions, as we compared notes with others, whether we could recognize a pup from the air in midwinter. Some captive-reared pups and some killed in the wild resembled adults in size and pelage to such an extent that even a group of dead animals could not be sorted readily. The fact that Isle Royale wolves were breeding with only a few, if any, young surviving to winter in some years of the 60s suggested high mortality in the litters. Food availability, both moose and beaver, was considerably lower in the early 60s than in the 70s, as will be explained in Chapter 15.

Beginning with 1962, a careful examination of photographs, plus field observations of behavior, convinced us that we could recognize some, but probably not all, young of the year. The state of nutrition in early life could make the difference in appearance between a wild wolf of the island and a well-fed young animal reared in captivity. The Isle Royale pups tended to be more slender, of a more uniform coloration from back to belly, and more fuzzy in appearance. They romped and played together and had a special relationship to the dominant adults.

That is how, at the time, we explained our identification of pups, and we had photographs that showed the pelage difference. Much later (1975) Van Vallenberghe and Mech published a paper on their trapping and marking of 73 wild pups, aged 8 to 28 weeks, in Minnesota. They found a wide variation in size and development in the same year and even in the same litter. Of particular interest is their statement,

> Pups of lower relative weight than their littermates sometimes also displayed retarded development of adult pelage. They retained a reddish, wooly-textured coat lacking guard hairs . . . for longer periods than their heavier siblings.

That wolves of a year class can be consistently malnourished in a particular range was shown by blood analyses made in connection with these studies. In 1972, 10 of 11 animals sampled showed an abnormal state of malnutrition (249). The plight of the wolves evidently reflected a major reduction of their deer prey as a result of forest maturation and winters of disastrously deep snow, beginning in 1968–69. The decline of wolves that followed reduction of the deer was characterized by a decrease in the production and survival of young (175, 178).

Studies have shown that in the summer pupping season wolves can piece out their diet in minor degree with small animals — in various areas hares, muskrats, marmots, ground squirrels, porcupines, raccoons, lemmings, mice, and even dead fish and other carrion. In August 1975 Rolf discovered that the wolves had joined the foxes in eating fruits of sarsaparilla, blueberries, and red raspberries. It affirms that, sooner or later, one finds nearly anything that is possible. However, as all investigators have emphasized, the wolf does not depend on small creatures. It must have the large ungulates to survive and prosper (cf. Olson, Cowan, D. Q. Thompson, Banfield, Pimlott, Haber, Kuyt, Kolenosky, Mech). Where scarcity occurs, survival of the young must be immediately affected.

Pups have behavioral privileges that are still evident in winter and which are steadily lost, totally by the second year. A young animal takes liberties with the dominants that would not be tolerated in an adult. They frolic and jostle and grab their elders by the jowls. They get rough and disrespectful and have to be seized by the head and pinned to the snow, which subdues them momentarily.

In this process they are learning the arts of submission, both active and passive, according to Schenkel terminology. The first, as practiced by accepted subordinates, is a deliberate obeisance and solicitation of favor. It involves a halting, tail-wagging, head-low, ears-down approach and then an upward nuzzling and licking at the sides of the dominant animal's mouth, a holdover of the food-begging of early puphood. In fact most gestures of subordinance originate in the interactions between the young wolf and its mother in the natal den.[5] While the above importunity is in progress the dominant animal may stand grandly with head up, ears cocked, and tail straight out. Sometimes the head will be stretched upward, evidently to minimize the intimacy.

Submission behaviorisms tend to appease the display of dominance and block aggression, not always successfully, depending on degrees of status and issues of the moment that a nonwolf is unlikely to understand. "Passive" submission involves an abject abandonment of dignity readily accepted by subordinate members of the pack. It is seen frequently when wolves play about near a kill or stop to socialize in their travels.

The direct stare of a dominant usually causes a subordinate to shape up quickly in whatever impropriety it is guilty of. For example, a low-ranking male may be paying too much attention to an estrous female. The compliant subordinate turns its head aside, flattens its ears, tucks its tail, and perhaps moves off in a half crouch; this is essentially the "guilt" reaction one sees in a dog.

If a dominant makes a sudden direct approach — the extreme is a growling lunge with fangs bared, ears and tail erect, and hackles bristling — the inferior wolf flops over on its side or back, head stretched out, hind leg raised, and genitals exposed. Then the superior animal is likely to "stand over" the head end of the supine individual and may inspect its genital area.

The cubs must learn the fine points of deportment, for by the second winter they will find the dominants more demanding, snappish, and ready to punish for overt infractions of the code.

Within reasonable limits, pups have feeding privileges around a kill. However, there appear to be adults that regularly associate but are not allowed to eat at the first table. The clearest example of this I recall was in 1969, when a calf was killed by the west pack of eight on Harvey Lake (p. 185). Only six wolves fed at first, while two others lay out on the ice perhaps 20 yards apart, obviously waiting until the others had finished. No doubt they fed later, without benefit of the giblet course. This pack had no evident trailing subordinate at the time. It is of interest that Haber found no feeding order at all in his Savage pack at McKinley Park, even when as many as 19 wolves were "crowded around a carcass."

A change in activity by the pack — after sleeping off the effects of a feed, leaving a kill, or simply starting off again after a rest — often is signalized by a bout of unmistakable revelry.

The characteristic invitation to a romp by one animal to another is an arching of the back, tail switching, front legs flat to the ground (or snow, as we know it), an artful "play face," with tongue dangling, and then often a turning and running away.[6] Except for a dignified dominant or two, most of the wolves join in to chase one another, play keep-away with a stick or bone, and engage in mock combat (including the "ambush"), scruff biting, sidewise body checking, and other roughhouse tactics. One or more animals will slide and roll down a steep embankment with snow flying in all directions. Of course, much of this is pup activity, if any are present.

Before it is over we will have a greeting ceremony, with all gathered around the dominant male to pay their fond respects, paw-poking, nosing his head, and licking his jaws. This must help maintain the solidarity of the pack; all field workers see it frequently. In his sympathetic feel for the personal side of things, Don commented that the poor wolves had to have a social hour without any booze!

Even after a rousing demonstration, the pack may not head away if some inspired impresario raises his (or her) head and begins to howl. Then the entire group will stand about and sing in glee club exuberance. Around pens of confined wolves a human voice, train whistle, or fire siren may start the animals going. In her book Lois Crisler told of having wonderful howls with her young wolves, the animals being obviously carried away in the joy of it all. On the island we have heard chorus howling both day and night, winter and summer. It is not always a major musical event; it can be a few wolves whooping it up, quite likely to tell where they are, for they may be answered by singles off in the brush somewhere. In summer the pups are often heard at such times.

As an example of summer play, Joe Scheidler described what he properly called a "once in a lifetime" observation at the upper falls of the Little Siskiwit River (above Hay Bay) on 4 June 1975. He and Jim Woolington had camped upstream and were coming down to the falls for breakfast — four wolves got there first:

> Jim happened to glance over at the river and saw 4 wolves romping, wrestling and engaged in a water fight. We stood and admired them for probably a minute before they moved downstream. There was . . . whining, and one wolf was throwing water on the others using his paw as a paddle. Two others were playfully grabbing each other around the neck with their paws.

Joe described the wolves in detail, from about 60 yards away.

Time of day does not seem to be a controlling factor on many activities, although the wolf is basically a nocturnal animal. Night watches of the Brookfield wolves have indicated little sexual activity at that time. Wolves of the island often travel and kill (or at least finish off kills) at night. In the work by Kolenosky and Johnston at Algonquin Park, more than half of the radio fixes from 11 A.M. to 3 P.M. showed the animals to be stationary. We often see them resting in the midportion of the day. On clear days this is readily understandable, since wolves are heavily insulated by their woolly underfur, and snow traveling in the sun would quickly lead to overheating. On 24 February 1976, with the temperature soaring to about 45°F., Rolf saw a wolf resting in a pool of water that lay over the ice. He remarked that the animal "must be hot!"

After the winter breeding period, it is evident that social tensions abate considerably. Especially if there are cubs to rear, a great deal of friendly concourse takes place among pack members from spring to fall. This is true among confined animals, and Haber saw it clearly in his wild packs at McKinley Park. Relationships tended to be peaceful, and the alpha male usually took the initiative in starting and leading hunting expeditions. He also was in the forefront in attacks on moose. Haber said subordinate members of the pack obviously became active in anticipation of a hunt, but they waited for the alpha to make the first move. When he was disabled for a couple of days, he stayed near the den and the beta male became master of the chase (95, 97).

Whelping dens used by wolves may be of several types. In McKinley Park, Murie, and later Haber,[7] found breeders using burrows from 10 to more than 20 feet long, excavated in sandy soil, and sloping upward to an inner chamber. It appeared that in all cases the wolves had remodeled a fox den. Various pack members, including a pup, helped to clean out dens — often several different ones — before whelping time. Not all the fox tunnels at a den were enlarged. The unimproved ones were used by the small pups later, according to Haber. An entrance used by adult wolves and entered by Murie to extract a pup measured 16 inches high by 25 inches wide. Dens were usually on high ground, with a view, and fronted on south or southeast slopes (95).

Both Murie and Haber recorded the production of litters by two females in a pack, who got along well. A young black female had helped care for the single litter in 1940, and Murie said she mated and had her own litter of six the next year. Her den was not found, but on 30 June she moved her brood into the den of her friend the gray female, who had four pups of her own. Nine days later the gray female, her mate, and the cubs had moved to a rendezvous a third of a mile away, but the black female and her family were still using the den. Later they joined the others at the rendezvous, and all of them probably constituted a pack of 15 that Murie saw later in the year.

On the Toklat River to the west, Haber observed two females with pups in the same pack in 1966 and again in 1967. They used a common den with two nursery chambers. After they vacated, Haber entered to investigate the burrow. He found the chambers bedded with underfur. They measured 50 inches by 14 inches (high) and 45 inches by 15 inches, respectively. The den was on a sandy peninsula extending 40 feet onto the gravel bar of the river (95). It is worth recalling that Young, Pimlott, and others have found no lining in wolf dens.

On Isle Royale a den was cleaned out on snow on the south side of a sandy slope and was seen by Jordan and Murray from the air on 13 March 1966. In the spring Jordan and Dunmire examined it and found the usual bone litter left in former years by foxes. The wolves evidently had not returned to use it.

For many years we had only one breeding pack on the largely forested island, and by 1972 we had practically given up the hope of finding a den. However, that winter saw the establishment of a new pack at the east end. Rolf took his cue from the work at Algonquin Park (Pimlott, Joslin, Theberge) and got a bull horn.[8] By howling from various vantage points he got responses and triangulated both denning and rendezvous areas. He examined the first den after it was vacated by the east pack

After this drained beaver pond becomes an alder thicket, the old lodge may serve as a wolf whelping den.

in July 1973. It was a grown-over beaver lodge in an old washed-out pond. Nearby he found another lodge that appeared to have been used in a former year. It seems to be invariable that both dens and rendezvous are located near water.

Under conditions comparable to those on Isle Royale, in Algonquin Park Joslin found three dens in burrows, one under the base of a tree, one in a hollow log, and one in a rock cave. In subsequent seasons on the island Rolf and his helpers found at least two hollow logs that had been used as dens, one an old charred pine trunk and the other a large cedar.

This suggests the validity of something I have long suspected. In the primitive forests where large hollow logs on the ground were plentiful, these must have been common denning retreats for bears and wolves, as well as lesser creatures. On occasion they were overnighting places for men.

On Isle Royale pups of a typical wolf litter probably are born in late April. Interpreting from other sources, we would expect them to be about six in number and to weigh a pound each. Their eyes would begin to open on the 12th day. The iris is gray-blue, which probably will change to tawny gold in the adult, although some adults have grayish eyes.

The bitch wolf grooms her young with the tongue, and especially licks the genital and anal areas. Soon after birth the cubs begin to nurse, and the licking stimulates urination; without it they would not develop the urinary function and would die. All excretory products are consumed by the mother until the pups are five to six weeks old (86). Pups begin to leave the den at three to four weeks of age, and soon they are exploring the surrounding area and feeding on meat regurgitated by the adults. According to Kuyt, the mother wolf no longer eats scats after the meat diet begins; she eats only "milk scat." By this time the pups are in control of their elimination, and this is kept outside the den.

It comes naturally for all wolves to love children — wolf children, of course — and commonly the entire pack cares for the litter. Socially the woolly moppets can get away with anything, and their presumption is colossal. They climb over, chew, and harass the long-suffering elders, who often have to move away from the den to get any sleep.

The mother of the pups stays with them for the first few days while the others are out hunting, but later she will (with obvious joy and relief) go off to hunt with the pack and leave one of the other adults, sometimes a young female, in charge of them. Haber said wolves left the den for a hunt in the evening and were out all night, often straggling in at various times the following day. He and Murie both noted that the hunters came back exhausted, often sleeping for hours without moving.

When adults return from the hunt they are swarmed by the hungry litter, who eagerly lick and nibble at their jowls and lips. This is the releaser that causes an adult, male or female, to regurgitate undigested meat carried in the stomach, a ration avidly consumed by the cubs.

That these contacts excite physiological responses in adult wolves there can be no doubt. In the Wolf Park two yearlings with no previous contact with young quickly took a solicitous interest in a litter of strange pups, regurgitating food in the normal manner.[9] When a litter was born to an alpha female, the dominant male was not permitted to enter the natal bower, but an unrelated yearling female had full privileges of the den. She moved the pups about in the usual way, by gripping them loosely around the back, and stayed with them while the mother was out.

In this connection, Klinghammer made a momentous discovery. He found that this accessory female, who had never been pregnant, was stimulated to lactation. It suggested strongly that if the mother of a wild litter were destroyed, another female might be able to take her place.

By the end of the second month, the young animals are ranging out from the den area and may have been led to feed on nearby kills. At this stage both Murie and Haber saw them hunting mice, sometimes successfully.

They are also being weaned, which is a gradual process. Relative to five lactation periods of which he had record, Kuyt said that one

> . . . lasted more than 34 days in the 2-year-old wolf, 43 days for her mother when 3 years old, 50 days for the same wolf, then 5 years old and 51 days for the same wolf, then 6 years old. These dates are similar to those given by Ognev (*in* Pullianen, 1965) who states that the female wolf suckles her young for 35 to 45 days.
>
> If these dates are comparable to those for wild wolves, it may be

assumed that cubs at the early age of 1½ months, depend largely or entirely on wild prey.

At about two months of age, which means by early July on Isle Royale, the young wolves are ready to abandon the den and move to their first rendezvous site: terminology first used by Murie. From July to October the litter is likely to be moved to three or four of these open-air kindergartens. They are places where the cubs can be left while adults — often split into several groups — are away hunting. Gradually the young accompany the pack members on more of their activities, visiting kills and traveling. In describing rendezvous of 1973 (two packs), Rolf said in his thesis:

> All five were located at or adjoining an abandoned beaver pond, with water still available nearby. They varied greatly in size, from an area of about an acre to a drainage one-half mile long. Most had a prominent open area where the vegetation had been matted, and holes had often been dug in nearby banks. A small den was found beneath the roots of a cedar tree at one area, and a beaver lodge had been excavated and used at another area, although it is not known if it had been used as a whelping den in previous years. Both dens and rendezvous are frequently re-used, with former rendezvous possibly serving as den sites at a later date, and vice versa.

The length of time a site is occupied may vary widely, but descriptions of summer home sites in other parts of the continent conform generally to these observations on Isle Royale.

The communication system of wolves seems much involved with their habit of occupying defended territories. Territorialism is characteristic of breeding packs, but not of loners or the several nonbreeders that form units often stable from year to year.[10] The record of pack formation on Isle Royale is given in Table 1. In compiling such a summary certain "rules" must be followed for consistency, and my standard was to give the maximum count for each pack. Seldom was that maximum maintained for a long period, incomplete counts and fragmentation being common.[11] However, it is evident that after 1971 we had two and then three breeding packs to deal with. This offered opportunities for studying territorial relationships, which had

Table 1. Summary of Wolf Packs and Numbers, 1959–76

Winter	Breeding Packs (in order of size)			Packs — Nonbreeders				Total Numbers	
Jan–Mar	First	Second	Third	First	Second	Third	Loners	Min.	BE*
1959	16				3			19	20
	(15 + 1)								
1960	16				3	2	1	19	22
	(15 + 1)								
1961	15				3	2	1	20	22
	(14 + 1)								
1962	17					2, 2	2	22	23
1963	16					2, 2	1	20	20
	(14 + 2)								
1964	22				3		2	25	25
1965	18	5				2, 2	1	25	28
1966	15 (−1)			4		2, 2	1	21	23
1967	7	7	6	4		2	3	19	26
	(6 + 1)								
1968	7	(−6)		4		2	2	15	16
1969	9			4 (−1)		2	3	14	17
1970	8			4	3	2	1	17	18
1971	10 →					2, 2, 2	4	16	20
1972	10	8			3			22	23
1973	13	8				2	1	23	24
1974	16	12 (−1)				2	1	30	30
1975	18	10	7		3	2	1	41	41
1976	17	9	4/3	4 (−2)	3	2	2	42	42

* Best Estimate

been minimal previously. The conditions that brought about this proliferation will be examined in Chapter 15.

Both summer and winter, wolves establish scent stations that they check carefully on their rounds. These usually are marked by urination on conspicuous objects along trails: rocks, moose bones, the base of a tree. Feces also are used to mark scent stations. I noted an unusual case on the Moskey Basin–Lake Richie Trail on a June morning in 1960. From the ground at the left of the trail I heard the hum of a swarm of flies, and under my feet was the scratching of a wolf. The flies were gathered on a strongly scented mound of moss, which was obviously soaked with urine. Precisely in the center of the mound were the scolex and numerous segments of an adult tapeworm. A wolf had tried to deposit a scat on the scent post, but all he could muster was the tapeworm, later identified as probably *Taenia hydatigena,* the cysts (larval stage) of which we find in moose livers.

In their traveling, wolves stop to examine scent stations with

Breeding packs of wolves are territorial and will kill any alien wolf with whom they come in contact (with Michael Wolfe).

a great deal of interest, the dominant male being seen frequently to "freshen up" the signal post. A pack becomes especially preoccupied with a station established by a different pack. In mid-February 1974, the east pack (numbering 16) moved west of their accustomed range into the south Siskiwit Bay and Houghton Point region. They killed a member of the west pack that had lingered in this area (a 64-pound young female, aut. 967) and then moved on southwest along the south shore until they came to a major scent post of the west pack. Rolf and Don were watching them. "After much examination and scent-marking of their own, they turned around and headed back toward the east end of the island."

We obtained our first record of a wolf being killed in an inter-pack altercation in 1969. The west pack (nine) had ventured beyond their usual range and met the nonbreeding group of four on the north side of Chippewa Harbor. There was a chase, and one of the four was killed. On 1 February the west pack was moving away from the scene. The next day Mike and Don landed in slush to retrieve the partially frozen specimen (adult male, weight 81 pounds, aut. 436).

Both of the dead wolves referred to above were frozen and taken back for autopsy in the Animal Disease Diagnostic Laboratory (courtesy Harvey L. Olander). Not unusually, the older animal showed two healed rib fractures.

Rolf described two additional killings — evidently stemming from the territorial nature of the wolf — in the winter of 1976. As in the previous year, there were three breeding packs on the island: the west pack (9), the middle pack (4 + 3: probably a split of the 7 in 1975), and the east pack (numbering 17).

On 12 February a new group of four was seen at the extreme northwest corner of the island (McGinty Cove area). This was west pack territory, and there was a clash. Bloody snow and tracks showed that one of the new pack had been killed and slipped down a bank and disappeared into the water. Two of the others were wounded, and one of them (adult female, 62 pounds, aut. 1216) died in the trail about a mile away. This wolf was retrieved fresh. The remaining two survived in west pack territory, although one was bleeding five days after the confrontation. It appeared that the four wolves (a female and three pups?) probably originated with the east pack, since three of them (not examined) were "thin tails," an evident genetic coat characteristic (see photograph on pages 280-281) that we had seen only in the east pack.

In 1976 the total known wolf population prior to the two losses had been 44, an all-time high. Density problems obviously were becoming acute.

The case of the four wolves at McGinty Cove is suggestive in several ways. A subordinate female of the east pack could have bred and then left or been driven out to find a range of her own. Unfortunately, on the island there was no space to spare. It is also possible that a thin-tailed male (a loner?) that had previously originated in the east pack could have sired the presumed pups of the normal-coated female.

Peters and Mech say that most of the young animals reared in a pack disperse and travel, and in their progress they are told by scent stations whether a range is occupied. "If the lone wolves can find a suitable vacant area and a member of the opposite sex, they may mate and produce their own pack."

From 1971 to 1974 they studied the scent marking of 13 contiguous territorial packs in northeastern Minnesota, aided by the radio tagging of 96 individual wolves and ground tracking to follow up on aerial observations. They found the raised-leg urination of dominant males or females to be the most significant kind of marking. Scats placed on prominent objects were also obvious scent posts, especially when combined with the ground scratching with all four feet that usually follows such behavior. Only high-ranking wolves scratch.

In summary, the findings were that a pack keeps its territory well marked throughout, but on the periphery, where it encounters the signs of neighboring packs, the scent stations tend to accumulate and overlap with those of the adjacent pack. Thus there is a well-delineated borderline, with scent stations constantly renewed, that tells a pack when it is on home ground and also tells would-be interlopers what to expect. In the region studied, territories were relatively stable and ranged in size from 48 to 120 square miles.

According to the most recent findings, there is even more to it than this. Scent stations convey a more complete message than anyone had previously suggested: As a follow-up on the work reported by Peters and Mech, a study by Rothman explored the scent-marking habits of lone wolves — usually the 2-year-olds traveling to find mates and ranges where they can

Members of the east pack in 1974. One of the "thin-tailed" animals is at left. This evidently was a genetic characteristic of some members of this pack. The coat lacked normal growth of guard hairs.

settle down. Wolves that are alone do not scent mark, but when two of opposite sex associate in the fall they travel until they find a vacant range. Then they begin *double marking* intensively, both male and female on the same spot. Thus they get their territory established rapidly, and the double marking indicates to potential invaders that a breeding pair is in charge. Older wolves in established packs double mark less intensively, but for the same evident purpose.

On the south border of the Superior National Forest, along the Minnesota shore of Lake Superior, Van Ballanberghe, Erickson, and Byman used radio tracking to determine the ranges of six adjacent packs. The ranges were discrete in the May-to-October season and varied in area from approximately 19 to 43 square miles. They included denning and rendezvous sites. Three of the wolves could be monitored through most of February and continued to use their respective pack ranges. However, they confined their activity largely to elongated areas (including deer yards) along the Lake Superior shore.

While scent stations may be the most important way in which wolves exchange information, howling obviously is another. The somewhat informal chorus howling that goes on around a rendezvous seems to say, "We are at home and hungry and waiting for someone to bring in the rations." However, it may have a more routine function of breaking in the pups to social activities. Frequently single wolves call in the course of their movements, and it seems beyond doubt that these are location signals. We have had packs break up temporarily about Windigo, possibly because of our activities, and it has happened at times elsewhere in the field. In each case the scattered wolves emit howls that tell us, and their pack mates, where they are. When Pete and Don were autopsying an abandoned moose on the north side of Siskiwit Lake, they howled and received answers from the pack down the shore. Of course, pack howling could have a function of warning off intruders, and on that no one has much information.

Several times we have heard wolves bark in the course of, or at pauses in, a howling session. However, the usual function of a bark evidently is similar to what it is in the dog: an alarm, warning, or challenge. Joslin describes it thus, as does Mech (126, 174). We have seldom heard it in this context because of minimal contacts evoking such reactions. On 10 July 1974

Scheidler and Woolington found themselves under inspection by a disapproving pup that stood 75 feet away on a rock outcrop near Pickett Bay. It barked at them four times, and they quickly retreated. It was their first proof that there was, as suspected, a new "middle pack" with pups.

In Mech's first two years of observation, the big pack used practically the entire island; after that it reduced its range to about the southwestern two thirds. The duos and trios had fairly consistent holdings in the eastern portions, and loners roamed widely. By 1966 we were beginning to suspect that this island of 210 square miles could support only one breeding pack. Actually, that winter was to initiate a new departure in what had been an amazingly stable situation.

On 24 February Pete and Don saw the big pack (of 15) traveling in two groups a mile apart, each group led by a courting pair. The dominant male, now known as Big Daddy, was trailing the second of these groups; he had been getting lighter in color for at least three years. That day he was limping, an impairment first noticed 20 February.

After another day a period of bad weather set in (I took advantage of a break and left the island 1 March). It was not until 12 March that Pete and Don could fly again and locate eight wolves on Feldtmann Lake. It was the big pack, but the alpha male was not with them.

Don did some expert back-tracking and they found two calf kills and an area of great activity south of Lily Lake. Don thought he could see a piece of wolf hide there, and Pete snowshoed 7 miles to investigate. There was no question that Big Daddy had found his last resting places. Pete got the piece of hide, leg bones, and several vertebrae, but no skull.[12] Under a windfall, there had been a great battle.

Don said the wolves had held an election. We would not again see the big pack as we had known them for so many years.

On 2 February 1967 Wendel Johnson and I flew to the island for the winter study. Wendel had joined the project in September to do a more intensive job on the smaller mammals, essentially the fox and its prey. Jordan had left in August, and Mike Wolfe would be coming to the island on 11 February to begin his 3-year post-doctoral stint.

It would be the coldest winter we had known, and with little wind the north channel was freezing into the most level and extensive ice field yet seen. The 23 inches of snow was light and fluffy clear to the ground — hard traveling, even for an adult and healthy moose. The weather was characterized by frequent drifting snows and poor visibility.

At the west end we had two breeding packs, the "six pack" and the "Lily Lake seven," evidently successors to the big pack of years past. We had our old foursome back again in the Tobin Harbor area, a duo, and several loners. But there was much breaking up, the snowing-in of tracks, and unusual movement; both the six pack and the Lily Lake seven made journeys clear to the east end.[13]

The six pack (probably) had a kill near Lake Ahmik (67-8), which is across the island almost directly west of Rock Harbor Lodge. Then a wolf appeared to be bloody and dead on the north side of Tobin Harbor. When Don and I went to retrieve it on 25 February, we found something new and startling on outer

Two black wolves, appearing to be pups, of the new "black pack" that came across the ice from Canada in February 1967.

Amygdaloid Channel: a pack of seven wolves, and four of them were black! We had not seen a black wolf before this on Isle Royale.

A wolf with a bloody head and neck was out on the ice running toward Canada, stopping at times to look back. One of the gray wolves in the new pack, a large square-jawed male, had blood on its face. The "dead" wolf on Tobin Harbor got up and left when I snowshoed in to it.

The fast-moving events of the remainder of that period had us mystified if not confused. The six-pack evidently disintegrated; it was not seen again. The Lily Lake seven, after a trip to the east end, doubled back and made a kill (69-7) south of Conglomerate Bay, which is south of the Middle Islands Passage into Rock Harbor. Whereupon, the new "black pack" left the Lake Ahmik area and crossed the island to the south, and at least 11 wolves were getting along amicably at or near the kill on 3 March. The black pack reassembled and followed their old tracks back to the north side, near where we first saw them.

Neither the Lily Lake seven nor the black pack was seen intact again. The wolves were in a state of social chaos, or at least complete reorganization. Three packs of six or seven each, at least two containing breeding pairs, had broken up, and loners and duos were scattered around to complicate any count. But at the end of the winter study (21 March) we knew there were three or four black wolves on Isle Royale, and they would help in identity problems to come. Mike Wolfe had reported for duty in the midst of the worst confusion and uncertainty we had seen thus far.

It appeared that the new pack had come across from Canada, possibly from Sibley Provincial Park, where wolves are protected (they were not afraid of the aircraft). Also, there were black wolves in that area.[14] Fighting with the six pack was understandable, but why could the new wolves get along with the Lily Lake seven, and why did the black pack itself split up? Had there been more interchange with Canadian wolves than we knew about? In former years the big pack had twice been seen as much as 2.5 miles off the northeast end of the island, and individual wolves could have gone or come without our knowing it. Our counts were not that accurate.

In 1968 we had more to think about along these lines. It was a winter of little snow, when wolves were trekking about inland.

There was much wind and we were often grounded. We accounted for a pack of six containing two black wolves at the west end — after two observations, they disappeared. It was about the time the north channel was freezing solid again. Another pack, of seven, also at the west end, had a dominant pair of gray wolves. The male was large and the female small with a pointed nose and a crook in her left front leg. Closely associated with them was a large black male. This became our new "west pack," and its history has continued with interesting changes.

In 1968 we were back to one group of breeding wolves and into a 3-year period of minimum numbers. The single black wolf would be well known to us through 1972 — more of that in Chapter 15. Moose and beavers were building up. I had a subversive feeling that some of our former ideas were coming apart. As things turned out, it was not a bad guess.

CHAPTER THIRTEEN

The Followers

ON MY EARLY TRIPS to northern Michigan or southern Canada I occasionally heard a cowbell in the woods. It seemed that forest-edge farmers were letting their stock wander pretty widely. In boyhood ramblings I had heard cowbells in Indiana woodlands, so I accepted the sound with little more than mild curiosity.

Years later my curiosity became more than mild, because I was hearing cowbells on Isle Royale; there were no cows in this national park.

Anyone familiar with the north woods will recognize my problem. Amid a widely variable repertoire of hoots, honks, bongs, twangs, and gurgles, the raven has a note that, from a distance, passes quite well for a cowbell. When you hear something in the forest that is unexplainable, it usually is a raven. Roy Stamey told us one evening that he heard a raven open a bottle and pour himself a drink.

"Only one thing he did wrong," said Roy. "He pulled the cork last."

The raven is the largest member of a family (the *Corvidae*) that includes the crows, jays, and magpies — all showing a high degree of "intelligence," as the word applies to birds; and some like the crow and raven are excellent mimics. Four members of this distinguished family nest on Isle Royale: raven, crow, whiskeyjack (Canada jay or gray jay), and blue jay. Crows go elsewhere for the winter; we have seen them along the Minnesota shores.

The raven is easily distinguished from the crow, being considerably larger, with a trowel-shaped tail and heavy bill. It can pry the eyes out of a dead moose, and often does. It is agile on the wing, and in spring these birds engage in elaborate nuptial flights that are a delight to watch. From Ojibway tower I saw two of them putting on such a performance in a high wind above Greenstone Ridge on 2 May 1974. By that time their nesting would be well along.

On 12 June 1965 Erik Stauber was on the high cliff above Feldtmann Lake and saw the "beautiful aerial display" of a raven over the valley. The birds tower and dive and turn over in flight. However, the raven meets more than its match when it approaches the nest of that elegant small falcon, the pigeon hawk (merlin). On 12 June 1964 over Rock Harbor Stauber saw a pigeon hawk hurrying a raven out of the vicinity of the Bangsund Cabin. It had its nest 100 yards behind the cabin. Erik said the raven turned over on its back in flight to meet the darting attack of the merlin.[1]

Ravens had a nest on a shelf of the Feldtmann precipice; young were still there in mid-June 1965, but Erik said they were a-wing on 4 July and were returning to the nest at night to roost. These birds also nest on the cliffs on the south side of Chippewa Harbor and at the Palisades, the steep rock face on the north side of Blake Point. All of these probably were sites of former duck hawk eyries. However, the raven does not require cliffs; it also builds bulky stick nests, lined with moose hair, in tall trees. These have been observed numerous times by our field people.

The raven is a versatile scavenger, eating nearly any kind of animal matter, including the rewards of beach combing. Like its near relatives, it catches small mammals and raids bird nests on occasion. Our principal interest in the species is its particular relationship to the wolf.

In winter a traveling wolf pack is normally accompanied by half a dozen or more ravens. When a kill is made the word seems to go out, and more birds gather. They sit around in trees, and share the wealth as opportunity permits.[2] Often we locate a kill by the ravens that fly up at any disturbance. I have tried to photograph them from a blind, but with little success; they are too keen-eyed and canny.

When a dead moose has been dismembered, the ravens and whiskeyjacks insert their bills through the foramen magnum at the base of the skull and extract brains that (in an adult) are unavailable to a wolf. When the end is chewed off a bone (usually calf), they pick marrow out of the cavity. They pry into crevices and help clean up the skeleton. On 29 January 1973 Rolf counted 27 ravens on an autopsy carcass across the harbor, which is the most we have seen together.

Ravens exploit the wolf in a more personal way. They follow the pack out onto the ice and gobble scats as fast as they are produced. Presumably they get fragments of skin, tendon, and even bone that pass through the digestive tract of a well-fed wolf. Often when we follow a wolf trail on lakes or bays, there are wing prints and bird tracks at places where scats have been picked to pieces. In fall and spring the birds do not seem to have this habit, since there are intact scats on trails with no evidence of such morbid feeding.

Until 1974 we had no indication that the scats of foxes were attractive to ravens. Then, on the south shore, Rolf found unmistakable signs that fox scats were being eaten. The foxes had been feeding on mountain ash fruit, which the ravens also like.

Fruit eating by these birds first became evident in the winter of 1971, when Don and I were finding kills in the Feldtmann area bedaubed with red-stained raven whitewash that contained fragments of mountain ash berries. We made a stop on the lake

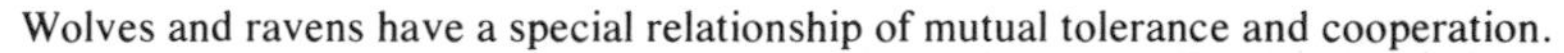
Wolves and ravens have a special relationship of mutual tolerance and cooperation.

to check what appeared to be blood and found that a raven had carried a head of mountain ash fruit to the spot and eaten most of it. We were somewhat mystified that a habit showing up so clearly this winter had not been noticed before.

Wolves are amazingly tolerant of the big black birds; there is a mutual understanding of some sort. On the ice, when wolves are resting or cavorting about, the ravens stand among them, often appearing to enjoy what is going on. At times they will be out somewhere with the wolves when they obviously could be feeding on a nearby kill.

On 7 February 1961 we had the young antlered bull carcass on Washington Creek (p. 209), and the big pack was on the harbor. I spent most of two days in one of the shelters at the campground hoping to photograph the wolves when they came up the creek to the bait. Two wolves and two ravens made it to the carcass, but the rest of the pack killed a calf on the north shore of the harbor during the night and came only part way up the creek to sleep on the ice. They were about 100 yards downstream from me.

For several hours in the morning nothing happened. Then a wolf stretched and got up, going around to the others and rousing them too. There were ravens about, and three of them came down onto the ice with the wolves, which began to romp and play and chase one another. A wolf would leap at a raven and the bird would fly up about 10 feet only to drop back down again. They actually seemed to enter into the "play" of the wolves. Don told me that on one occasion he had seen a raven alight momentarily on the back of a wolf.

One of the most impressive episodes Don and I witnessed was near the south shore of Thompson Island (at the mouth of Washington Harbor) on 26 February 1971. The west pack of eight had a kill on the island and were out on the ice. The black male was a quarter-mile to the east, trotting toward the others along the shore. Several ravens were around, and one of them flew to the big male, who stopped on the ice while the raven walked around it within 3 feet. The wolf raised its head and howled twice without alarming the raven. The bird stopped at the nether end of the wolf and obviously inspected its anal area. Don and I were slowly circling; he turned his head back toward me.

"Not much question what he's interested in, is there?"

Eventually the wolf walked to the others, with the raven hovering overhead. We had an impression that the raven preferred wolf scats to what was left of the kill a short distance away.

Usually ravens are prudent enough not to feed at a carcass when a wolf is close at hand, and we have sometimes seen a wolf deliberately flush birds off autopsy remains across the harbor. However, in 18 years of winter watching, we have only one record of violence done to a raven by a wolf. On 16 February 1974 Rolf and Don were circling five wolves of the west pack near a Francis Point kill (74-20, Houghton Peninsula). One of the wolves made a couple of quick bounds and caught something: a raven. The wolf shook it vigorously, then went and lay down shaking it some more. It took the bird up the bank and buried it in the snow, then soon retrieved it. Rolf said the wolf was still lying on its belly probably chewing on the raven when the plane left.

In February 1963 the big pack made a kill (aut. 171) on the northeast side of Wright Island, and later tracks indicated that a loner was picking the bones. The pack was miles to the west at Halloran Lake on the 23d, and Don and I decided to set up a bait station on the ice outside the Holte fish house on Wright Island. There was a possibility that one of us could make observations from the fish house.

We had autopsied a tick-bitten young bull on Washington Creek, and we chopped off most of a hind leg and took it to Wright Island. There we cut a hole in the ice and inserted the lower part of the leg so it would freeze in. Nothing but a fox or two came to the bait before I left on 26 February.

The next morning Phil and Don found the bait gone. In making a low pass before landing they saw a wolf in the shadow of trees on the northwest corner of the island. Carrying what was left of the moose leg, it headed across Malone Bay. Phil said the bones looked nearly clean, but the leg was all the wolf could manage.

Following the wolf were nine ravens and a fox. The wolf had to lay the leg down frequently on its way across the bay. Three to six of the birds constantly hovered within a few feet, diving repeatedly. They would alight at each rest stop and approach closely only to be driven off by a rush — at which other ravens would come in behind the wolf for a peck at the bones. Then the harassed animal would pick up its heavy load and start out at a walk, gradually breaking into a trot for a short way.

Escorted by about three ravens, the wolf made it to the north shore and lay down under a leaning cedar. Half a dozen birds lingered at one of the stopping places on the bay (scats?). Then the watchers overhead saw the fox emerge from the woods at Wright Island and follow the trail out across the ice directly to the group of ravens. Phil said:

> The fox trotted right in among the ravens with its nose to the ground [ice] on the wolf's trail. The ravens paid no attention to the fox, except when it ran directly at one, which it did but once. The raven merely hopped out of the way, the same as if the fox had been a wolf. It looked to me as if the fox could have caught a raven easily had it wanted to.

Phil and Don checked the wolf three more times that day and found it guarding its prize. Since this appeared to be the local loner, it may have been the best meal the animal had in some time.

A point of significance is that ravens seem to have a truce with foxes as they do with wolves. Don and I found the first dead moose of the '71 winter season (2 February) when we saw a fox and a raven feeding on the open ice southeast of the Belle Isle campground. We landed and found a calf that had been floating and was now frozen under the ice with only a side of ribs showing above the surface. This exposed part had been trimmed of meat. The fox droppings contained mountain ash fruits.

Foxes and ravens frequently associate on our bait stations across the harbor. Many times a fox could have killed a raven, but it has not happened. We have seen examples of intolerance by foxes, but the birds sense this quickly. On 13 February 1971, when Don and I flew off the harbor, there were three foxes on the autopsy remains, and 30 feet away two ravens had something they were working at. A fox ran at the ravens, and they took off. Later we saw a fox and two ravens feeding within 10 feet of one another.

As elsewhere, the whiskeyjacks of Isle Royale are well adjusted to people. They frequent campgrounds in summer, and in winter they have been everyday callers at our back door. Like other corvids, they eat nearly anything, and what they cannot eat they carry away and store in the woods. When foxes are not around and a handout is available, they get the message from afar and float in like puffs of gray down to a gentle landing on the snow or a low branch. Usually we can account for three to five at a time, when rations are plentiful, and dominance relationships are evident among them. The first-comer, or the next in line, will quickly stuff his elastic throat to bulging and fly away to stash the boodle in trees somewhere. The whiskeyjack is more plentiful and sociable than the blue jay, and there is no cordiality between them. From some chasing I have seen in the tops of nearby balsams, I suspect the blue jays of pilfering caches not their own.

The whiskeyjack seems to understand the other fellow's problem. I once saw a red squirrel ascend the yellow birch beside our entranceway carrying a piece of pancake. It was obviously looking for a place to put the morsel, since it was already overfull of such things. A couple of feet away a whiskeyjack moved from branch to branch as the squirrel went up, stopping at short intervals to look around. Where could he put this choice confection so it would be safe from that dratted bird? The higher they got, the more frantic the squirrel became, and finally it was

Whiskeyjack, or Canada jay.

running out of tree. The frustrated rodent made its last stop, and from inside the kitchen window I could clearly hear it say, "Oh, the hell with it!" It socked the pancake into a fork and ran down the tree. The whiskeyjack accepted the gift and flew away.

Woodsmen of the north sometimes call the Canada jay the moose bird, because supposedly it hangs around "moose yards" for some reason. It has been observed picking up the ticks that one finds in moose beds, according to Anderson and Lankester. We saw ticks in beds most frequently in the early 60s; often the beds contained tufts of hair and oozings of blood on the snow.

Whiskeyjacks quite evidently like moose — dead ones the wolves have killed. When we pick up specimens at a kill one or more whiskeyjacks frequently are there. I suspect that by the time a carcass is used up the nooks and crannies in surrounding trees are well stocked with the whiskeyjack equivalent of jerky. Visiting foxes may stall this program at times, since there is no trust between the fox and the jay.

By the time we close our camp in mid-March whiskeyjacks probably are pairing off, and it is likely they have young in the nest by late April. In mid-June family groups may be seen in the woods with two or three of the dark-colored young. They reputedly feed on insects, birds eggs, and all manner of fruit. I have seen them eat the yellow flesh of skunk cabbage heads in fall, and they make the most of mountain ash fruit when the crop is plentiful in winter.

As noted previously, from June 1966 to October 1968, Wendel Johnson carried out a doctoral study of the island's most significant smaller mammals. In August 1969 he submitted his thesis, "Food habits of the Isle Royale red fox and population aspects of three of its principal prey species." Those potentially important prey species — as judged by studies elsewhere and our preliminary observations — were the snowshoe hare, red squirrel, and woodmouse.

In terms of available time and money, we started Wendel's study when we could. In a biological perspective he was at a disadvantage, because both snowshoe hare and fox populations were down. Hares and foxes (the latter following the former) participate in the northern 10-year "game cycle" that is widespread across Canada, where it also includes the grouse. Since the 1940s, and possibly before, the ups and downs have fitted roughly into calendar decades.

In the first years of the decade the trend is up, and in some northern habitats hares reach a spectacular peak of numbers. This appears to bring prosperity to certain predators, notably the lynx, fox, goshawk, and others locally. Then in the midpart of the decade there is a plunge to scarcity by the hares, followed by a reduction of some predators. Late in the decade recovery is evident, and there is a build-up to another peak. The fluctuation is not precise; variations are evident from one 10-year period to another.

Sketchy reports indicate such a pattern in the past history of Isle Royale, although hares probably have not reached the extreme high and low levels sometimes seen in Canada. Nonetheless, changes have been drastic enough to be remembered or recorded by observers on the island, and they probably indicate the major peaks in hare numbers.[3]

Relatively little is known about the 10-year cycle, affecting in its classic form such distantly related species as hares and grouse. However, recent studies by Lloyd B. Keith and his students in Alberta are helping to analyze the mechanisms. Meslow and Keith reported on their work from 1961 to 1967, a provincewide period of abundance, decline, and partial recovery. They estimated a maximum population of 622 adult hares per square mile on a study area in April 1962 as compared with a minimum of 3 in the summer of 1965.

> The decline in the hare population resulted from a decrease in adult survival, a juvenile survival of only about 3 percent, and a halving of the reproductive rate. Recovery from the low was a function of doubled adult survival and reproductive rates, and marked increases in juvenile survival.

Johnson's best evidence of the amplitude of the hare fluctuation in the 60s came from the operation of a 30-acre trapping quadrat near Chippewa Harbor. This had been established in 1962, when hares were high, by Larry Roop, who caught 8.9 hares per 100 trap-nights. The corresponding figure for the summer of 1968 was 1.9 hares per 100 trap-nights. This probably is a rough index of relative numbers.

Wendel found that during their population low hares did not occupy marginal habitats; they were found only in the best coverts. Rolf made a similar observation in the 70s. As might be expected, the largest area of good habitat on the island is the 1936 burn. The wide range of topography and cover produces a diversity of ground vegetation that favors the hares. However, the shading by maturing trees has a progressive adverse effect that will reduce carrying capacity.

In its food habits, the hare joins the moose and beaver as a major influent on forest growth. It lives largely on herbaceous plants in summer, and in winter it prunes and barks a wide variety of woody brush and seedlings. From notes of our field men one could compile a long list of hare foods, but that is not necessary. Our records show, as did those of Dodds in Newfoundland, that hares eat practically all things taken by the moose and some besides, most notably spruce. The greatest hare-moose competition is in the liking of both for aspen, birch, red osier, Juneberry, mountain ash, mountain maple, and green alder.

Principal prey of the fox and large feathered predators, the snowshoe hare in winter pelage.

It is fair to assume that the red squirrel has been an inhabitant of Isle Royale for some thousands of years, since this isolated population has changed enough to be classified as a distinct subspecies. Logically, it was the food mainstay of the marten before the tree-inhabiting weasel disappeared early in the century.[4] Since then, as Wendel remarked, the squirrel has had no efficient predator to contend with. Live trapping in the major forest types has indicated a preference for conifers; the maple forest does not support many "pineys," as this squirrel of the north woods is sometimes called.

Squirrels on the island produce only one litter of three a year, on the average. In a more productive population it might be expected that females would often bear both spring and summer litters, averaging about four young. Breeding appeared to be strongly influenced by food crops. In 1966, a year of cone, mast, and fruit failure, the squirrels failed to breed.[5]

In general these results are consonant with findings in various studies elsewhere. A notable discovery from the trapping data was the high survival of individual animals from one year to another. Of squirrels of all ages trapped in the summer of 1967, the known survival rate for one year was 55.2 percent, remarkably high for a small prey species.

It is predictable that more intensive predation would immediately bring about a higher (compensatory) rate of increase in Isle Royale squirrels. One might almost consider this a "stagnating" population. Thus, there is no question that the island could again support martens if the National Park Service decided to restore them.[6]

When Wendel and his crew began their surveys of possible foods of the fox, we were quite sure the hare was important, as indeed it proved to be. On the other hand the squirrel and the woodmouse (or deermouse) were open questions. It developed that nothing much was exerting pressure on the squirrels, and we were surprised to learn that the only small (mouse-size) mammal on the island also got little attention from the fox. Avian predators take both squirrels and mice on opportunity, although the deermouse is abroad only at night.

This study revealed that, despite the low level of competition, the woodmouse was no more plentiful on the island than it was reported to be in mainland habitats. It occupies all areas where woody vegetation occurs. For some reason the Siskiwit trail cabin is overrun with them, even after careful disposition of things edible. During the night a light sleeper is constantly aware of their pattering and dragging around at all levels.

For three years our summer research crew operated a live-trapping grid of 81 traps south of Conglomerate Bay. In addition, using snap traps, they sampled the island extensively with a total of 98 lines. In these and other snap traps they caught more than a thousand mice, which were used for studies of breeding and productivity as well as food habits. It was found that the breeding season lasts from late March to early October in years of favorable food production. In his thesis, Johnson stated:

> In summary, it appears that mature females in April have the capabilities to raise three and possibly four litters [averaging 5] if they survive the entire breeding season. First-litter females may have two broods, and females born from mid-June to mid-July may have one litter.

The woodmouse is the only mammal on the island that can breed the same year it is born. It is short-lived, and in two years of trapping at Conglomerate Bay only a single individual was carried over from one year to the next. This could be due to both movement and mortality.

Of 151 stomachs analyzed for food content, 66.8 percent contained remains of insects and other arthropods, and 30.9 percent contained plant material. The latter comprised various seeds, fleshy fruits, and seeds of fleshy fruits, which undoubtedly were a small sample of the foods taken year-round. An inventory of the winter stores probably would show additional mast and fruits that ripen in the autumn. Signs indicate that mice rapidly garner nuts of the beaked hazel, and they probably store the pits of many fleshy fruits (cherry, dogwood, etc.), as do their relatives farther south.

The limited value of deermice to the fox seems to be a matter of palatability, although they are frequently killed and sometimes eaten. In widely scattered records, Shelton, Johnson, Peterson, and I found deermice dead in the trail where they evidently had been left uneaten by foxes. Fred Montague, who studied the fox in Indiana (and spent two weeks observing foxes at our camp in 1972) found that the prairie deermouse and white-footed mouse — both closely related to the northern deermouse

— were eaten occasionally, though obviously less preferred than meadow-mice. He found them left uneaten and marked with urine (189). At Bangsund cabin Johnson tried to feed a visiting fox the deermice they caught in snap-trap inventories. The animal turned them down and sometimes left its mark on them. In contrast, as Wendel stated in his thesis,

> The ermine is the most specialized for hunting mice and may take the heaviest toll of any Isle Royale mammals. A shorttail weasel that took up residence in the Bangsund cabin in October 1969 became very bold about accepting freshly-killed deermice.

Information on the foods of Isle Royale foxes is drawn largely from collections of scats made by various investigators since Mech began the work in 1958. Dave gathered and analyzed 295 scats over a 3-year period. The results showed hares to be the major item, increasing as a percentage of food occurrences from 17 to 46 to 52 in the years 1958–60 respectively. Shelton gathered 127 scats during 1961, and in Keeler's analysis of these, hares comprised 38 percent. In 1962 Shelton and Long collected and analyzed 107 scats in which they found 57 percent of food occurrences to be hare. These figures reflect the increase of hares that took place over the series of years in question. From 1961 to 1963 foxes observed per 100 hours of winter flying numbered 22, 37, and 45 — another indication that foxes too were increasing (251).

Johnson's work on fox foods is representative of a period of snowshoe hare scarcity and also primarily of the warm season, as were other collections. Lumping the three years, 1966–68, he found that the 448 scats contained 618 food items, of which 19 were from winter.

Following is my summary, taken from Wendel's more elaborate breakdown, of items of primary interest for present purposes (120, 121):

Food items	*Percentage of occurrences*
Hare	14.3
Muskrat	8.3
Squirrel	7.9
Deermouse	7.1
Birds	9.2
Cold bloods	5.0
Insects	5.8
Plant matter (esp. fruit)	40.0

Some significant seasonal trends do not show up in this summary: In winter, moose carrion would be a locally important item, although harès are the primary prey, supplemented by a few deermice and squirrels. In August and September the great dependence on fruit is beyond question. Relative to what I have called "cold bloods" (vertebrates), foxes seem to regard these as marginal foods. Shelton saw a freshly caught sucker dry up uneaten beside an active fox den. In October on the Feldtmann Trail I found a recently chewed green frog within an inch of a large fox scat.

We think of the fox as a "follower" of the wolf because the wolf is so obviously a winter provider. Seldom do we locate a moose carcass that is not, sooner or later, visited by foxes. A kill site seems to be a zone of relative tolerance among local foxes, and 3 to 5 can often be counted at once. The most anyone ever saw was 10 (72-10, Lake Whittlesey).

Wolves have only a casual interest in foxes. It is sometimes observed that a couple of wolves using a kill will take no notice of a fox or two that will be curled up nearby, day after day, awaiting a chance to feed. When wolves are hunting, the chance flushing of a fox is likely to mean a sudden sporting chase and a kill. In 1960 Dave and Don saw the big pack snatch up a fox that bounded off an old kill as the wolves arrived. The fox was practically shaken to pieces, as is usual.

This shaking of medium-size animals by big carnivores has a practical purpose. Not only is it damaging to the victim, but it prevents the attacked animal from doubling back and inflicting wounds on the head of its assailant. A small dog shakes a rat for the same reason.

In some chases the fox has an advantage; it can sail away over soft snow at times when the wolves are floundering. Rolf saw a chase abandoned after only 75 yards on 28 January 1975. On 2 February at the east end Rolf and Don saw part of the east pack in hot pursuit of a fox across Hidden Lake. But when another group of the pack appeared, the chase was forgotten in favor of a greeting ceremony.

On 21 February 1971 Don and I saw a dead fox in the middle of Lake Mason. We landed to look the situation over. The fox had gone onto the lake from the west as two wolves had given chase, on opposite sides and each about 40 feet from the trail of the fox. The three trails converged at the center of the lake. The fox had been seized by the back and its viscera shaken out. The wolves had moved on, leaving the remains to the ravens. We have two records of foxes probably eaten by wolves, although the evidence is circumstantial. In another case, the pack of 10 killed a fox on Malone Bay, and Rolf and Don collected the carcass (aut. 650). A day later they saw a trailing wolf come along and carefully investigate the bloody snow. Before the plane left the loner was seen to roll in the spot three times. Evidently there is something about a fox . . .

In 1971 I wrote in my notes that foxes were more plentiful than we had ever seen them, although hares appeared to have dropped off some from a peak in 1970. On 2 February in 5:15 hours of flying we saw 16 foxes, a record for the winter studies.[7]

In the early 60s, when we first began to see "tame" foxes around the Windigo camp, we did not recognize the beginning of an era that would afford unparalleled opportunities to observe the behavior of this species. Foxes were breeding in the area, and young animals were fed by help at the lodge kitchen (until the closing in 1973) and by the families of other summer workers. Some of these people were of Finnish extraction, as are many in the upper lakes region. Thus, young foxes sometimes were passed on to us with Finnish names.

In 1962 we saw "the fox" around occasionally, and in the next two years Sleepy became a part of the establishment. He was succeeded from 1966 to 1969 by Fido, alias "Heikki" (Henry, as I was told), as he was known in summer. In 1967 Wendel placed a tag in Fido's ear. Other vulpine visitors were around, whom we did not try to identify, except that one was a distinctive grayish female.[8] From physical characteristics, we thought this pair founded the fox dynasty that began in 1970 and was so much a part of our winter scene for six years.

In the 60s we learned that foxes would have to be taken into account in our daily operations. Snowshoes stuck in the snow by the back door had their bindings chewed off. On 5 February 1966 I made a large kettle of chili — up to the top of a 3-gallon pot. I set it out in the snow to cool and then forgot it for a couple of hours. On going out the back door I found the cover 20 feet away and about an inch of chili gone. Chili was splashed on the snow in a ring around the pot, which had melted down even with the surface. Fox tracks left no doubt about the culprit. I brought the kettle in to freeze in the back room. It was the way we kept our soup fresh.

From 1970 on, we saw many more foxes around camp than I shall mention. Probably some of the animals that stopped in were young of the year from litters born in dens of long standing in nearby ravines. In 1970 there were five individuals we got to know well:

Heikki (Dominant Male): This young animal must be distinguished from the first Heikki (Fido), who disappeared in 1969. In the winter of '70 we carelessly (in terms of records) used the Finnish name, but in future years this "dominant male" would be called just that. He was light in color, without a white tail tip, and with little marking on the hind legs. A large animal, he was socially dominant over all the others.

Scarnose: Another male, age unknown. Slightly smaller and usually subordinate to Heikki, although he challenged frequently. Duller in coloration but with little black marking.

Vixen: Only resident female this year. A young, short-coupled, handsome animal, not as tame as foregoing. Dark shading on throat (similar to Heikki, Sr., in the late 60s) and underparts. Black strap across hock.

Blackie: A young wild male silver fox, slender, seemingly starved, subordinate to Heikki and Scarnose. Stayed on outskirts of our area, frequently under a building.

Cross fox: A partially melanistic, heavy-bodied, and mature alien male that came in to challenge Heikki, occasionally with some success. The cross fox was not tame, and Heikki (DM) may have survived in rank because of his preferential treatment at headquarters.

Unlike wolves, foxes do not socialize well with people.[9] Our familiar animals could not be touched, and their chief interest in us was as a source of food, which meant practically anything of organic origin except uncooked vegetables.

The above-named were our regulars, but in addition an elusive light red fox sometimes surveyed the environs (for buried food?) when others were absent. This one always got chased away, even by Blackie, when they were present. It probably had a range to the east somewhere.

No chivalry was involved in food competition. When something was thrown onto the snow the dominant males usually got it.[10] The strategy was always the same; the successful fox grabbed the morsel and presented its back to the others, body checking sidewise any animal that tried to get at the prize. Obviously, in these routines there are rules at work; a fox does not attack another (superior) individual from behind. In a real fight no one turns his back. Actual fights are rare, but when they occur the blood flows, with each animal grabbing the other by the snout or jaw.

The system of food monopoly that went along with dominance, regardless of sex or condition, got under Don's skin and gave rise to sessions of instruction in the back lot. Don would fare forth with a pan of autopsy leavings (which he always brought back for his friends) in one hand and a stick in the other in an attempt to bring some justice and morality into the system. Actually he seemed at times to make his point, to the extent of a faint indication that the foxes were taking turns. However, from one day to the next Don always had to start over with basic principles.

It is the nature of foxes to curl up on the highest perch

Heikki, alias Dominant Male, alias DM.

available where the view around is advantageous. The turned-up roots of a windthrow, a fallen log, or any high mound of snow would be used as a resting place by our local animals. After snow was dug off the roof, it piled up in the southeast angle of our building nearly to the eaves, giving the foxes ready access to the roof. Frequently one or more of them could be seen looking down through our kitchen window as they sought their rooftop napping places. Sometimes these habits resulted in the marking of chimneys or the anemometer and degraded the quality of icicles used for domestic purposes.

In 1971 Scarnose was gone, and the cross fox visited only once. Two young animals joined our winter group, obviously tame as a result of summer conditioning. The female was with us only one season and needs no further notice. The male was light in color, like our dominant male, and his character was unmistakable. He was totally subordinate to DM, whose sufferance he cultivated by a perpetual caterwauling from a crouching posture of submission. We were to see much more of this fox, who soon earned the name of Whiney.

Vixen produced a litter of six in 1971, according to Windigo ranger Frank Deckert. One of them was still around the following winter. Our new female, yclept Bitch Kitty, was a gracile and winsome adolescent with the dark throat and general appearance of Vixen. Ingenuous and trusting, she became a favorite with all. She would be a member of the fox community through 1975.

By the second year it was evident that Vixen was the *grande dame* of local society and, in an undemonstrative way, the chosen one of DM. Modest and withdrawn, she did not participate in the mad scrambles for food. As someone said, Vixen was the kind of fox you would like to take home to mother. When we arrived each winter she was in noticeably better condition than the others, and her droppings attested to honest endeavor. They contained the fur of hares caught in fair chase. In a winter when cage traps were set for hares in the campground, someone got to running them and pulling the animals out through the wire. But that could not have been Vixen.

There were many more fox characters, but only one is essential to our story. In 1973, probably from a litter produced by Bitch Kitty, we acquired a glossy light red, piquant, and impish female that we called, simply, the little female or Little One.

After Scarnose, we never had two truly competing males. We got a distinct impression that in any aggregation of foxes a single male is the quota. DM efficiently repelled all nonfemale joiners except Whiney, who got along passably through his debasing tactics. He was an annoyance, but no threat to the system.

The original authority on fox behavior is Tembrock, who sketched many of the postures we saw almost daily in our free-living animals. His work was helpful as an introduction, but present interpretations are based on our own observations.

When foxes convene at a food source there is aggressive testing, and the same applies to disputes over land tenure. Evidently the males are territorial and the females more inclined toward sharing. However, our females sometimes ran off aliens of the same sex — their sex being judged by the fact that DM took little interest in the matter.

In the typical interaction two foxes would stand on their hind feet pawing at one another with the forefeet. Ears would be

Our dominant fox (right) gives the treatment (stylized aggression) to an interloper. The individual on his home ground has an obvious advantage.

flattened, tails kinked in an S-curve, and jaws gaping widely. Both uttered a staccato snarling "Kuc-kuc-kuc-kuc-kuc-kuc-kuc-kuc," in obvious threat of bodily harm. Usually the known subordinate would quickly back down, which meant he crouched and sometimes cried in a high-pitched wavering yowl. This was the treatment DM gave to any strange male that moved in to challenge him, like the husky cross fox of 1970.

After an encounter, the cross fox would be further humiliated as our head animal deliberately turned his back and went into a full-dress dominance display. This consisted of a magnificent piloerection — head, shoulders, back, and tail — that increased his size impressively. He dropped his tail straight down and humped his back like a Halloween cat. Then DM would stalk away stiff-legged to fresh-up his nearest scent post. Effects on the would-be interloper were devastating.

Challenges and dominance testing were frequent and elaborate between males, but often they were male-female and female-female. Young animals sometimes were left crouching in the snow, noses a few inches apart, both crying in a seeming uncertainty of threat and submission.

The grand termination of the behavioral episode had a devastating effect on the subordinate male (left).

Through February, social tensions mounted, and more foxes would drop in at times to complicate things. By the end of the month there was increasing absenteeism, and it was suggestive when DM and Vixen were gone at the same time.[11]

At home DM displayed a grave interest in the scent marks of all females, and any of the latter might be involved in coquettish tail switchings and slitherings-about on the snow. Sometimes DM slithered too. From 1971 to 1974 Whiney was not much involved. With rare exceptions, he kept shop at the shack while others were off on more interesting missions.

When Whiney was the only male around, we heard little caterwauling, and there was less emphasis on status. Things were much different when DM got back from "a business trip," and it took a few days to reduce Whiney to his usual size. This was especially so on an afternoon in 1973, when DM returned, an ominous kink in his tail. Like Odysseus of old, he found things in what he considered a state, and we were exposed to a degrading spectacle.

Whiney crouched in the snow and cried for quarter, then took refuge behind a screen of small birches. DM was not to be appeased. Swelled out in awesome display, he approached humped and glowering, hissing and snarling. Repeatedly he struck the bushes with a sidewise slam of the hips. The stricken Whiney was reduced to a psychological shambles, as the great one turned his back, then eventually moved away to lift a leg at appropriate places. It was the most vulgar exhibition of temper we ever saw, and DM seemed to realize he had achieved something extra. For the rest of the day he sat aside, squint-eyed and flat-eared, aloof and grand, gazing off into the woods, exuding charisma.

Occasional activity of this kind was seen among the wild foxes we observed from the air. On 13 February 1971 Don and I saw two on their hind legs interacting near the Sivertson dock on Washington Island. On the 18th I made a note about it:

This is no doubt a territorial type of activity; a dominance decision is involved.

Yesterday Don and Jim saw two foxes in a prolonged action about 1.5 miles east of Long Point on the south shore. One fox began to chase the other across the open hillsides to the east. It was a determined pursuit that ended a quarter-mile farther on with the chased animal at bay against a tree. The other stopped, swelled out its fur,

DM carefully investigated the scent marking of any female.

dropped its tail, and humped its back, turning completely around with its back to the first fox. It was the typical dominance display we see behind the shack.

The exact status of each female was difficult to determine. In degree DM responded to them all, but the degree was minor at times. On 21 February 1974 I saw Bitch Kitty wriggling around on her back, evidently with some orientation toward the big male, who stood 20 feet away and looked on. That afternoon I saw DM on *his* back with the little female not far away. Speculatively I threw Vixen a chicken neck and she ran under the building next door pursued by DM. There was a thumping under the building, and the little female went in to investigate. Then Vixen came out, followed by the male, who was munching the bones. Lastly Little One emerged, and both she and the male had blood on their faces. In my notes I said,

Several times lately we have given food to Vixen and she has stood off the DM and also the LF. This has not been usual for her — she sits back and has been deprived by the rest of them. This afternoon Tolley and I have thrown food to different foxes and it was immediately possessed and not contested by the others. There are situations in which possession has priority.

I never was able to define the criteria for many changes in status and attitude that took place almost on a day-to-day basis in the second half of February and early March — the mating season.

After these observations DM and Vixen disappeared for nearly a week. On 26 February they were back again, and Whiney got the treatment; he "raised a continuing racket." I fed DM some bones, which he crunched up and swallowed.[12] Little One was crouching about 20 feet away, and as I stopped at the door, something happened. Suddenly she blasted off like vengeance incarnate, straight at the big male.

DM knew instantly that his fate hinged on a quick getaway. In total abandonment of alpha dignity, he fled over the snow yammering like the most abject subordinate. A red streak of terror, he circled the premises, Little One hot on his tail, in deadly resolve to eat him alive. Three times the male sped around the lot, up the ramp, and over the roof. On the west side he launched himself into the blue to land in a snowdrift with all four spinning. As he emerged in a flurry of white, the young female plumped into the snow behind him.

This was Little One's finest hour. It was no stylized attack, but honest murder in the making, and the boss fox had no doubts. In a wild burst of energy, they skirted the edge of our compound, passed behind the trail-crew cabin, and disappeared in the woods.

Half an hour later they were behind the shack as though nothing had happened. Experimentally I gave DM a worn-out soup bone. He walked away and buried it. Vixen dug it up, and he appeared not to notice — nor did Little One, although I had a feeling she had come up in the world.

On 14 February 1972 four of us were in the kitchen eating supper. There were five of our neighborhood foxes in the back lot, all well fed from the proceeds of a field trip Rolf had made to a moose carcass; he had brought in about 60 pounds of head, organs, and bones. The foxes had been burying food[13] against harder times.

There was a thump and a slight commotion at our door. Don

went to the window and looked out. "Would you believe there's a fox dying on our back step?"

We found a strange fox breathing its last in the shoveled-out entranceway. Whiney had been perched on the snowbank at one side, and the others were scattered about taking it easy. None of them appeared to have attacked the dying fox. We injected the body cavity with formalin and froze the carcass, later taking it back for autopsy in the veterinary lab. The fox was an old 8-pound male that had lost some of its teeth. However, there was no wound and no evident cause of death.

My personal diagnosis was "shock," obviously a noncommittal term. I suspected that this fox, driven by hunger, had invaded strange territory occupied in force by the rightful proprietors. The adventure could have produced a psychosocial strain of colossal proportions. By the time he had pursued the enticing aromas to our door, was the build-up of stress just too much for an individual well past his prime? Exactly what occurred we will never know.

A couple of years later I felt a bit better about my guessing in reading Barnett's *Instinct and intelligence*. He was discussing crowding and social stress (p. 60):

> An extreme example, which I have studied in laboratory groups of wild rats, *Rattus norvegicus,* is sudden, unexplained death. An adult male, intruding in the territory of others is attacked by a resident: the attack consists of a threat posture, leaping and brief biting. The biting is usually harmless. The Intruder may nevertheless die — indeed, often does so; there are no post mortem changes to account for the collapse.

I am led to believe that social sufferings are among the worst that befall an animal, including the most sentient of them all, the human. Especially tragic and punishing are frustrations long continued. They lead to desperation and physical breakdown.

The terrors of misfitism and insecurity were evident in the plight of Blackie, although some of this is part of the usual trials endured by young animals. In particular, the potentially territorial males must travel during their first fall and winter. They wander through unfamiliar fox-inhabited country in an attempt to settle down somewhere. Unless there is a high death rate in the population, some will never make it, but will remain as unestablished opportunistic vagrants. Or they can accept the status exemplified by Whiney, which would not always work.

Blackie needed food, but we had difficulty getting any to him when other foxes, including females, were present. When threatened by another of our boarders — which was nearly always — he would flatten his ears, half crouch, tremble, urinate, defecate, and appear to be on the brink of nervous breakdown. Then he would dash under the building next door.[14] Blackie was inferiority personified.

After observing these social and physical incapacities in the winter of 1970, I suspected that here was one young animal who could not survive. However we had a surprise the following year. At Whittlesey Lake there were two kills, and on 14 February Dietz and Murray saw several foxes in the area.

One of them, keeping well apart from the others, was black. Don and I checked it again in the afternoon, coming in low for a good look. It was curled on a mound, and as it stood up, we

We saw a cross fox frequently in the winter of 1970. This color phase represents partial melanism.

were convinced. It had the familiar silvery color and white tail tip of Blackie. The location was about 25 miles from Windigo.

We were still more impressed the following winter (27 January 1972) when this same animal (as we assumed) appeared on Wood Lake only a short distance to the west. This time Blackie was freely associating with other foxes. Evidently he was now socially assimilated and localized in a range.

It was good to have such information on an individual three years in a row; we knew there would not be two such animals at one time. Then, three days after the above observation, Don and Rolf came in with some news: There was *another* black fox on the small islet east of Hawk Island near the mouth of Mc-

Blackie, a wild silver fox, was inferiority personified — perhaps a common characteristic in wandering juveniles.

Cargoe Cove, 10 miles north of Wood Lake. A day later Don and I saw them both in one flight.

Now all our knowledge was guesswork again. When I assembled the records on black foxes, I found more than expected. Mech had been told that Bill Lively, the enterprising Michigan conservation officer, raised black foxes in the early 30s. Had some got away?

In March 1963 Phil and Don had seen either a black or very dark cross fox running across Siskiwit Lake. In September 1969 Pete Edisen encountered a black fox on the lighthouse trail. On 3 February 1973 another black fox — this one with less silver and a larger white tail tip — appeared at Windigo and was chased off by local animals. After a few days, I saw it come through the lot once more when other foxes were absent. In the summer of '73 a black fox frequented the Belle Isle campground, and "it" appeared on Amygdaloid Island the following winter. These could be repeats of the one of 1972. We had other, secondhand, reports of black foxes. The genes for melanism definitely were being carried by the red foxes of Isle Royale.

As further proof of this we had seen at least a dozen "cross" foxes — at any rate, animals showing partial melanism — at various times in our field work. These records make quite an impression, but they must be measured against hundreds of observations of animals exhibiting no aberrant coat color. Rounding off fur-return figures given by Peterson and Crichton (1941-46) for the Chapleau District of Ontario, about 250 miles east of Isle Royale, we can represent the proportions of the three principal color phases of the fox as 90 percent red, 9 percent cross, and 1 percent silver (including black).

Per E. C. Cross, records of more than 100,000 fox pelts sold to the Hudson's Bay Company in the period 1916-36 can be compared with 880,000 recorded from 1850 to 1909:

	20 years	*59 years*
Red	74.7	74.3
Cross	21.6	20.4
Silver, black	3.2	5.1
Bastard	0.7	

In such a perspective, what we have seen on Isle Royale may be according to expectation in this northern population of the red fox.

Dry statistics on the percentage of various foods taken do no justice to the fox as a hunter. The creature is exquisitely adapted for its role as a minor predator in the food chain. Its senses are keen, and it is incredibly quick and agile.

The environmental awareness of a fox must be constantly whetted by odors brought on every breeze. The animals readily locate cached food anywhere, and twice I have seen them detect an approaching fox that was out of sight in the woods between 200 and 300 yards away. After one of our autopsies in the early 60s, I bunched up some visceral fat of a moose in a wire holder and hung it about 6 feet up on a tree trunk for the birds. Our familiar Sleepy was trotting through the woods about 30 yards away when he got downwind of the suet. He instantly whirled, nose in the air, came directly to the tree, leaped up to jerk off the feeder, and sped away with it.

On an afternoon in February 1972 Rolf hung part of a moose head in a tree for the benefit of whiskeyjacks. It soon became evident that Bitch Kitty was interested. She pointed her nose up and circled the tree. Rolf had left the ladder leaning against the trunk, and soon we saw the fox at the top of it chewing on the meat. When we went out the door she came down *head first,* the pads of her small feet hitting the rungs of the ladder with dainty precision!

Another time we saw her climb 6 feet up the trunk of a 10-inch yellow birch in an attempt to get suspended food. She clasped the trunk with her forelegs and pushed with her back feet, using them alternately. Red foxes, in contrast to grays, are not supposed to be able to climb trees.

On the latter occasion the food in question was a hard loaf of bread I forgot to take out of the oven. I tied it with a cord and hung it up for the squirrels. The first squirrel nearly made a mistake. It clung to the loaf, then severed the string with one bite. Bread and squirrel dropped within 3 feet of a fox. Both of the animals moved instantly, but the squirrel got up the tree.

On 6 February 1966 we had 3 feet of snow, and the weasel had made deep tracks behind the cabin. Don saw one of the squirrels running on the packed snow in the weasel trail. The animal was too low to see Fido standing nearby, but the fox was aware of the squirrel. As it passed, Fido leaped into the trail and snapped it up. Don said the bones crunched as it began eating the squirrel at the head end, and the last thing to disappear was the tip of the tail.

I watched both Fido and a wilder fox hunt squirrels that winter. The snow was sufficiently firm so that squirrels were active. On 12 February the wild fox came in for a handout, and later I saw it curled up on a log about 60 yards behind the cabin. It was watching squirrels and whiskeyjacks intently. It lowered its head flat to the log as our tame (and very fat) squirrel worked closer in, traveling over the snow. Finally the squirrel ran between trees only 20 feet away, and the fox made its move. With several bounds it nearly caught the squirrel, which kept about 2 feet ahead as it made two circles, each time bouncing off trees to keep from getting caught. Finally the squirrel hit the side of a tree (rather than the middle) and circled behind and up. The fox walked away without a look back.

Fido used the same technique, crouching on the snow at a vantage point with head up and ears cocked. As a maneuvering squirrel got nearer, the head of the fox hugged the snow between its forepaws. I spoiled the hunt by taking pictures, and Fido walked off in disgust. A few days later I found tracks where a squirrel and a fox had rounded a high mound of snow from opposite directions. A patch of blood told the story. There was more blood with fox tracks on top of the mound.

On 14 October 1976, William Kohtala, Windigo maintenance mechanic, told me he saw a fox catch a squirrel by lying quietly near the base of a tree until the squirrel worked its way down the trunk. Then the fox leaped up, knocked it off with a paw, and caught it on the ground.

Some other hunting methods of the fox will be mentioned in the next chapter.

Throughout the warm season, fox usage of the foot trails is evident. There is less sign in the maple forest than anywhere else, but it picks up noticeably in burns and openings around the edges. Reynard does not waste much; he places his scats where their scent-marking value is greatest. Commonly they are on top of rocks (the "double dip" recorded by summer help), or near anything conspicuous, such as a pile of bleached wolf droppings or a moose skull.

Trails above Daisy Farm get much use by both hikers and

foxes. Campers cast aside plastic bags that are not degraded by weather and must be picked up by conscientious rangers. By the time they are carried out, these bags get to smelling very high indeed, for the foxes mark them regularly.

In winter the urine marks of foxes along trails are plentiful and conspicuous, becoming especially so in late February. Fox trails across openings lead from one bush or weed to another, each sprinkled with yellow drops on the snow. On 22 February 1974 Rolf counted 74 marks on about 600 yards of trail between the bunkhouse and the powerhouse. Male foxes being highly territorial, we ascribed much of this activity to our dominant animal. Farther out, the "wild" foxes were doing it too.

We would have liked to know more about Bitch Kitty's social life. She enjoyed economic security at the bunkhouse. But she knew about the breeding season, and in a group of variously

Anything unusual along the trails became a part of the territorial scent-marking system.

adjusted females, it was evident that her outlook on sex was quietly contemporary. We were told that she produced one of two local litters in 1972, when she was a year old.

This was explainable by what Don told me later. He said that after I left the island, he saw a "happy wanderer" come through the lot and, with little ceremony, subvert BK in the trail to the backhouse. Don was pensive about this, but I reminded him that we did not know where Kitty went on her time off. This male could have been an old acquaintance not forgot, and in addition, who were we to judge the problem of an overdue female in the Ides of March.

At times BK practiced her arts on DM, usually with little response, but there was increasing evidence of her more serious liaison with a male we seldom saw. It could have been the same one each year. Rolf made notes on one of these amorous interludes on an evening when our regular males were absent. The scallop-tailed male was chasing BK around. She was burying food, and the (evidently hungry) visitor was digging it up.

> She ran onto the roof and pounced on an old cache, throwing it 3 feet into the air over her; she did the same on the ground. Vixen looked on from the other bunkhouse. She whined submissively when the male came near, and she and Kitty squared off verbally once. Kitty often hid, head low, ambushing the new male, then running away with a bouncing gait. Soon Toivo [Finnish = devil: one of Don's names for DM] showed up, and he immediately took after the new male, playing the ritualized dominant role, with the new fox whining submissively. When he had the chance, however, the strange male took after Kitty again.

By that time it was getting dark, and Rolf could not follow the further proceedings.

After we had been in camp for a few days in 1975, it became evident that something was missing; we had not seen Vixen, and DM obviously was not there. The back lot was quiet, and it appeared that the days of Don's "reform school" were over. We were under the burden of a front-office mandate not to feed foxes. Each summer the prospering litters had been a nuisance around the campground at Windigo, and our top pair could have been the victims of a deportation program of some sort. We understood the food embargo and did our best to cooperate, although there were occasional emergencies.

BK, a favorite of all, at age of 10 months.

I never tried to find out what happened to DM. The truth was, of course, that his churlish cant was not just personal cussedness, but a part of the system. If you overlooked occasional tantrums he was a forceful personality, one not likely to be replaced.

We did have a dominant fox in 1975 — Whiney, no less, and he presided over the remaining females (BK and Little One) according to the rules. His star was risen and his underprivileged past forgotten. By default he had become a dominant male. Now he was first at the food, first on the mound, and first in the hearts of his ladies.

The change in social status had obvious physiological effects on this fox. He openly courted BK, his testes enlarged, he began to scent mark, and he was in breeding condition for the first time in his life. It was another example of those stop-gap adaptabilities that so often appear among wild creatures in times of crisis.

We had difficulty adjusting to all this. Perhaps Whiney showed to disadvantage against the high standard of leadership we had known in previous years, but for me his alpha-ism never quite came off. He had been a career subordinate, and his exaltation in one easy lesson was a bit much. Admittedly, inadequacies could have been in the eye of the beholder, since his supporting cast did not seem to know the difference.

In particular, he and BK seemed to have much in common. However, she was gone for a time, and on her return it appeared she had been in the wrong company. I heard something in the backyard and went to the window. Ten feet from the door, BK and Whiney were touching noses, something rarely seen. Kitty was in a deplorable state.

Her coat was rough, she looked thin, her left front leg was swollen and useless. I thought it was broken. There was a 2-inch gash at the elbow, and when she put her paw on the ground she closed her eyes in a grimace of pain. Pieces of lip were missing and there were cuts on her nose. Spots of blood marked her tracks. She seemed so weak she could hardly get around.

With visible misgivings on the part of Little One and Whiney, we got some food to BK, but we had scant hope she could come out of it. In the next few days she spent much time off in the brush, where she was haunted by Little One, who seemed to be taking dominance liberties. Rolf tracked them and kept up on the situation. By that time I had left the island; it was the last of my winter work. Surprisingly, Rolf reported that in another week BK had nearly recovered, and she seemed her old self when they closed the camp.

In one starkly realistic way Whiney was a success; he lived to be at least five years old. Rolf saw him last at Windigo in May 1975. By 1976 the no-feeding policy had dispersed the fox community and returned the winter camp to the squirrels and whiskeyjacks. An era was at an end, and I was glad to have been a part of it.

CHAPTER FOURTEEN

The Whole of Nature

IN PONDERING THAT BAFFLING ORGANISM we call the ecosystem, we are stopped on the one hand by frustrations of partial knowledge. On the other, vistas of understanding seem to open. As a way of muddling ahead, we collect vignettes of fact and theory that we try to fit together in a logical pattern. For many of the gaps we find no pieces.

All of the world's pioneer ecologists have perceived something ever more evident and convincing: that the organized living things on any area of the earth's surface are in the present configuration because of (1) the influence of climate, (2) the nature of the substrate and topography, and (3) the length of time a relative stability has endured. Thus there is one of the most finely tuned stabilities on earth in the Amazonian rain forest, which has been little changed for some 70 million years.

In contrast, a much less stable condition is evident on Isle Royale, which emerged from under the ice less than 10 millenniums ago. This does not imply a raw beginning for the island so recently, because there was an adapted biota that had been pushed southward by the glacial influence. It was ready to reoccupy lost ground when the climate changed. Of course, reconstruction is still in progress. With only one species of mouse and no poison ivy on the island, there is quite a way to go.

We have considered certain niches — jobs in the community — and there are more to be discussed. What a plant or animal *does* is a clue to its role, and also the roles of associated things. We have a general concept of how the system works, how green plants synthesize the nutrients for everything else. Some of the annual yield is fed out in the form of summer forage, winter browse, and fruits and seeds. Only a small portion of the total warm-season production is used by the vertebrate animals with which we have been primarily concerned. Eventually all of it gets back to the soil, which we have neglected.

Soil is a complex organism in its own right. It is the cracking plant, a subassembly of the ecosystem, devoted to degrading every kind of organic substance into more simple nutrient forms. These are started again through the many pathways of energy flow that enable plants and animals to live. Most plants of the soil are not green; they live on residues of living things. They are bacteria, yeasts, molds, fungi, and those enigmatic filamentous mycorhizae that envelop or penetrate the rootlets of trees and other plants. These have a symbiotic relationship with the higher plant, which means a mutually beneficial exchange of services. They help to break down soil minerals and other compounds into usable forms and do much to promote the health and vigor of forest trees. As part of the total process, there are animal components at work in the soil, including many kinds of invertebrate life.

As for fungi that rise above the ground, Isle Royale is a mushroom hunter's Eden throughout the season. A wide variety of edible species are there, and a few that are fatally poisonous. An interesting saprophytic (living on decaying matter) plant of August and September is the translucent white Indian pipe that springs up, like an artist's dream in alabaster, from the spongy mold of mist-dampened forests along the north shore.

The further one goes in studying the habits of animals and plants the more perplexing the details and variations become. Nothing seems to hold still long enough to be understood. Niches are seasonal, and the seasons are constantly changing. No year is like any other year because of weather, that day-to-day manifestation of climate. How long do you have to stay at this to see the same conditions come back again? Likely enough, they never do come back.

Sometimes we see animals doing things that spell individual destruction. How do they get away with it? How could *this* have survival value? If we pause to reflect, these are not questions to be asked of individuals, but of species. Maybe it was for the common good that the individual did die. Then too, everything dies; it is only a question of when.

It can be difficult also to explain why animals do certain things that simply seem to be wasted effort. A stumper that often challenged me is why the beaver girdles or half cuts many food trees without finishing the job. He spends untold hours and energy chiseling away and gets nothing at once to show for it. Where lies the incentive?

I may have got a bit more insight into this situation on a hiking trip of about 40 miles over the center portion of the island in early May 1976. As mentioned before, I found many aspen windfalls blocking the trails that spring (p. 199). It was evident the trail crew would have a big clean-up job.

Around beaver ponds girdled aspens and birches were budding out and the sap was running. The trees were still alive, even though some of the girdling was from the previous summer. As evidence of age, the cuts had begun to weather and turn gray.

High winds had blown down the partially cut trees far and wide. I began to realize that the half-job of the beaver was effective after all. With a minimum expenditure of energy, the animal now had fallen aspens to feed upon, and so did the moose

Often the beaver does much work for little immediate gain. This 19-inch aspen hung up in adjacent trees and had to be cut four times.

and hare. The removal or killing of large trees would open up the stands and accelerate plant succession.

As another observation of that spring field trip, the pileated woodpeckers were hacking their rectangular holes in the bases of certain trees, just above ground level, where roots taper into the trunk. Three that I noted in particular were all balsams; they had rotted wood inside the outer shell, and the two to four holes undoubtedly weakened the base of the tree. One that had been chiseled perhaps a year before was a windthrow; more moose food.

It is profitable intellectual enterprise to ask why things happen and to study the alternatives; good logic may take several directions. Cutting or half-cutting trees could sometimes represent a wasted effort that has no benefit for the individual beaver. However, this animal does not work by judgment, but by a compulsive innate behavior pattern.[1] He must cut trees because they are there, and in the long run it gets beavers ahead, even though any one piece of business could be a loser.

It came to me vividly how animals do these things on an evening of beaver watching along the Lane Cove Trail 12 September 1973. A beaver flooding had inundated the trail in the valley north of Greenstone Ridge, and I crossed on the dam. I climbed the low ridge ahead just before sundown. Tomorrow I would follow this ridge westward (it was the wrong one), and the weather outlook being good, I simply laid out my sleeping bag, using the sleeve tent as a dew cloth.

Then I went back to the pond for some late-evening watching. Earlier I had noticed that a dead aspen stub had been freshly girdled by a beaver. The wood seemed hard, but the stub was limbless and so far gone it had broken off 20 feet up. Strewn about was better material for a dam; the stub was not food; why cut it? By chance I located myself about 30 feet away on some logs over shallow water.

I soon accounted for at least three beavers in a 200-yard stretch of flooded bottom where visibility was fairly good. One large one came within 20 feet of me and sat on a log. It was on its haunches and scratched and groomed itself on the head and elsewhere. I could not see that it used the double claw that is on the third toe of the hind foot, supposedly for this purpose.

On the other side of the pond a small beaver swam within a dozen feet of a green-winged teal; neither showed any alarm. It

was nearly dark when this beaver came out of the pond on my side and moved among the logs and thimbleberries to that stub. As I sat there I could hear him gnawing at it. The old dead stub had to go.

The animal returned to the pond twice and then went back to work. After a while, things were quiet. It was 6:50 and shadows were deep around me. It was getting cold; time to head for camp. As I stopped to look at the stub, the small beaver broke from the ground cover and ran past me to the water, where it turned and lay, regarding me suspiciously. I went back to my sack with a lingering impression that catching beavers should not be particularly difficult for a capable wolf.[2]

I mused at length on why a beaver would spend its energy on that useless stub. The explanation may be an extension of the half-cutting idea proposed earlier. The cutting of anything and everything is generalized behavior that provides for individual and colony survival. But it goes beyond that; it is community service. It contributes to the pond-meadow-forest rotation that brings diversity to bottomlands and benefits all living things in this association. The beaver is a good ecological citizen.

In this vein, what seems like wasted effort is not wasted in the long run. For the beaver to survive, its life community must survive. In these self-operating organizations of living things nothing actually is wasted: not substance, not energy. Possibly this is what George Perkins Marsh meant when he said, ". . . in the husbandry of nature there are no fallows."

To follow through, one must assume the existence of logical reasons for everything he sees. Some reasons seem obvious; many more are a mystery. All about us are adaptations we take for granted, but which might be understood if we looked long enough.

For example, how does one explain the unique bark of the birch; the bark that, once peeled, will never grow back? It became a magnificent cultural material for the Indian. In the forest, that tough thin bark lasts for many years after the tree is dead. Why?

Again, we may be dealing with community benefits. When beavers girdle an old birch, or when a stand of trees is flooded behind a new dam, the trees are killed and the largely limbless stubs continue to stand. The birches will be there after other species have rotted away, protruding from the water for dec-

Old beaver works have turned the Tobin Creek drainage into a succession of ponds, meadows, and thickets.

ades. The wood softens, and woodpeckers dig out nesting cavities. The downies make small holes, the hairies and flickers a size larger.

These cavities become secondhand nesting places for tree swallows, chickadees, nuthatches, bluebirds, grackles, starlings (!), and other small birds. Some of them become large with use and house the wood duck, hooded merganser, and goldeneye. The wood will get so weak the stub eventually falls, but the old tree has served for a long time.

Is there some connection between this community usefulness and the amazing durability of birch bark? Whatever the relationship, it had its beginnings many millions of years before the first beaver ever flooded a bottomland. Birds, mammals, and trees underwent a slow and coordinated development. We can not assume a simple answer to the question, How did the birch tree get its bark? But somewhere there is a bio-logic that accounts for such distinctive qualities. The bark did not evolve as a covering for canoes.

Amid the complexities of nature we can ask endless questions, both practical and academic, that could be profitable to explore. For now, perhaps, they have value in pointing up our ignorance. They seem to say, keep some country in its natural state so we will have time to learn what it is really like. Our questions and ignorance suggest a go-slow policy in destroying anything irreplaceable.

In all realism, it is not our nature to slow down on anything. As the master species, we are hard at it. I lie in my tent and listen to the singing mosquitos. How great it would be, I think, to be an all-powerful benevolent biological dictator for just a short spell. What good one could do.

Then it occurs to me that someone said mosquitos pollinate the habenaria orchids.

The changing seasons were discussed in Chapter 7, and it was evident that migrant birds, either in passing through or in spending a part of the year on the island, have a well-established part in annual events.

I have been impressed that, in fall, as the fruit season is tapering off, foxes get some of their support from flocks of ground-feeding sparrows and other birds that are on their way south. This first came to my notice in the second week of

The birch stubs persist long after most woody growth has rotted away. Hole-nesting birds are benefited.

October 1972 in the vicinity of Windigo. There were snow buntings, juncos, and many species of sparrows on the move, and I found at least two places where it appeared a bird had been caught and eaten.

The case became more convincing when I started a field trip at Siskiwit camp 22 September 1974. I reached the cabin in early afternoon and spent my time following out local trails, since my plan was to head for Lake Desor the next morning.

From Siskiwit dock for half a mile westward, where the trail traverses old fields dating back to CCC days, many birds were feeding on the ground, including various sparrows: song, field, chipping, fox, vesper, white-crowned, white-throated, and possibly others in fall plumages I did not readily recognize. They were mostly in the grass, but on the trail I found 14 places where a bird had been eaten, the primaries and tail feathers being left in a pile. Fox droppings on the trail sometimes contained bird feathers, and one fox had left its marker right with the feathers. White tail feathers identified the kill as a junco.

Isle Royale's most important fleshy fruit in winter is the mountain ash.

John Wiehe, towerman on Feldtmann Ridge, had excellent opportunities to watch foxes hunting. In August and September he saw them catching grasshoppers, squirrels, and birds. He described the technique of pouncing in the grass:

> The fox walked carefully through the grasses lifting its paws high and looking from spot to spot in a somewhat jerky fashion seeking out the insects. When an insect was spotted, the fox would slowly lower its body to a crouching position and then jump forward with both front legs extended side by side in an effort to cover the prey with its paws. I saw the fox capture and consume grasshoppers . . . but there may have been other types of insects being eaten.

On a midsummer morning John watched a fox approach with something in its mouth. He descended the tower to investigate, and the fox dropped its prey on the trail. The catch proved to be a female indigo bunting, which the fox retrieved later. In a similar episode at Windigo, John found that a fox had killed a white-throated sparrow. When he laid the bird down and re-treated, the fox came back and got it. Foxes do not confine their birding to summer and fall. Ron Bell and Steve Ruckel saw one trot up the trail to Ojibway tower dangling a couple of ducklings from its mouth.

Aside from being eaten, migrant birds provide other benefits to the fox. In the Feldtmann area in October 1974, I saw flocks of robins taking a small crop of mountain ash fruit, some of which fell to the ground for the benefit of foraging foxes. The following winter no mountain ash fruit was to be seen. I noted that when robins were cleaning up the crop of chokecherry — black, juicy, and inviting — I found none of it on the ground, and the scats of local foxes did not contain cherry stones.

On Feldtmann Ridge foxes also feed on the hard pomes of hawthorn; the bony nutlets from this fruit may remain on the ground for several years before sprouting. Many seeds of fleshy fruits probably require scarification in the digestive tracts of mammals or birds, plus a winter of freezing, before they are ready to germinate. Lest we miss the ecological implication, the only evident reason for a plant to produce a fleshy fruit is that some creature will eat it and distribute the seeds. The only reason for a fruit to be bright colored is that birds are not color blind.

Winter birds of Isle Royale are subject to great variation. In the mid-60s, after several cold seasons in the field, I began to realize what made them vary. In this chapter I must repeatedly come back to weather and the matrix of events that is mediated by year-to-year conditions. A principal effect is on the crops of fruits and seeds that some animals use or require.

Spring rains may come at the wrong time for effective wind pollination, or a late frost may kill the blossoms. Such happenings can explain why we have fewer good seasons of bird food production than poor ones. Also, a bountiful harvest of seeds and fruit may be on the way, but a dry summer will shrivel it to nothing. Even when abundance is evident in fall, an early winter ice storm sometimes spoils the program. However, when mountain ash berries are stripped from the trees in this manner, they keep fresh in the snow, to emerge for the delectation of foxes and whiskeyjacks (and ravens?) later on.

The most significant winter bird foods are birch seed, the seeds of spruce and balsam, and the fruits of mountain ash. Conifer cones are first cut when green, by squirrels, in August

Pine siskin (left) and redpoll, small fringillids that flock together and appear to have identical habits in winter.

and September. They ripen in fall, and most of the seed is windblown to its destination on the ground, where it lies "stratifying" (preparing for germination) all winter. Not all of the cones open, and an abundant supply may be on hand in winter for small groups of white-winged or red crossbills to open for the seeds.

When we arrive on the island I look carefully at the birches, white and yellow. The abundance or scarcity of seed decides whether or not we are to have the flocks of small fringillids (finchlike birds) that characterize the winter season. Birch seed is shed all during winter as a result of bird and wind action; it blows over the surface of the snow and then becomes layered within the pack as more snow falls. Anyone who melts snow for water will have birch seeds in his coffee.[3]

Birds using the birch are mainly the redpoll, pine siskin, and goldfinch. These small finches are closely related, and they occupy the same ecological niche in winter. Incidentally, animals of different species are not supposed to fit into the same niche in the same community. However, this is legal, since it is only for the winter. As far as the community is concerned, these *are* the same species. But when spring comes the goldfinches go south and are found almost the length and breadth of the United States. They are the summer-nesting "wild canary" of farm boys. The siskin breeds in a wide range across Canada and a northern fringe of the states; some of them stay on Isle Royale. Before the warm season the redpoll migrates far to the north where verges of the boreal forest give way to tundra. It nests in scattered spruces of the "taiga," where barren ground caribou have wintered.

These birds appear on Isle Royale in the fall in mixed flocks, sometimes mostly redpolls, sometimes predominantly siskins, with here and there an olive-colored (winter plumage) goldfinch thrown in. One winter (1971) we got a surprise because our small fringillids were nearly all goldfinches. Maybe the makeup of our flocks, usually 20 to 100, depends on what happens on the breeding grounds. At any rate, if there is a crop of birch seed they stay for the winter, and if there is little seed they move on somewhere, no doubt to better feeding areas.

We got interested in these feathered miniatures during the early 60s in seeing them gather on the snow behind our shack. We had only an occasional visit from a fox in those days,[4] so birds and squirrels had the run of the place. The attraction in our back lot was a strange one: The birds were avidly eating mineralized snow around the urine holes that collected in a certain area — it was said the Brownies did this at night, but I never caught them at it. For various reasons Don dubbed the flocks with a generic term, which seemed ecologically sound. Our small winter birds became, forevermore, "piskins."

In winters when we had few birds and practically no birch seed, we would sometimes see a small flock feeding on the seed-bearing catkins of green or speckled alder, which are much like the birch. This seemed a back-up food that never was in great abundance.

Birch seeds collect in windrows on the snow, and I watched for the piskins to make use of this supply — never. They would pick up nothing off the snow, except the yellow snow. One year I took a bag of commercial bird seed to the island and scattered it where the birds were feeding. It might as well have been gravel; they did not touch it.

These feeding habits got to be an issue. I reached the McCargoe Cove campground on a sunny day in the first week of

Birch seeds, yellow (left) and white, are the main dependence of wintering siskins and redpolls.

May just as the first dandelion seeds were ripening around the grassy dock area. Siskins hopped about on the ground eating seed directly from the plant. I watched as close as 4 feet, and also through binoculars, but not once did I see a bird pick up something from the ground.

All of which does not agree with anything written about the

siskin. They are fed at feeders; they pick up grass seeds off the ground. But in winter on the island they will have no seeds except what they get directly from the plant itself. Incidentally those windrows of birch seed do not go to waste. If snow texture is right, the woodmice will be abroad at night, and tracks show that they patronize these sources of food. No doubt such feeding helps to spare the caches of other things they have laid by for the cold season.

Other birds we look for, and which mix in some degree with the smaller finches, are the purple finch, pine grosbeak, and less frequently the evening grosbeak. Grosbeaks are mainly dependent on fruit of the mountain ash. When this tree bears abundantly and is laden in midwinter, we have the best of bird watching. The islands bordering Siskiwit Bay on the south have many mountain ash, and the species is frequent on edges and cliffsides. As I have mentioned, foxes seek out these sites on the south shore and elsewhere and track up the snow under the trees. A thaw brings the bright red fruit to the surface and gives an impression of blood on the snow. We have been fooled by it many times.

Small flocks of up to a dozen white-winged and red crossbills are seen when we have a good cone crop. The snow is littered with debris under heavy bearing spruces or balsams. These birds are prominent among feeders on the yellow snow. That is the way we found it in 1968, our winter of least snow depth, and the first winter of red crossbills. Birch seed was abundant, as were the cones of spruce, fir, and cedar. Redpolls were missing — maybe they stayed farther north because of good food conditions. Flocks of 20 to 60 siskins, with occasional goldfinches, were common. Mixed in were a few purple finches. We saw scattered pine grosbeaks — many on the south shore — and flocks of a dozen or more evening grosbeaks, keeping mostly to themselves, were budding on the conifers.

To a crew interested in practically everything, it was always a letdown when we arrived on the island to find the trees bare of the bounty that we, somehow, thought should be there. Something was lacking when we waited long to see any of the small creatures that livened the woods in our "best" winters. Discussions to follow, as well as the notes in Appendix IV, will attest that the periods of silence and untracked snow are part of the big plan also.

As I see an osprey in the heat of noon shading its nestlings under drooping wings, I realize that we creatures of the wild know and react to common problems. As I lie under my shelter on a night of rain, I snug down in my sack and know that the squirrels and woodmice are likewise in dry cavities. If they got wet, it would mean pneumonia and death. As I retreat from a stinging wind on a cold day and find comfort in dead air among lowland cedars, I understand why foxes are curled up under drifts and windthrows and moose are bedded in thick conifers. This is a suspense period, and they will earn their living under pleasanter conditions.

Sometimes in February the weeks go by and we see few, if any, tracks of squirrels or mice. In our experience, this inactivity has always been tied to soft fluffy snow. Both species are garrisoned against want by caches of food, but this does not mean they eat regularly. They are capable of spending many days curled up in a furry ball, enjoying a low-metabolism sleep.

Around logs and the bases of trees, where mice have their nests and food storage, there are tunnels under the snow, and it is not possible to judge reliably the extent of woodmouse activity. Their foraging on the surface of the snow is easy to appraise, and the conditions they favor (fairly solid footing) correlate well with requirements of the red squirrel.

Long periods of holing-up are evident in the squirrel. In the winter of 1967 I began to suspect some disaster had overtaken the species. This proved not to be so; the prudent creatures were riding out a long period when foraging for cones, birch and alder seeds, or dried-up thimbleberries would have been hazardous and unprofitable. Wendel told the story in a statement he wrote for the annual report:

> In 1967, no squirrels were observed in the immediate vicinity of the camp for the entire month of February, and few signs of this species were seen elsewhere. Our first impression was that the population had nearly died out — logically, as a result of the marked shortage of seeds and fruits of woody plants during the past two winters (1965–66, 1966–67).
>
> In March, however, this species became active, and trapping was fairly successful . . .

In winters of deep or soft snow that were to follow, we became more familiar with this habit. In 1969 we did not see or hear a squirrel around Windigo until 17 February. On that date

I wrote also that I had seen less than half a dozen woodmouse tracks. In 1970, signs of squirrels were a rarity during February, and it was not until 5 March, on a new crust, that we finally saw one of the animals at the camp.

The use of food stores permits a squirrel to localize its forays abroad, even when snow conditions are not critical. I saw an example of this on 12 February 1963.

> On the floodplain by Washington Creek I found a tame red squirrel sitting on a stubby branch of balsam about 5 feet above snow level. It had a midden of shucks on the snow under the feeding perch. It would go down into the snow at the base of the tree and bring up a spruce or balsam cone, then sit there and neatly pare off the layers and eat the seeds. Handled five cones in a few minutes while I watched.

I cut down through a couple of feet of snow at the base of the tree and found the cone debris layered clear to ground level. The animal had been feeding here all winter.

Wind is one of the important influences governing the daily activity of animals. At times of low temperature and high wind, the woods seem devoid of bird life, squirrels are silent, foxes do little moving about, the nocturnal hares leave few tracks, and moose remain in dense cover.

It is evident in winter that moose move from highlands to low ground and in and out of thick conifers in response to changes in snow conditions, temperature, wind, and sunshine. Like anyone else, they must maintain their body temperatures within

Crossing of the trails: moose and hare.

some middle range of comfort. Much depends on the wind-chill factor: It can be conducive to more activity when the day is warm, and it slows things down in cold weather.

One might not realize that, even in a northern winter, the moose, with its great bulk and relatively low rate of heat loss, can easily become overheated. Thus, on a sunny windless day, they stay in the shade and move little. Under such conditions we have difficulty seeing them from the air because of the high contrast between snow and shadow. Correspondingly, tracks show up to good advantage.

As a sample of our observations, on 26 February 1971 we had experienced six consecutive days of predominantly clear weather, with daily maximum temperatures ranging into the 20s and 30s. That day we did little flying, but I recorded that

> yesterday I flew 2 hours and 50 minutes, and Jim flew 1 hour and 35 minutes, but all of us saw only one moose, the presumably injured one on Houghton Point. Don proposes that it is too warm in the sun and moose are seeking heavy cover. This seems a plausible explanation for the complete failure to see moose . . .

Actually, it does not need to be so warm; almost any temperature above zero may have minor influence if there is little wind and no cloud cover. Under these conditions it is also quite possible, as indicated by tracks we see, that moose do much of their foraging on clear nights and are correspondingly inactive by day.[5]

Many flights have illustrated the contrasts. On 18 February 1975 Don and I flew for 2 hours, 18 minutes, covering all the major portions of the island, as usual. During the previous 24 hours it had been cloudy and warm, 25 to 33 degrees, with much moisture in the air. Our flight began at 2:20 P.M., after it had brightened somewhat at Windigo, but there was cloud cover over the rest of the island. Moose obviously were active: During the flight we saw 22, with no special effort made to search for them. Of the 22, 13 were standing up and moving about, usually in open areas of short brush and sparse tree growth where browse is found.

A somewhat comparable trip was made three days later, 21 February, after a night of quiet, bright moon, and (on track evidence) much moose activity. The morning was sunny and delightful for flying but warm for a moose. The thermometer read +22° when we took off at 10 A.M. for a flight of 2 hours, 2 minutes.

On this flight we saw only six moose, three of which were bedded in heavy cover. Since moose were not in the open, I was looking hard at these shaded areas where tracks were plentiful. In the conifers north of Lake Richie, Don saw two moose, one standing and the other bedded. He circled to show them to me, and we could not find them the second time.

Many contrasting records of this kind make us chary of counting moose for census purposes on any but obviously favorable days. Probably what is needed to carry out a moose census on the most reliable basis is an elaborate weather-recording system coupled with extensive reruns of specific areas under varying conditions. That is, of course, if you do not have that mechanical scanner we look forward to.

What has been seen provides the basis for a not-very-specific hypothesis on moose activity and movements in relation to weather:

1. At any given time, a moose can be doing nearly anything.
2. On cold windy days or nights, moose are largely inactive. Presumably they are bedded in sheltered areas.
3. On quiet clear nights moose move about and do much browsing. Following which, on sunny days there is little activity, and the animals stay in heavy cover.
4. Following a prolonged stormy period there may be unusual daytime activity.
5. On cloudy days moose usually move about and feed conspicuously. This could mean relatively little feeding on a cloudy night, in which poor visibility may be a factor.

I have referred to niches occupied by various animals in the life community, and have defined the term loosely as the function, role, or job that each performs. Such description is nonprecise, but it can help elucidate the gross structure of the life association. In the final resolution every species has a function all its own. The word *niche* probably has its greatest usefulness in comparing different communities and in identifying species that do the same thing in each — thus, the bird-of-prey niche, the seed-eating-rodent niche, or the dominant-carnivore niche. Subdivisions of these often are evident.[6]

A scavenger-predator of great efficiency in summer is the herring gull. It reached abundance around the island as a result of the plentiful offal at commercial fishing establishments. Today it nests on offshore rocks not accessible to foxes. Pete Edisen had a pet gull that would sit on his head while he drew his nets. Gulls line up in soberly watching ranks on fish houses when netting is in progress. The birds are a decimating factor for broods of young ducks that gravitate to inhabited sites for handouts of grain. At Rock Harbor Erik and Annette Stauber saw gulls kill a nearly half-grown goldeneye. Roop wrote this indictment in a report of 1965:

> I would regard the herring gull as the main predator of black duck, mallard, and American goldeneye ducklings. On two occasions I have seen gulls swallow ducklings whole. Peter Edisen has seen the gulls take nearly the whole brood of a black duck, and Art Johnson saw gulls take two of a brood of four from a female black duck. In all cases I have seen or heard about, the gulls did not attack singly, but several would intimidate the female at once. While the mother would chase one gull, another would dive at any duckling that lagged too far behind.

Feeding may have attracted these ducks out of their accustomed habitat. Ordinarily such dabbling ducks as the mallard, black, and teal, as well as the divers, goldeneye and ringneck, would be raising their young on marshy ponds and shallow lake margins not regularly frequented by gulls.

Birds more typical of the open lakes and bays are the American merganser and loon. These feed on small fish, leeches, insects, and other aquatic life, which they catch under water by diving. In the case of the merganser, I suspect predators are confused by the closely bunched and often fast-moving brood of young. We also see multiple broods (as many as 26 ducklings), which might be a baby-sitting operation while a parent is diving for food.

As broods of mergansers are thinned out by gulls and other perils, we frequently see several of the young riding on the mother's back. This is also observed regularly in the loon, which usually has only a couple of young. The habit may be an adaptation that could preserve several chicks in waters where weedbeds are home to that voracious predator, the northern pike.

Gulls obviously share the beachcombing kind of scavenging with crows and ravens, although the gulls are likely to get there first. When Dorothy and I were camped in one of the two shelters on Caribou Island, our neighbors had been fishing in Lake Richie. They gave us a 2.5-pound pike, from which I cut two fillets for chowder. Several gulls gathered around the remains, and the dominant bird ate the two strips of skin from the fillets. The next in line, no doubt in a hurry, grasped the head and gulped it down, with the rest of the skeleton attached. Evidently the backbone looped enough to get into the bird's gullet, but the gull sat on a rock looking foolish, with a portion of the tail protruding from one side of its mouth. Finally it waddled to the edge of the dock and spread its wings for a grand takeoff.

That bird had the aerodynamic characteristics of a watermelon. It plopped into the drink and, riding low in the water, taxied across the harbor.

Gull feathers around fox dens indicate that the species has its own predation problems, and a most unusual example was described to me by Larry Roop. In October 1964 he was on his way by boat to Pete Edisen's when he noticed a big half circle of gulls on the water around the beach. Pete motioned him away from the dock, so he cut the motor and rowed out a way. On the beach was a snowy owl feeding on a freshly killed gull. He and Pete could approach within a few yards, and the owl stood its ground, hanging on to the gull and feeding.

The bird-of-prey niche (or niches) is not one of striking abundance on Isle Royale. We mentioned three rare or unusual species in Chapter 7: the peregrine falcon, bald eagle, and osprey. Only the latter has produced any young in recent years.

Today the falcons are represented by the pigeon hawk, or merlin, and the sparrow hawk, or kestrel. As a result of his summer studies in 1963–65, Stauber thought each species numbered 10 to 15 pairs.

The agility of the stylish pigeon-size merlin is attested frequently by its catching dragonflies on the wing and its ability to harry much larger birds (raven, broad-winged hawk) from its nesting area. Our summer crew have nearly always made notes on a few nests of the pigeon hawk, most frequently near Bangsund Cabin, on Raspberry Island, or at Crystal Cove. This bird seems to prefer the vicinity of shorelines and open bogs. Stauber, Coble, and Johnson made food studies, collecting regurgitated pellets and other remains around nests and feeding perches. This evidence shows that many small birds are taken: chickadees, sparrows, shorebirds, crossbills, bluejays, flickers,[7] and downy and hairy woodpeckers. Johnson and Coble made a discovery of special interest in collecting four bat wings among food items beside the Raspberry Island bog. They found fur of the red bat in a third of the castings, and identification of two wings of this bat gave us a new mammal record for the island (122).

Stauber located two kestrel nests, both of which were in old woodpecker holes in pine stubs on open highlands. In the summer of 1963, from 7 June to 2 September, he saw from one to three of these birds at 13 locations on 17 different days. As he noted, the sparrow hawk is the most often observed of Isle Royale's birds of prey. It is seen catching dragonflies and grasshoppers, and Erik saw a kestrel carrying one of the small red-bellied snakes we find on the ridges.

Of the so-called blue darters, or accipiters, all three found in the Midwest are sparse summer residents of Isle Royale. In the order of size, they are the sharpshin, Cooper's hawk, and goshawk. The male in each species is considerably smaller than the female, the male sharpshin being about the size of a robin or kestrel. All three are forest-dwelling bird predators with short wings and long tails. They are secretive, and Stauber found only one nest of the Cooper's and none of the sharp-shinned hawk. However, he saw two nearly fledged sharpshins in the burn above Siskiwit Lake on 26 July 1965.

The largest of the accipiters, the goshawk, is a bird-feeder too, but it has no plentiful grouse to prey upon at Isle Royale, so it relies on smaller birds and an old mainstay, the hare. Shelton flushed one off a freshly killed hare (nearly white) about a quarter-mile south of Moskey Basin on 15 November 1962. These fall migrants appear to go on south before winter, since I recall only one observation (a hare kill and hawk seen by Shelton) during February or March. We have seen goshawks and recorded several other hare kills earlier in the fall.

Our only verified nesting of this species will be well remembered by Stauber, Jordan, and me. A quarter-mile west of Moskey Basin and 100 yards north of the trail, a female goshawk had a nest containing a single female young. The male parent was never observed.

Stauber found the nest on 28 June 1964, and the young bird was fledged on or about 26 July. Among food remains under the nest Erik identified Canada jay, blue jay, hairy woodpecker, snowshoe hare, and red squirrel. He remarked that the most outstanding characteristic of this female was her aggressiveness toward humans: something previously recorded for the European, but not the American, goshawk. I had an opportunity to observe this on 20 July, when Erik, Pete, and I hiked to Lake Richie. The female came at us repeatedly in a strafing attack that seemed calculated to tear us apart. Erik said the nest was found in this way. As he described it,

> The hawk did not avoid contact and did not hesitate to fly into a person's ribs with full force or to scratch over the back of a stooped person. On one such occasion the hawk proceeded to make a long scratch through a heavy khaki shirt, and at another time an eleven year old boy was forced to lie on the trail face down until someone came by to rescue him. At the beginning the goshawk announced its attacks by excited warning calls so that one could get ready and protect his eyes; but later the bird came without warning.

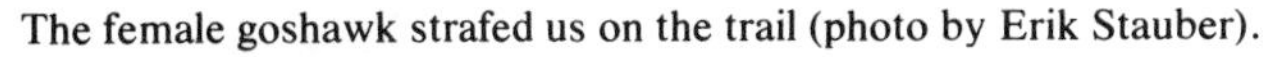

The female goshawk strafed us on the trail (photo by Erik Stauber).

Based on the appearance of birds in other localities, Stauber thought there might be three or four pairs of goshawks nesting on the island.

He estimated that the red-tailed hawk occurs in slightly larger numbers, possibly 7 to 10 pairs. He found two nests with only one young each. They were in high open forests next to large openings. As Stauber observed, the redtail's usual food (small mammals) is not plentiful, the most obvious prey being the hare and squirrel. This evidently results in small broods; only one young was reared in each of the nests.

Aside from the redtail, our other soaring hawk, or "buteo," is the broadwing, the most common large hawk on the island. It seems to prefer more dense forest and lower ground than the redtail. We have seen both of them decorate their nests with green leaf-bearing twigs. The broad-winged hawk feeds on squirrels and young rabbits, plus slow-moving birds, including the flicker and young waterfowl. But it is distinctive in liking such cold bloods as snakes and toads, for which it has little competition.[8]

All of the above species leave Isle Royale for the winter, except an occasional goshawk. The earliest migrant is the broadwing; Joe Scheidler saw seven of them milling about in a premigratory flight on 22 August. These birds must get an early start, for they have a long trip ahead to Central and South American wintering grounds.

In winter at Windigo I have sometimes heard three different horned owls, from as many directions. These probably indicate nesting pairs. Some winters we rarely heard one. We have never found a nest, but this capable hunter must be counted among resident birds of prey. Occasionally a marshhawk is seen, and the same can be said of the golden eagle and rough-legged hawk — all migrants, we think. Another migrant, which winters erratically on the island, is the snowy owl.

It probably is a fair appraisal that predatory birds impinge most heavily on the hare, but they probably are not a primary population limitation for any prey species on the island.

Biological systems — especially, self-perpetuating ecosystems — are the most complex entities we know anything about in the universe. The natural life community is an organism that functions as the sum of its parts. Its metabolism is the flow of energy through diverse forms that are held together by their common need and interlocking functions. The community is hedged against extremes and has seemingly endless feedback mechanisms to steer its fluctuations toward a midpoint. Just to review, an ecosystem is a complex of living things plus its physical environment (p. 447). Often it is the environment (e.g., weather) that produces instability.

We have seen such changes in the northern hardwoods and the Hudsonian spruce-fir. Such a wilderness is not just a stabilized monotype of protected climax. It includes all the successional stages of vegetation and animals leading to the climax. Of necessity, it also includes those natural disturbance mechanisms: the '36 burn, the wind that touseled the forest on Locke Point, the spruce budworm, the browsing by moose and hare. These initiate and help maintain productive sub-sets of the ecosystem.

To live with nature on an enduring basis, men must sense and believe in original harmonies that have to be there. Possibly this is more an attitude than a way of thinking. It requires a conceptual base many do not have. Most important, I suspect, is a realistic sense of time. Time is the only unlimited thing we know about or think we do. But even this is questionable, for without the bias of human interest time may not exist, only galactic rhythms of some sort. Where there is no eye there may be no light.

Which is a way of saying that evolution can wait indefinitely for some unlikely event. Anything remotely possible is going to happen. Such logic explains the genesis of life on earth. It explains how, in resolution of its tangled dependencies, the community has come to be programmed for long-term survival. What happens from day to day is inconsequential; it has all happened many times before.

We can conceive of evolution as integrated at three levels:

At the first level is the individual, who must find nourishment, survive, and breed.

A second is that of the species, an aggregate of societies and populations. Here the process involves competitive liquidation of the least fit and the propagation of an adapted breeding stock.

The third is the level of the community in its ecosystem. On this totality of functions, services, feedbacks, and compensations the fate of the whole depends.

At first inspection it is not possible to appraise properly the processes of the ecosystem. Each would inevitably be destructive and lead to catastrophe if it were not countered by some other trend. In natural events nothing is all right or all wrong. If the system survives, it has been successful. This should give us reassurance in preserving natural things we do not yet understand.

Only through such abstraction can one subordinate himself and regard with dispassion the timeless drama of death and transfiguration. In the structure and physiology of every creature there is contrived obsolescence, an essential part of the big plan. It insures the turnover of generations — experiments with new genetic combinations, on a schedule appropriate to the functions of the species.

In confronting the welter of relationships, one senses a perpetual progress toward a steady state. But the fossil record witnesses that such a state, if ever attained, must always be temporary because of planetary perturbations. Rapid changes find many species too specialized to adapt, and these must disappear. Slow changes, measured in thousands or millions of years, mean the gradual emergence of a new kind of ecosystem, one fitted to new conditions. Plant and animal species may have been dropped, added, or converted into something a taxonomist would classify as different.

In contrast to some other national parks, Isle Royale can be regarded as a complete ecosystem. It is a self-contained unit that probably can perpetuate itself in the present mode if sufficiently undisturbed. Nonetheless, as our brief inspection shows, seasonal dynamics of this life community are tied into events in areas far removed to the north and south. Migrant birds may or may not reach the island and be able to breed or feed there, depending on what has happened to them elsewhere.

In this light Isle Royale and its biota are part of a larger unity, a combination life zone that we have regarded as Hudsonian-Canadian transition (p. 445); such large zones or provinces, in turn, make up the biosphere of the earth. No unit is totally disconnected or independent of the others.

As expressed elsewhere, *E pluribus unum.*

CHAPTER FIFTEEN

New Rules

WHEN WE FIRST BEGAN this study, I speculated that it might take 10 years. That, it seemed, should give us essential information on how moose and wolves affect one another. But as the situation unfolded, my original thinking proved inadequate. Interactions of predator and prey were not always the same, and conditions governing their relationships could change for years at a time. We know now that weather is a critical factor.

As the work progressed, we saw it had begun with winters of exceptionally light snowfall. I have roughly categorized all 18 winters on this basis in Table 2, and there are further weather notes in Appendix IV. In the first decade only two winters, 1965 and 1966, stood out as having sustained conditions of deep (and sometimes soft and fluffy) snow. In most winters crusts formed at various times to make the going difficult for moose. High-level crusting was favorable to the wolf.

As a generalization, the first 10 years of the project were predominantly a period of little or average snowfall. The second 8 years comprise a contrasting period of heavy snowfall, with 1969, 1971, and 1972 particularly notable in this respect. Thus I have lumped the first decade and the second 8 years in making certain comparisons (Table 3, p. 347). Both periods include nonconforming seasons, but these do not obscure the major effects on both moose and wolves. The winter of 1969 introduced a new era in which we had to alter and broaden some of our well-established ideas.

Details of many findings were discussed in previous chapters. From the beginning of the work in 1959, for six years we rarely

Table 2. Snow Depth Summary*

First Period, 10 Years	Little	Average	Deep
	12–20″	20–30″	30–44″
1959	X X		
1960	X X		
1961		X X	
1962	X	X	
1963		X	X
1964		X X	
1965			X X
1966			X X
1967		X	X
1968	X X		
Second Period, 8 Years			
1969			X X
1970		X X	
1971			X X
1972			X X
1973		X X	
1974		X X	
1975		X X	
1976		X	X

* For notes on individual years, see Appendix IV. Light, fluffy, or heavily crusted snow may offer difficulties not shown by this classification.

recorded a dead moose that was in the 1-to-5 age group. We came to speak of these seemingly invulnerable animals as "prime." However, the consistency broke down somewhat in 1965 and 1966, when nine prime-age moose appeared in the kill, six of them in the second winter. In every succeeding winter some young animals were taken by the wolves. The summarized age distribution (Appendix III, p. 433) shows that young moose have been killed most frequently in times of deep snow or during winters following such conditions.

In early years we became accustomed to seeing moose high on the ridges and in the '36 burn above Siskiwit Lake. But in the two heavy-snow winters of the mid-60s this was not true.

Table 3. Summarized Age Composition of Known and Probable Moose Kills* (Winter)

First 10 years, 1959-68		*Number*	*Percent*
Calves		44	24.8
Ages 1-5		17	9.6
Ages 6-10		58	32.8
Ages 11-15		58	32.8
	Total	177	100.0
Second 8 years, 1969-76			
Calves		148	42.2
Ages 1-5		64	18.2
Ages 6-10		82	23.4
Ages 11-15+		57	16.2
	Total	351	100.0

* See details, Appendix III, p. 433.

The open ground was largely devoid of tracks, and moose were concentrated in lowland conifers, commonly along shorelines. The icy edges and frozen slush of lake fronts were travel routes for both moose and wolves, and the two species were brought together in greater degree as deep or fluffy snow made inland movements difficult. We would observe this unmistakably during 1969 and the hard winters to follow.

In 1971, during his two weeks of winter field work, Rolf saw that weather had become a major issue in our studies.[1] Hence he undertook systematic measurements of snow depth and condition. His findings applied constructively to what we already knew about severe winters.

As Des Meules and others have explained, a maximum snow pack collects on open ground and in sparse cover. For a moose this has the double disadvantage of buried food and increased difficulty in getting around. Long legs help,[2] but they do not solve all the problems. The high openings are exposed to the sweep of wind and its enervating chill on cold days. On warm quiet days there is no shade.

In contrast, as Rolf stated in his thesis (p. 218), "Snow-covered conifers are effective windbreaks and serve as thermal insulators, reducing the amount of upward infrared radiation at night . . . While conifers may provide a warmer habitat during cold periods, they may also be cooler in warm periods."

During severe winters moose tend to localize and do less moving about than is usual under more favorable conditions. They forage in the densely wooded areas where low-growing foods are already scarce. Along shoreline travelways the overhanging cedars have long been heavily used. Thus a food crisis may ensue.

A main attraction of lowland conifers is that snow under them is less deep. Pruitt cited an Eskimo word, *qali,* that describes the snow caught and held by trees. Correspondingly, the depressed area of little snow under a tree is designated *qaminiq*. Windbeaten snow is *upsik*. As we stumble awkwardly in describing conditions we see, it is evident that if we lived closer to nature our vocabulary would be richer in appropriate terms.

In the mid-60s, when moose were seldom seen on the ridge and in the big burn, we ascribed the change in distribution to maturation of the forest. Trees started in the early 40s were growing beyond reach of the moose, but we overestimated the effect. This became clear in 1968 when, with only a foot of snow, moose were again tracking up the high and open ground.

Moose swept aside the foot of snow with their noses to feed on bottom whorls of the small balsams.

In preliminary sampling for his census, Mike Wolfe said it appeared that moose were distributed about evenly over the island. Around Windigo they were moving more extensively than in the years of concentration, and we observed their feeding. With its nose, a moose would sweep aside the shallow snow and browse on seedlings and bottom whorls of the small balsam trees, sources of food that usually were protected by deeper snow (p. 456, note 6).

It is likely that the heavy runoff of 1965-66 had helped the beaver; wet years and beaver increase continued. Though the snow melt was minimum in 1968, the growing season brought heavy rains. Lake Superior reached the highest level in 95 years. Beavers had ponds well up on the sides of ridges (p. 256). This trend would persist well through the 70s, and it had much to do with what happened to the wolf.

The first decade ended as it had begun, with little snow and "moose all over the island." The winter of '68 also introduced a 4-year period when the wolf population would be 20 or under, the lowest level in our 18 years (Figure 14, p. 364). Conditions were propitious for further increase of both moose and beavers.

When we terminated a 4-day weather wait and flew to Windigo on 31 January 1969, new conditions and challenges were waiting. It was evident that Isle Royale, like northern Minnesota, had the most snow in many years: 44 inches on the level. In survey flights, we saw that moose were concentrated in lower parts of the island more than in any previous year.[3] The burn and high ridges were almost devoid of tracks.

No one could have designed a more effective experiment than the contrasting conditions of 1968 and 1969: a winter with about a foot of snow, and then one with nearly 4 feet. Such extremes bring cause and effect into clear focus.

Beginning with 1969, three of four winters would always be remembered as the times of deepest snow: 1969, 1971, and 1972. These three years had a cumulative effect that redefined some "rules." The drastic change in moose distribution would inevitably influence forest successions. When snow is deep, large areas inland are spared heavy browsing, after which young growth can add to the gains of the previous summer. In this way, some pines are becoming re-established on the ridges. Conversely, in years of little snow the lowlands are lightly browsed.

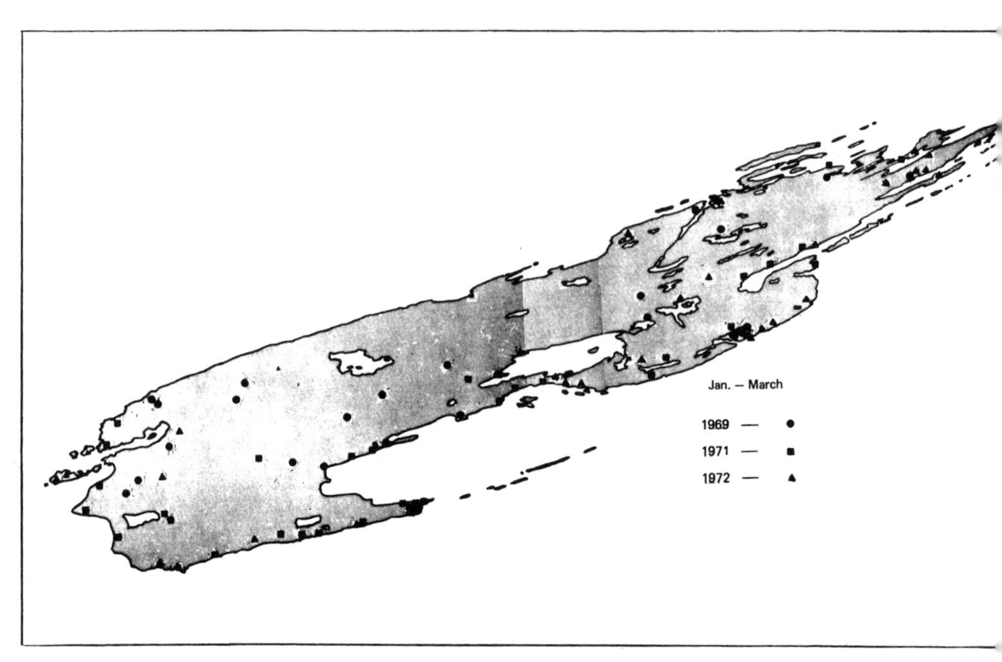

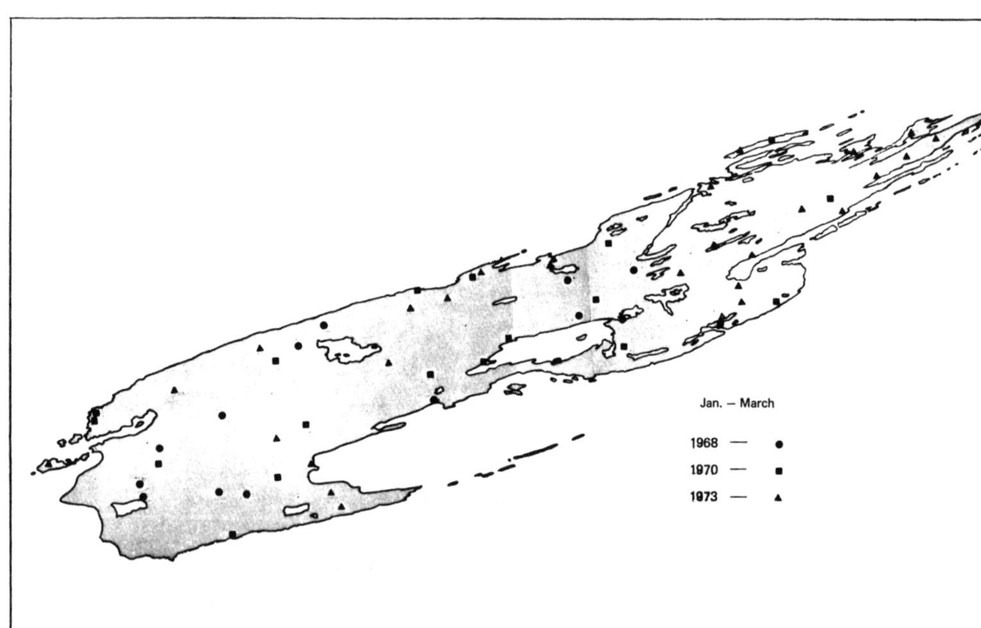

Top: Figure 11. In winters of deep snow, moose were commonly in lowland conifers, and kills were frequent near waterfronts.
Bottom: Figure 12. During winters of light snowfall, moose were widely scattered, as were the kills by wolves.

Once the balsams are too large to be broken over by moose, the trees are secure. In 1969 we frequently saw moose standing under large balsams with their heads stretched high, feeding on the lowest branches. They were standing on enough snow to enable them to raise their previously established browse line.

That year, it will be recalled that the moose population reached a peak for our period of study (p. 167). This intensified browsing pressure in the concentration areas and contributed to an increased kill by the wolves.

In Figures 11 and 12 I have compared the locations of known wolf kills in the three winters of deepest snow, beginning in 1969, with the three winters of little or average snow. The more common association of kills with shorelines is shown in the first map. In compiling locations of all winter kills for 1959–74, Rolf found that when snow depth was less than 20 inches, 47 percent of kills were within half a mile of Lake Superior or Siskiwit Lake. When snow was more than 20 inches deep, 79 percent were near shores. The figures also demonstrated a differential between calves and adults. Fifty-nine percent of the calves and 39 percent of adults were killed within ⅛ mile of shorelines (211).

The map of moose density strata prepared by Rolf for his winter "census" of 1972 is shown in Figure 13 (p. 352). He found that only three strata were necessary in this winter of deep snow and heavy lowland concentration of moose. In degree, this is comparable with Mike Wolfe's map (p. 166) for 1970, a winter of less than 2 feet of snow.

Records on moose aggregations, principally in hard winters, suggest what some have called "yarding." Moose were grouping more than usual in 1966; in 1969 we saw more of it. Mike and Don said some cedar swamps were so heavily browsed and so tramped up they resembled a deer yard. On 4 February we counted 12 moose within a 300-yard stretch along the south slope of the spine of Beaver Island.

On 4 March 1970 Don and I saw 12 moose on about 5 acres of the northeast point of Johns Island, at the mouth of the harbor. Here food was an attraction, since they were stripping a thrifty stand of young balsams. On 3 March it had rained all day, and the snow was slushy. Freezing produced a crust, after which there was less moose movement than we had ever seen.[4]

In the winters of deep snow we frequently saw groups of

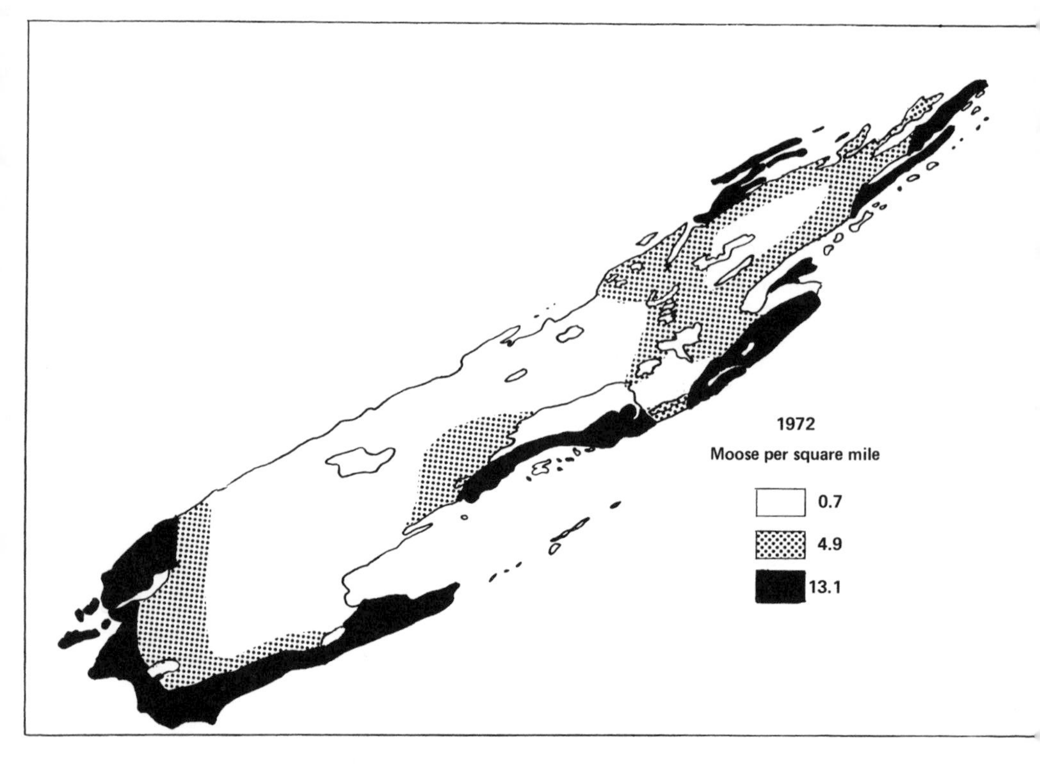

Figure 13. The three density strata used in the winter count represent the calculated distribution of moose in February 1972. At this time of deep snow, moose were concentrated in lower parts of the island. (From original by Rolf O. Peterson.)

three to nine moose — commonly on islands where feeding conditions were favorable. As we observed it, the close grouping of wintering moose seems to fall into what Randolph Peterson (209) called "voluntary" yarding, which means the animals were attracted to an area, but not confined by unfavorable conditions. Food did not always account for the aggregations. Whether easier walking in the trails of other moose was sufficient incentive to cause the coming together we are unable to say.

The way things run in streaks was never more impressively illustrated than in February 1971. On the morning of the 11th, Don and I tracked the pack of nine from the inner harbor to the north side of Thompson Island, where they had a fresh kill at the base of a 30-foot wall of rock. The calf (aut. 543) had fallen off the cliff and was buried in a deep snowdrift, now well packed by the wolves. On examining the kill three days later we were inclined to regard the event as an "accident" quickly exploited by the pack.

It was startling to discover that two nights after the death of the calf the same pack moved 2 miles to McGinty Cove, where tracks showed they had put a bull (aut. 545, age 5) off another 30-foot cliff.

The situation became still more interesting on 18 February. Jim Dietz and Don tracked the pack of nine around the south shore and across to the north side of Houghton Peninsula. There a 4-year-old bull (aut. 547) lay on the ice beneath a rock ledge about 15 feet high. It had obviously been pursued along the icy margin by the wolves.

That winter we recorded two additional "moose jumps" involving other groups of wolves. Our conclusion had to be that weather conditions made the footing favorable on these low ridges, and wolves were hunting where the moose were. As mentioned in a previous chapter (p. 226), moose killed in falls were not unknown to us before 1971, but it seemed remarkable that it took 13 years to discover that hunting by the wolves is sometimes aided in this way.

In the winter of '69, moose probably were more plentiful than we had seen them previously (p. 167). More kills were on the

Snow depth was 44 inches in early February 1969.

Wolves pursued this bull along the icy edge, and the animal fell to its death. It was one of five such events in the winter of 1971.

ice of shores than ever before, and more of them were calves than ever before. During the first 10 winters, wolf kills averaged about 25 percent calves. For the four years beginning in 1969, calves represented more than half of winter mortality.

The wolves had changed their feeding habits. Unlike what we had been accustomed to, carcasses were not being cleaned up. The west pack would kill a calf, eat the soft parts, and then move on to make another kill a night or two later (see photograph below). It was the highest predation rate yet seen. They were not even dismembering the skeleton. Several times young adult moose were neatly skinned out, the hide left in a compact bundle with leg bones attached. Incomplete utilization and excessive killing of starved deer by wolves were observed that same winter in the Boundary Waters Canoe Area of Min-

In the winters of deepest snow, moose carcasses were not cleaned up immediately, and a large kill of calves occurred on lake shores.

nesota and in east central Ontario.[5] In a hard winter 10 years earlier, similar conditions had been recorded in Algonquin Park.

We saw repetition of something first observed in 1965: When a large moose was fed upon by only a few wolves, the warm, heavy body melted down into the deep snow and the lower half became firmly frozen into a layer of ice. Only the upper side of the animal was eaten, and sometimes we had to dig and chop to collect tooth and bone specimens. Much carrion would be available on the range that spring, and we wondered whether this would contribute to higher pup production. It did not work out that way, since 1970 brought the island's lowest wolf population.

In summers immediately after the winters of deep snow, calf production was reduced. This appears in field records of cows and calves after 1 June (p. 169). In the following summary the female category includes yearlings; these had to be counted in because sometimes they are mistaken for adults:

	1969	1970	**1971**	**1972**	1973	1974	1975	1976
Calves/100 cows	32	39	26	28	49	37	34	22

In the above series, the years of deep snow are in boldface. The milder winters were followed by summers of higher calf productivity through 1974.[6] By then the wolf population was reaching its peak and bringing even heavier pressure to bear on calves. The reduced twinning rate after severe winters has already been referred to (p. 174), and a lower incidence of calves appeared in fall aerial counts (p. 453).

The smaller calf crops resulting from winters of malnutrition could begin with losses before birth.[7] This probably depends on the previous health of the cow. As a more immediate effect, calves dropped after such a winter are likely to be weak and may not survive. Also, as we were to find, a calf that lives through its first autumn and then experiences three months of deprivation may not develop normally. Thus there are both cryptic and visible environmental influences at work in affecting the productivity of animals, and some are prolonged over a series of years.

The hard winters brought a triple penalty to the new generation of moose: the young animals could not move about freely, they had to feed on shorelines heavily browsed by adults, and they were not well protected by their mothers. In 1971 the calf

On Carnelian Beach a cow guarded her dead calf for more than 30 hours.

kill was at a new high of 21, and we were able to examine 17 on the ground. Sixteen of the calves had fat-depleted bone marrow. In a later winter (1974) one of average snow and another substantial calf kill, we checked the marrow of 13 calves, only two of which showed obviously low fat content.

The problems of calves in the deep snow of 1971 were evident to Don and me on 10 February as we circled over Wood Lake at the southeast corner of Siskiwit Lake. On the west side is a small island, and a high peninsula extends into Wood Lake. A cow and calf were standing near the south shore of the island, and near them were two wolves we had seen on Moskey Basin the day before. The wolves were interested but not attacking. A large dark moose came off the island and stood 30 feet from the cow and calf. The two wolves started east across the lake. No action today.

Then one of the wolves made a wide circuit to the south and dashed up the steep slope onto the peninsula. A moose was there that we had not seen. The wolf immediately went after it, savagely attacking the rump as the animal floundered down through drifted snow toward the three moose on the ice. At the lake edge the calf — for that is what it was — nearly went down, with the wolf hanging on to its anal region. But it struggled out onto the better footing of the ice, as the dark-colored moose ran toward them to intercept the wolf. The wolf disengaged and ran away, while the cow led her calf north across the ice and through deep snow into the woods. Where the calf had stood and bled, a raven and a wolf came over to eat snow. The cow and calf still on the ice appeared to be uninvolved observers.

With the cow breaking trail, the injured calf and its mother worked slowly through the belly-deep drifts northward to the edge of Siskiwit Lake, about three quarters of a mile. The wolves followed at first but would not approach when the cow stood facing them on the alert. Then the wolf duo went eastward toward Intermediate Lake.

Don and I continued our flight. The air became rough, and we were numb with cold when the plane landed on Washington Harbor. We had seen a near-kill and accounted for 15 wolves. The astronauts had made a safe splashdown in the Pacific. Everything considered, there was justification for an enjoyable social hour.

The following day the cow and her badly injured calf were in the same spot. We fully expected that the wolves would be digging up calf remains somewhere in the locality during the next few weeks. To our best knowledge we never saw the calf again, but two days later an exceptionally spooky cow was on the shore of Wood Lake.

The attack on the calf appeared possible because the cow had not tended her offspring. We had other examples. On 22 February, Don and I saw what was apparently an abandoned calf on the Malone Bay shore. It was chased and nearly killed by a single wolf. We interrupted the work of the wolf and then had to finish the job, converting the calf into an autopsy (aut. 550, p. 210). At this time another lone calf joined us in the dock area at Windigo, using the well-packed weasel tracks for getting around. The animal grew tame and was still there when the camp closed. It was obviously underweight and lived on coarse browse, lichens, and other substandard provender.[8]

The following winter Rolf and Don saw a calf abandoned by two cows when seven wolves of the west pack approached. The adults ran away on shore ice, while the calf struggled inland through deep snow and was quickly killed (aut. 750).

In the above cases, we speculated that the responsible cows had left their calves to go to more favorable feeding areas. The young animals could not follow through the deep snow.

Weather influences may be favorable or otherwise for the current generation of calves. In 1973 we witnessed an all-time high wolf kill of 42 moose. But the winter was mild, and only 29 percent of the mortality were first-year animals. Unexpectedly, the kill of young adults kept on increasing. Our annual report, issued the following year, described this:

> Moose in the 1+ to 4+-year-old group comprised 53% of the total adult kill, compared to 18% for this age group for the years 1959 through 1972. All the moose in this age group that were killed in winter 1972–73 were either carried as a fetus through a hard winter (1969, 1971, or 1972) or experienced such a winter as a calf. This suggests that severe winter conditions may have long lasting effects on the viability of the youngest age groups . . .

Young animals had been physically impoverished as calves and were exhibiting in their early adulthood a *residual* vulnerability to wolves. Another stumper; how did this one work?

The 7th of February 1971 may have been the turning point in our realization of what was happening. On a bank above the north shore of Moskey Basin, Don and I found a calf (aut. 538) that was almost a dwarf. Thin and scruffy, it was hardly more than skin and a rack of bones. The two wolves had eaten little of it, and Don dragged the carcass out of the balsams with one hand.

Later, in the shack, the thought of that stunted calf plagued me. I realized it was not a product of conditions during the current winter. It failed to grow last summer; it was *in utero* through the cold season of 1970, and its mother had endured the previous winter (1969), the worst we had seen. We most urgently needed an objective method of comparing one generation of young moose with another.

There was no way to get weights; a standardized bone measurement seemed the best possibility. The bone most likely to be present and undamaged was the metatarsus, the hind leg from the hock down to the toe. As they were heading out for a flight, I asked Dietz and Murray to stop and bring in a leg of that calf. We collected the hind foot (it *is* the foot) of other calves the rest of that winter period, and the next year Rolf extended the collection to moose of all ages. He learned more from studying these bones than anyone could have hoped for.

A long bone of the leg consists of a shaft and the two end pieces, or epiphyses. In young animals the epiphysis is joined to the shaft at a suture containing the bone-forming cartilage. As long as this cartilage persists, the bone can continue to grow. When growth stops, the suture disappears through ossification, and the bone becomes one solid piece. Rolf cleaned up the specimens by boiling, and if an epiphysis did not come off after several hours of boiling, he considered it to be fused. Metatarsal length was used, as he stated, "as an index of body size and an indicator of the nutritional status of the mother before and immediately after birth of her calf" (211).

Based on these criteria, Rolf's eventual conclusions were highly significant: The generation of young moose that were physically smallest were those born in 1970. Animals in the next three generations were progressively larger. The small size of those in the 1970 cohort still was evident when they were killed later as 2- or 3-year-olds. As previously noted, food scarcity in 1970 probably was a partial result of heavy browsing in 1969.

Physical effects on the cows during that hard winter could also have been involved. Calves born in 1973, after a relatively open winter, were the largest produced in the 5-year period.

Epiphysial closure was complete in only two of nine moose killed in their third winter, the earliest for any examined. It may be that the third summer is the "normal" time for the completion of metatarsal growth. In two of eight animals killed in their fourth winter, fusion still had not occurred. Evidently these young moose would not have attained maturity until their fifth year.

To sum it up, concentrated browsing during hard times by an increasing moose population was propagating food deficits into more favorable winters. In addition, the weaknesses of a deprived breeding stock and their offspring were showing up in succeeding years, and the effects were compounded when severe conditions recurred.

Retarded physical development, probably originating from fetal malnutrition, was associated with that lag vulnerability in the 1-to-5 age group. Why this should be true — when yearlings of former years were able to protect themselves successfully — only a wolf would know.

Wolves killing malnourished animals fitted logically into the vulnerability pattern we had known for so long. Starved moose dying without attention by the wolves were new to us. We did not immediately identify the first examples of this.

A review of records indicates at least one such case in 1966.[9] Possibly the next we saw was in 1970 when Don and I examined a "kill" (aut. 508) a few yards inland from Spruce Point (Siskiwit Bay). On 19 February we backtracked the west pack (six) to this site, and they had obviously found the carcass already frozen. I recorded that the moose had "died on its hunkers," which is to say upright with legs folded under and frozen into the 2 feet of snow. The wolves had cleaned off most of the flesh and pulled viscera out from behind the rib cage. How had they done that on a frozen carcass? The 15-year-old cow had worn teeth, a severe jaw necrosis, and fat-depleted bone marrow.

That same day, Don had dropped John Keeler off to pick up a kill north of Lake Desor. What John found doubled my perplexity. His note, like my own, described a cow (aut. 509) that was frozen "in an upright kneeling position" with the rumen still in place below the backbone. The skeleton was intact, the bone marrow without fat. We learned later that the cow was 16 years old. Its mandible showed an extensive infection below the roots of both rows of molars.

Things were running in streaks again. A year later (1971), another high kill of calves and their poor nutritional state helped to fill in the picture of stresses that were affecting the moose population. Since a food deficiency frequently catches up with animals in late March or April, such specimens often were first discovered during field work of May and June. Rolf found a male calf (aut. 604), still in the upright position but leaning against a tree, as late as 8 July in 1971. Our spring field work was inadequate for two years, and Rolf's first full time in the field began with the winter and spring of 1972. By then — the second of two unfavorable years — evidence of nutritional problems among the moose was piling up. Our annual report for 1972-73[10] stated:

Evidence of the severity of the winter of 1972 was found the following spring when we recovered 10 carcasses of moose that evidently died

This moose died of malnutrition and then was fed upon by the wolves. Typical of such cases, the carcass remained in an upright position with legs folded under (photo by John C. Keeler).

of malnutrition. When found, these animals were untouched by wolves, and usually they died in stands of balsam fir that had been browsed severely. This suggests that these moose had been restricted to small areas by snow and had completely consumed the food supply available to them . . . All but one of these carcasses were located less than half a mile from the Lake Superior shore of the island, reflecting the concentration of moose in conifer areas in response to deep snow conditions.

Rolf said all of these moose showed complete or partial fat depletion in leg bones. A cow (aut. 680), with head turned back, still resting on the sternum and a bed of snow on 3 June, contained a fetus indicating death sometime in April. Live ticks were all about.

In the relatively favorable winter of 1973, no moose mortalities ascribed to starvation were found, but more appeared in subsequent years. There were seven in 1975. As further evidence of nutritional stress that year, Rolf and his assistants saw moose still feeding on balsam in May.

During this work I finally realized how the wolves could pull the viscera out of a supposedly frozen carcass. It was the result of that heat-producing decomposition process sometimes seen in dead animals not killed by wolves (notes, p. 450).

If we were surprised by developments in the moose population in the 70s, it is proper to say we were astounded at what happened to the wolves. In Figure 14 our counts of pack numbers and maximum populations are shown from 1959 through 1976 (see also Table 1, p. 276). Beginning with their productive shoreline hunting of calves in winter 1969 and the increasing use of multiplying beavers in summer, wolves entered a period of unprecedented prosperity.

In 1968 and 1969 the new west pack, using the western half of the island, was led by a large gray male and a small sharp-nosed female with a crook in her left front leg (p. 286). By 1970 it was evident the black male had beta privileges; he was constantly close to the bonded pair. In still another year this too had changed: The black male was now alpha (no sign of the old gray male), and *he* was now bonded to the foxy-headed female.

This situation lasted for two years. In the second winter the little female — now at least 6+ years old — was limping. On 26 May 1972 Rolf made his last observation of this pair, an occasion to be remembered:

He and Carolyn had camped on Red Oak Ridge northeast of Lily Lake. They were awakened early by the howling of wolves to the west, including the deep-voiced animal heard the winter before. A loon on Lily Lake seemed to respond.

Rolf and Carolyn left the tent and at 5:15 were standing 50 yards from the bog mat bordering the lake. They saw movement and then became aware of the black male and small female walking northeast on the bog about 75 yards away. "The black male stopped, followed shortly by the female. He stood quietly, apparently seeing us behind the trees and knowing that all was not right. With 9X binocs I could see much white on his belly and face as he looked at us for several seconds."

The wolves turned and slowly walked onto the ridge and out of view. Rolf said the small female was much browner than she appeared in winter and was limping noticeably, favoring her left front foot.

After 14 years, I had begun to wonder whether Isle Royale was big enough for two breeding packs. We had hoped to study the

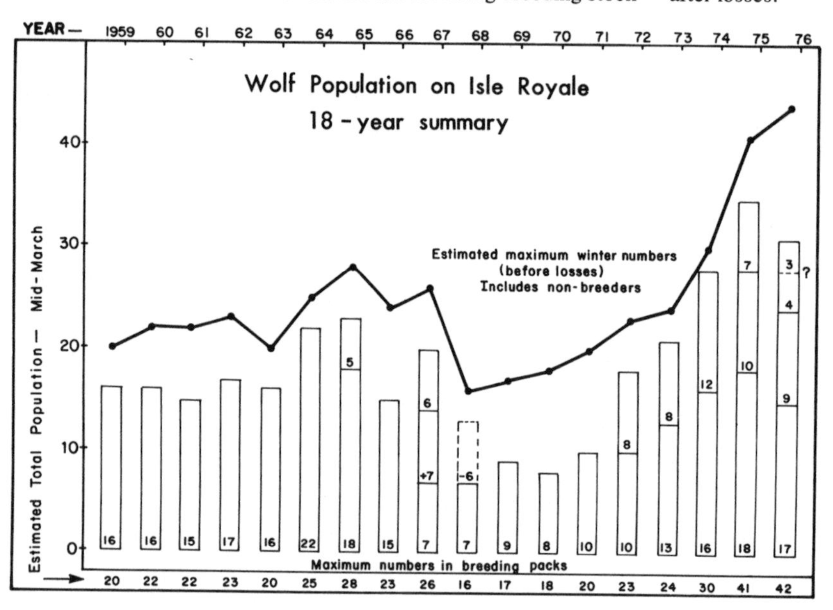

Figure 14. A figurative history of Isle Royale's wolf population. Bars indicate numbers in breeding packs. The −6 in 1968 represents a pack that disappeared and presumably left the island. Numbers at bottom are the surviving breeding stock — after losses.

relations of territorial packs, but it seemed not to be. Then, in 1971, we witnessed strange events that proved significant:

On 7 February Jim Dietz and Don saw that the west pack, then on a south shore kill (71-8), had lost a member and now numbered nine. Four miles northeast, on the north side of the tip of Houghton Peninsula, three more wolves and a couple of ravens seemed localized around a clump of brush at the shoreline, suggesting kill remains.

In the afternoon Don and I checked the pack (nine) and then the trio. Occasionally one of the three would enter the thicket by the shore, but something else was taking place. We circled and watched. Previously we had accounted for three different duos, probably all males. Two of those animals appeared to be at Houghton Point. The other was a dark-backed female that we suspected was the tenth member of the west pack.

When one of the males approached the female, she tucked her tail and sometimes sat down on the ice. When the two males walked toward the point, she followed at a distance of 25–50 yards. Only one male showed any interest in her. Twice he turned back and jostled and hip-slammed her from the side; she snapped and probably growled at him. After watching for 15 minutes, Don and I flew on.

The next morning the west pack was 8 miles from Houghton Point, still numbering nine. The three wolves were still there, and courtship activity was still in progress. The female appeared ready to join this duo, but she was not yet ready to breed and had not formed a pair bond with the interested male.

As the three wolves started north across Siskiwit Bay, the female was trailing. We left them, since, when out on the ice, they seemed skittish of the aircraft. We did not see them again, although tracks indicated they may have revisited the point. The west pack was nine for the rest of the winter. Don and I thought the dark-backed female had come into estrus and had been run out of the pack.[11]

We were not especially surprised the next winter to find a new pack, numbering 10, at the east end. There were as many as five pups, and if this figure was correct, then two additional adults had been accepted into the group. From 1972 to 1976, yearly numbers of the tightly knit east pack were 10, 13, 16, 18, and 17. The dominant female survived the entire period, al-

though she had at least three different dominant males, and in the last year she alone seemed to control the pack. At that time she was at least 7-plus years old.

Pup production had attained a new level of efficiency. In the summer of 1973 more than half of 540 wolf scats contained beaver: a plausible key to the greater survival of young. Beginning that year, there was increasing evidence of fruit eating, mainly blueberries and sarsaparilla.

In 1974 a duo that turned out to be a mated pair reared a litter and established the middle pack (seven), first seen the following winter. In 1976 this pack appeared to split, forming two groups (Table 1, p. 276). The apparent effort of a lone female and her cubs to start still another pack has been described, as have the territorial incursions and disputes that led to at least four killings (p. 278). In the last two years of this record, the west pack had divided management: two dominant pairs of uncertain relationship. Strange social doings.

Wolves of the island had reached an unheard-of population density. On a flight of 2.5 hours, 27 January 1976, Rolf and Don counted 33 wolves: "Must be a record for first flight." It certainly was; 44 wolves were now competing for vulnerable individuals of a reduced moose herd (p. 168). In the 7-week study period 50 fresh kills were made, more than one each day. How far would the moose decline go, and how long could the wolves maintain anything like this density?

Problems for Rolf. My own field work was at an end; 1976 was the first winter I had missed since 1961. It was time for my personal reporting, via this book — by that time well along.

When Dave Mech brought in the first moose jaws for aging, he and I discussed the possibilities for a life table illustrating the population dynamics of this species on Isle Royale. We knew it could be no part of his project, but at some time in the future enough specimens would be on hand. The moose work by Jordan and Wolfe had a life table as a major aim.

The point in such an exercise is to use the number of specimens in each age group to develop a table of statistics on the age composition of the population. From the table a smoothed curve is constructed showing the decline and disappearance of a theoretical generation beginning with 1000 animals. The table of statistics (not used here) and the curve derived from it give

The east pack at its maximum number of 18 (1975). In contrast to others, this pack seldom split.

a picture of recruitment, mortality, survival rates, and turnover in the range where the collections were made.

Using known ages of 374 specimens, Mike Wolfe constructed a survivorship curve in 1969. His curve postulated a generation of 1000 calves and showed the year-by-year reduction in numbers until the last animal disappeared at 17.5 years, our oldest moose at the time. Mike's curve showed the essential things we needed to know, although it was based principally on the age distribution in moose mortality during and immediately preceding the first 10 years of our work.[12]

Rolf compiled a second life table and survivorship series in 1974 using 532 known-age specimens. By that time our autopsy record was more representative, since it included the years of deep snow and increased young-adult mortality. Also, since 1969 the number of random pickups had more than doubled as a result of intensified summer field work.

Figure 15. Rolf Peterson's survivorship curve for the Isle Royale moose herd. The decline and disappearance of a theoretical generation of 1000 yearlings are depicted, with males (♂♂) and females (♀♀) shown separately. The oldest moose in the "autopsy" series was a cow of 19.5 years (211).

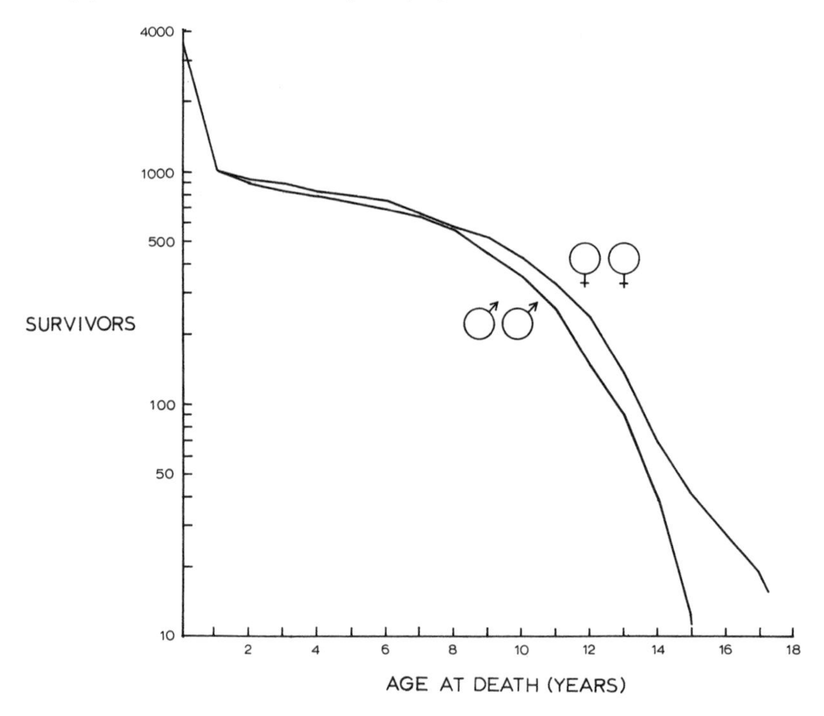

Rolf separated the records of males and females and constructed two curves that show the higher life expectancy of cows (Figure 15). Two cows of 17.5 years and another (aut. 978) of 19.5 years now were the oldest animals in the collection. Information to date indicates that bulls have disappeared from the population by the time they are 15.5 years old.

Calf remains, being highly perishable, are inadequately represented in any field sample, but yearlings presumably appear in proportion to the number that are dying. Thus Rolf used a generation of 1000 yearlings as a starting point. The curves, plotted on a logarithmic scale, would be a straight line if moose of all ages died at the same rate. The age-specific differential killing by wolves is well expressed in this treatment. Despite the inclusion of data from hard winters, a high average survival rate in animals in the 1-to-5 age group is evident. At age 1 year the life expectancy of a male is 7.02 years; of a female it is 7.81 years. The increasing liability to being killed, or dying otherwise, appears in the steep downward slope of the curves for adults.

With the continuation of this work and the changes in conditions and mortality rates that probably will come, it may be useful to redo the life table at appropriate intervals. However, the base of numbers and the spread of years are now sufficiently large so that any major change in results is unlikely.

A main problem of scientific studies in natural communities is that important changes can sometimes be witnessed only at long intervals. They happen under conditions that represent a continuum of variations; they involve causes that are unknown or even unsuspected. Our only approach in appraising such things is to observe and record what we see. As records accumulate, they reveal aspects that are repeated or which seem to conform to some pattern. Eventually we find ways to measure what is going on, which means to prove or disprove theory. Often a final step is to devise experiments that can assess the precise nature, limits, and effects of various phenomena.

A weakness of ecological research in its first 50 years has been the short-term approach. Those time-consuming weather trends, plant successions, and things we do not know about are poorly evaluated in the three or four years it commonly takes to produce a doctoral thesis. Thesis jobs are immensely valuable, because they can be financed as convenient units, and the students must give their all. Degree candidates are inspired by an immediate goal. For a period that cannot be stretched too far, they live at subsistence level and gamble their wives, their fortunes, and their sacred honor.

In this project we had it both ways by running students in relays on manageable segments of the long-range study. At intermission time there was always a good reason for taking off on a new start. It is for the best that the program worked out as it did.

If I had called a halt at 10 years and done my writing according to that period of remarkable stability, the story would have been convincing. We had proved wolves to be amazingly selective and consistent in their killing and feeding habits. It was evident that they *almost never* killed a moose in that 1-to-5 age class. Of course, any generation included a few weaklings, but the wolves got most of those as calves. Up to that time, the information from other studies, usually concerning wolves that fed on deer, seemed to bear out our Isle Royale findings.

To plump up the account, we could have asserted with confidence that wolves cleaned up one kill before they went on to another; they did not indulge in excess killing. Our findings proved that wolves were almost entirely carnivorous. Unlike foxes, they did not eat fruit. Relative to another issue, wolves had strict social controls on their own numbers. When the population spurted up a bit, it quickly dropped back to that magic number, a wolf per 10 square miles.

By way of contrast, what would have been the conclusions if we had begun this work in 1969, the advent of years of deep snow and many changes? What if we had followed it through to 1976 and then stopped? Again we would have had a consistent record.

Aside from a heavy calf kill, we would have said there was little age discrimination in the predation on moose by wolves. They took young animals about in proportion as these occurred in the population, just as they did the older less numerous moose. A habitat condition — heavy overbrowsing of the winter range — resulted in wolves killing malnourished calves and other moose beyond their needs and only half cleaning up the remains. In a summer of plenty, they ate fruit. Around rendezvous in 1975, more than a quarter of the scats contained fruit.

It was plain poppycock that wolves would not build up beyond one animal per 10 square miles. Give them enough food and they would increase to twice that number.

In some respects the two accounts would have been enough alike to attest that we were studying the same area in both periods. But the answers to some vital questions would have been different. And if the two studies had been done by two groups of people, some eyebrows might have been lifted. Understandably, public confusion could have resulted (11).

The research on Isle Royale exemplifies a kind of scientific inquiry that should be carried on through many decades and many changes in weather and vegetation. In reality it is one of those continuous searches into the unknown that has no foreseeable end.

CHAPTER SIXTEEN

Contemplations on Wilderness

ON WINDLESS WINTER NIGHTS a quiet of dramatic purity pervades the forests of Isle Royale. In the looming presence of ancient ridges there is come-what-may stability that makes one stand and listen, and smile to himself. Possibly he smiles because, for a time, he has outmaneuvered the world of staring lights and unresting motors.

I have been privileged to live at this time and to know the privacy and intimacy of the truly wild. The rising tide of human numbers inexorably closes in on enclaves of nature and infiltrates the great spaces that once were security to those who must run away. Like me.

On tours afield it is good to sense the yield of new snow under webbed footgear and see ahead the expanse of white that I must blemish with my strange fish-shaped tracks. I would know my own trail anywhere, because the left foot is toed-in a bit. No doubt congenital.

I have seen no clarity so clear as the rounded ice crystals we found among jumbled chunks on the outer shore of Johns Island when Don and I were traveling afoot from a landing place to find a kill. These evanescent gems I tried vainly to photograph. Like shapes of life, they were beauty for a day, due for early destruction but fated for limitless incarnations.

Winter gales from the southwest shatter the locked-in pack ice between Isle Royale and Canada and sweep it out of the channel with a grinding roar. Then the driven swells explode against cliffsides, points, and islands, icing them over in contorted masses of taffy white and glacial blue. On abrupt shoreline faces the glassy overlay assumes strange animalistic forms that whisk by the wing-tip of our exploring plane. We leave behind ranks of maddened gargoyles, fat improbable monsters, and giant plasmic worms that crept out of the gray depths to freeze against the rocks. For hundreds of feet inland the forest is rimed with a hoary sleet of spindrift. Through streaming clouds the sun models the frosty glaze in changing contrasts of dazzling light and sullen shade. One views in wholesome awe the ageless violence of thrashing waters and the chastened land, feeling the presence of powers he should not confront.

Winter is perhaps the best time, for all things that move on the surface leave a story of their passing. Deeds of carnage are witnessed by bloody evidence. Day by day the snow settles and grows old with icy imprints and windrows of blown debris. Finally there comes the time of welcome change, a sunless morning of lowering sky when the storm hare startles and sheds its immaculate down to wipe away the record of the past. Hour after hour the great flakes drift to earth, covering all outdoors in an amnesty of white.

Intrusive sounds are muted. Except for a few persistent woodpeckers, birds have retired to security in thick conifers where each clings, a motionless puff of repose, among the laden sprays. In the monastic quiet no squirrel chatters, for they are curled away in dry hollows, perhaps in half-sleep, waiting for days — weeks if need be — until the snow packs again and will bear their weight for further foraging.

The night comes early, and we wonder about the wolves seen yesterday, less than an hour off at the mouth of the harbor. Did they come our way? Could they, even now, be feeding on the bait across from the ranger station?

Long after supper we find that the gentle snowfall has spent itself, and a star shows through thinning clouds. Another hour and the moonless sky is clear; it is turning cold again. A crackling aurora sprays tinted shafts of flickering cloudlight toward the zenith, and the land glows under its new coverlet of white. It is a rare time of impeccable calm, when one must be out and savor the frosty air. At some hazard the slick soles of bunny boots squeak and slide as they feel through new snow for the packed surface of the hillside trail. Speaking in low tones we stand and enjoy a beatified idleness.

Then we hear what we might have hoped for — a low moaning call that rises, lingers, and dies away.[1] Like the keening of a lorn soul, it resounds across the ice-bound harbor and ridges beyond. It comes again, this time with accompaniment on a higher key, a soaring note that breaks downward in rich contralto. The spirit catches them, and a medley of many voices join in, each on its own scale and tempo. The volume swells in joyous yodels, organlike harmonics, and wailing dissonance; skyward from the ice it reaches to make the welkin ring.

For a moment the chorus lags, and echoes reverberate from the far shore. Then it is renewed, more inspired than ever, building up from the shadowy void that is Beaver Island. In wild exaltation the music mounts in a crescendo of tonal splendor mixed with yips and ki-yis that we take to be puppish sharing of lofty artistry.

Another minute or more the concert flourishes, then gradually subsides. In mournful threnody the soft basso is alone again. His finale trails off (diminuendo) to empty stillness.

Around us the forest, the harbor, and the brushy hillside. Above us the darkened sky resplendent with jewels of the galaxy. We shrink to nothingness under the inscrutable regard of Sirius, Aldebaran, and Betelgeuse. These were watching when Isle Royale's ridges were formed, before earth's highest animal life had learned to crawl across the ocean floor. They will still be watching when . . .

No words are needed as we take the bend in the trail toward the lighted windows of the shack. Nature has shown us many moods, but this is the fulfillment of it all. Actually, what we experienced tonight has been known to men in every generation before us. Except for uncertainties ahead it would not be such a big thing. But we have at once a sense of privilege and achievement. Not entirely by chance, both we and the pack were in the right place at the right time.

We listened for a voice crying in the wilderness. And we heard the jubilation of the wolves!

When one has sojourned in the wild outdoors he tends to tell about the great adventures. But behind each idyllic moment in the field there is a hard-earned back-up of routine unexciting days less well remembered because there were so many of them. And one must understand that the closer he gets to the primitive

When a storm swept pack ice out of the north channel, the shoreline forest was coated with ice.

system, the more he must live by its unforgiving tenets.

In our winters on the island there were sobering interludes, and the worst of the burdens was almost entirely my own. When I went flying, it was frequently early in the day, and this gave me the afternoon in the shack, alone with the typewriter, the sourdough, and visiting foxes. Between stints of more serious work, I made preparations for the evening meal.

Ordinarily, between 5:00 and 5:30 the plane came in and was gassed up before being tied down for the night. Our ranger — perhaps on a mission in the power house — might join the crew at the dock, and the trio would labor up the hill, thoroughly chilled and ready to thaw out by the bunkroom stove.

Once or twice a winter I found myself with supper in the oven, after sundown, the cold of nightfall settling on the land — no plane. In rigid discipline I sat at the typewriter, fidgeting and making no sense. I paced the floor, then stood outside and listened to the silent woods.

Inevitably I lost composure, got into boots and parka, and hurried down the trail. I had to do something — it made no difference what. At the dock it was no better. Shadows deepened; drifted snow around the harbor lost its contours in the

Gassing the aircraft after a late return. A full tank accumulates little moisture.

waning light. Could they still land safely? Not really a problem; I was convinced Don could feel his way down in the dark. But what could they be seeing at this hour? And their gas was surely gone. Was it the day I had long dreaded, when the plane would not come in and I must head for the radio? Were they stuck in slush on some remote lake, or had something worse happened? Damn the suspense!

My ears were betraying me with false vibrations. But now there was a real hum in the distance, could be the Champ. Again, it sounded to the north, an aircraft over Canada. Then I heard it for sure, that familiar tone, building up in sudden glorious volume as the plane bored down the creek from Lake Desor. Momentarily the motor idled, and I could hear the sibilance of wind in the struts as Don cut power and settled in over the trees.

Two minutes later the boys were describing some wonderful event at the other end that had caused them to hang on — usually a kill that did not quite come off. Yeah, they were sorry, Doc, if they forgot to call in; finally did over Little Todd, but couldn't raise the shack.

At the social hour all was forgiven, but my evening potion was likely to be strong. Don was especially understanding and always vowed they would not again be guilty of such negligence. Then, for the benefit of new personnel, he would tell of the first and most memorable time this had happened.

It was 1960, a year before my first winter on the island. The field party was Dave, Don, and ranger Dave Stimson. Mech and Murray had lost track of time in watching a chase. They had come in for gas in late afternoon — when the wind is right you can do this without being heard — and they neglected the usual procedure of buzzing the shack. Long after sundown they landed and taxied in to the dock. There they found Stimson, fuming with rage.

As Don described it, Dave got them out of the plane and lined them up on the ice. The air was red, then blue, then purple, as Stimson gave them the most elaborate and competent "chewing" Don had experienced since Korea. I do not now recall what Stimson's role had been in the military, but Don judged that only long practice could produce the sophistication Dave displayed on that occasion. He was completely right, and the troops slunk up the hill with tails tucked and no arguments. Don had great respect for sturdy characters and good craftsmanship. He always spoke of Dave in terms of admiration, as indeed did all of us who knew him.

One cannot realize the stresses of this daily life until he gets away from them. My five-day loner trips on the island have impressed this on me in a particular way. At home, in confrontation with my wife's culinary products, I am in a state of chronic self-discipline to hold my weight at a decided-upon level. I have an appetite that could easily get the best of me.

But I found I was taking too much food on my hiking trips. Carrying a pack over those ridges should make one ravenously hungry, but it did not. A bowl of quickie cereal, hot coffee (if practical), and several prunes were all I needed for breakfast. A piece of bread (a small loaf of home-baked whole-grain bread lasts the five days), a handful of pecans, and several strips of jerkey are enough lunch. For supper a hot drink featuring dry milk, a small dehydrated meal of some kind (preferably protein, like beans), a thin peanut butter sandwich, and more prunes. Usually I lose only a pound (or three) on one of these trips, and seldom have any keen sense of hunger. Nuts are my most satisfying food.

I have concluded that complete freedom to do as I choose on the trail — no commitments, no interruptions, no demands that must be responded to — begets a mental relaxation that is not experienced elsewhere. In other words, at home my eating is in part a compensation for stress of which I am not even aware.

In being totally alone I have a sense of security, of knowing where I stand. It seems likely that some of the impacts of social stress result from the uncertainty about what the other fellow is going to do. A natural disaster can overtake you, but you are most likely to get into it through calculated risk for which you alone are responsible. Natural events are reasonably predictable and never malicious, only indifferent.

In degree each of us has become the hyperkinetic victim of an overpopulated range, overused living space, competition for increasingly scarce resources, and the harassment of sounds that invade every attempt to be alone. We are confused and intimidated by a cultural complexity we do not even recognize. In isolation and self-examination the insight might come that would make one honest and whole. Of course, a person could talk to himself too long and become tedious company.[2]

Man is the tragic figure of the organic world; he is the only one who worries. He wonders how it is going to be; he finds out how it is; then he remembers how it was. He is haunted by the past and threatened by the future. Lacking big things, he worries about small things. It is the penalty for what we call intelligence.

In a sense, trusting oneself to the caprice of natural events is a retreat from responsibility. You cannot plan far ahead because you do not know how the sky will look in the morning. You know what the job is, but you do not know just how you will get it done, or when. One cannot work on Isle Royale or any comparable place unless he can accept, even enjoy, this kind of uncertainty. After all, who is to blame if it rains?

I have learned for myself, and seen in others, that there is a particular satisfaction in meeting unforeseen happenings and then finding that you have done the right thing. Don and I shared such an experience.

On 14 February 1967, we were flying in cloudy weather over the east end. Westward the sky had darkened, and we suddenly realized that a storm was moving in over Washington Harbor. Don decided we should get back, and we flew in a light snow southwest along the shore of Siskiwit Bay. We were soon under heavy clouds, and as we reached the end of the bay at Island Mine Trail, I could see that Don was in a hurry. Ordinarily we would have followed shorelines and waterways. It was routine safety practice, and these always were good sites for wolf hunting. This time the plane headed directly west over the swamp toward Greenstone Ridge and Windigo. The weather was thickening ahead.

In fact it got too thick. As we reached the edge of the swamp, we could not see the ridge in front of us. Instead we were flying directly into the solid white wall of a snowstorm — zero visibility from the sky to the ground. Don turned his head back toward me with a half smile as I shouted above the noise of the motor:

"Better pick up Feldtmann Ridge!"

Don nodded; he was already banking the plane. There still was scenery to the south, and we headed across Siskiwit Swamp. We both knew that, by flying southwest with the ridge off our wing-tip, we could follow landmarks without much visibility. For those accustomed to the casual exploits of the big transports, it should be explained that our small plane has no instruments and flies everywhere by contact. If you got out of sight of the ground, you would be downside up in short order.

As we approached the ridge, the snow was enveloping us. With the steep slope on our left, Don came down to a level just above the trees. There was nothing ahead to hit, but there is uneasiness in flying into a white void.

When we passed the high cliffs we knew we had lost the ridge, but now the shore of Feldtmann Lake was below, and we followed it to the west end. Then a hop over the old beach lines for three quarters of a mile. When the Lake Superior shore appeared, we knew we were in Rainbow Cove.

Now we felt better. We were nearly down on the deck, and Don followed the winding shoreline around the west end into Washington Harbor. He let down at the right spot and taxied up to the dock.

As we tied the plane down I was aware that Don and I both had a vast sense of relief and almost exhilaration. I pointed to the thermometer on the wing strut.

"It's ten above. Anyway, you can't ice-up at that temperature."

Don grinned. "Doc, we did it just right this time. But it's something we won't try again!"

On a sunny day in February Don and I were flying over Lake Desor when we encountered a circling red-tailed hawk, one of only two that I recall seeing in winter. We got above the bird and likewise circled, the wing tipped down so we could see it comfortably. Its wings and tail were broadly spread and showed richly colorful against the snow. I could see that Don was fascinated. He continued around and around, glancing back at me occasionally.

Here it should be said that one of the things giving me a sense of kinship with this man is his never-failing appreciation of the beauties of the natural world. He is a tough Irishman who did his stint in Korea running a motor pool. In case of need he can draw upon the most exquisitely debauched thesaurus of unprintable verbiage devised by the GIs of two wars. Yet he has a streak of gentleness that shows through at every turn.

Don's reform movement among our backyard foxes (p. 303) always was a source of amusement to the rest of us. His patient efforts to remodel that social system were invariably in vain. Despite food incentives and the respected stick, millions of years of evolution had produced something far above his poor power to add or detract.

Don is an accomplished woodsman and hunter, a sportsman of the highest type. In the 60s he was constantly plotting with a few trusted friends to give more protection to the wolves of northeastern Minnesota. I gave him my cull slides taken on the island, and he used them in lectures to local school groups. He has taught his sons and daughters to hunt and fish — for fun. If a kid misses half a dozen grouse, great! He has had a wonderful time. If he hits one, they have had a successful hunt and are ready to do something else. Don chuckles often about the fishing trips. When they have caught some fish, the young ones demand the right to build a campfire and cook them in the field. Any greenhorn can take them home to the kitchen. In spring two or three of this family will be sitting quietly on a hillside watching the play of young foxes around a den, and keeping the whole thing secret to guard against what might happen if the wrong people found out.

Surely it says something for the origin of man that we usually see beauty in life forms and vistas of the natural world. It is nothing new. During a period of some 20,000 years, especially in Les Eyzies region near the French Pyrenees, men of our kind, hunters of the magnificent European megafauna, established the first great art movement. In such caves as Lascaux in the valley of the Vézère and others in the Dordogne, by the flickering light of grease lamps, the Cro-Magnons painted dynamic scenes of animals and men. What drove them to artistic expression deep in caverns of the earth no one knows. But driven they were, and their portrayals of bulls and bison, reindeer and horses, are moving reminders that man's earliest stirrings of artistic impulse must have long pre-dated these triumphs of animal portraiture (216).

The refinements of artistic appreciation and talent are cultural, though based on innate capacities that frequently are handed down. One's personal background may make all the difference in his perception of nature. A coral reef can be a delight to the senses, a symphony in color and form. Or it could be a den of horrors where millions of creatures are being eaten alive.

In the right situation the artifacts of human industry take their place in consonance with naturalistic things: I linger in my inspection of a long-abandoned fishing boat on the shore of Chippewa Harbor. I wish its many stories could be told. As we stop to look at an old mine pit overgrown with the redeeming forest, we experience a moment in history and become a part of the progression of events. Possibly such things imply the transient nature of man and his works. But in the presence of dissolution and healing I derive a strange comfort. There are processes basic and substantial that transcend the last slow-moving million years and also the speeding century ahead.

Does it mean I am an unbeliever in man? Not necessarily, but certainly a believer in an ultimate justice in which man will get about what he, in all honesty, deserves. And how does one deserve? Perhaps by being modest enough to learn, courageous enough to face reality, and frugal enough to work hard.

The full enjoyment of wild things and wild places requires some input of curiosity and understanding. The more you know and the closer you get to elemental living the more you come away with. Your experiences will be memorable, if not always comfortable. It is the unknown that both men and creatures fear. A stranger in the forest may not enjoy its nights and its violent tempers. And what one knows sometimes will be no comfort; I admit readily that I do not sleep well in bear country.

On Isle Royale one can rest with a greater sense of security. A stick snapping in the woods is, ten-to-one, a bumbling moose. I am alert and listening, but not disturbed. An animal trotting lightly through dry leaves is almost certainly a fox; rarely it might be a wolf. We know relatively little about what goes on at night. Mostly one interprets in seeing the evidence later.

I have spent most of my recent birthdays on the island, and on 11 October 1976 I had the Feldtmann Lake campground to myself. I was enjoying plenty of room in my pop-up tent; the trip was a one-nighter out and back to Windigo. Before 4 A.M. I heard bushes shaking and the deep breathing of an animal close beside me. Certainly a moose browsing. I was glad my tent lines were tied close in and not across any trail in the surrounding brush. That area at the west end of the lake has long been kept open and in the shrub stage by moose feeding. I went back to sleep.

The next morning I found a fresh wolf track in the wet beach sand 100 feet from my tent. I savored the thought of the wolf passing by and paying no attention to me or the moose. I vaguely recalled something that, on my return home, I looked up in Rolf Peterson's notes for 1970.

On 22 June — his first month on the island — he, Ron Bell, and John Vanada were camped in that same spot. On the previous day, starting at Siskiwit camp, they had covered 13.5 miles in their wanderings. In his notes Rolf said that Ron and John slept in the tent and he rigged a piece of plastic as a shelter outside. His remarks on the night's rest were laconic and expressive:

At 1:00 I heard a moose browsing grass by my head. In the morning I found that he was about five feet away — funny munching sound.

3:15 — Strong wind off the lake blew down my shelter, so I slept in the open next to it.

5:15 — The same two bulls (adult and yearling) were back. They stared at me from 10 feet away as I lifted my head to look at them. The yearling was browsing, coughing and wheezing. The sun rose over a golden mist.

7:30 — An adult bull appeared on the shore by the sign. We finally got up, as we couldn't sleep too well with all these moose running around anyway.

Shortly after, Ron got within 20 feet of a cow moose, eating grass on shore. She remained undisturbed.

A mayfly hatch was on this morning in front of our camp.

We left camp at 8:45 after cold cereal breakfast.

The stories I read in early boyhood of adventures in the "north woods" commonly featured some lucky stripling who was conducted through the wilderness by a didactic and knowledgeable woodsman. The rustic techniques of living close to nature were explained in detail. In my case they effectively charmed the imagination of a city-bound kid who had to enjoy the sights, smells, and discomforts of wild country by proxy.

As the tale was told, one picked his campsite in the virgin forest, cleared and ditched his tent, cut saplings for poles, a green backlog for the fire, huge armloads of fragrant balsam branches for his bed . . . He might even be partially living off the land — legally, of course — with meals of tender grouse or succulent venison. The hunting was always good and the fishing pristine and fabulous. Thoreau's journals of his rambling in the Maine woods are replete with similar allusions.

Those times and those feckless backwoods habits are now largely gone. During the warm season some two thousand people hike the 40-mile length of the Greenstone Trail or some major portion of it. They are in a wilderness formally designated by act of Congress.[3] Whereas Thoreau regarded the campfire as spiritually essential, today's backpacker can experience its genial warmth only at strictly limited sites, often in company with many others. He must range far afield to pick up dead wood, for his predecessors had similar needs. He cuts nothing that grows and endeavors to leave the habitat with minimal signs of his passing. Whatever he sleeps under or on must be carried and, often enough, a portable stove as well. He has no bottles or cans and packs out whatever trash he brought in with him (him = him or her).

It has been reassuring to observe how often people cheerfully accept the necessity for these restrictions and manage to achieve a quality experience. They live as simply as possible and learn some of the historic crafts of the backwoods. At least that is an accepted ideal, and I have had great admiration for the do-it-yourself skills of many who come north to use Isle Royale's hiking trails. The island gets some of the best, and on the average its visitors are a cut higher in knowledge and habits than the clientele of most parks. There is less slobbery, less littering, less peeling of birch trees than one would expect.

For the more sophisticated user of the out-of-doors there is a primary value in discovering how, through a simplified approach, he can take care of himself in natural environments. Seeking and finding essential comforts and ignoring the frills produces a satisfaction that brings him out in the weather and makes him, for a time, a part of the wild community of life.

This kind of person is a supporter of wilderness preservation, in part because he likes the idea that there is always a more primitive level of lore and accomplishment to beckon him on. If he cannot attain it in his lifetime, it will be there for his children or someone equally deserving.

Obviously, not everyone is physically able or has the inclination to do his outdooring the hard way. In fact, simple things do not attract some people, and they carry along the modern manufactured gear they understand. They are confirmed in doing this because the industrial promotion of gadgets assures them that this is the way for a true progressive to overcome the privations that are taken for granted in the out-of-doors.

On a day in June I came out on the trail from Lake Richie to the Moskey Basin dock. An elaborately equipped powerboat was tied up there. In the small campground were two men sitting on chunks of firewood in front of a radio. It was tuned in on some juke-box masterpiece. They stared at it, somewhat stoically, I thought.

There was music in the woods that day, although if one knew nothing about birds it probably had small meaning. These men bothered me in one way: They had come to the north country, and maybe they expected to find something there. If so, they did not find it and would decide henceforth just to stick with familiar things. Or, the open spaces might be just a place to use their equipment. We have new generations of people well isolated in large cities who grow up having no real opportunity to get personally acquainted with the natural scene. If this goes on long enough, there will be no natural scene.

Actually, too much gear is a common problem with all of us. I first learned to cut down and be selective when I had to carry it over portages. Today, I have to backpack it all day, and I still take too much. Almost invariably I come back with things I did not use. When my wife is along, I am automatically committed to dealing with her cherished possessions. She has not had even my level of conditioning on such things, and she plans at length on the contingencies she might have to face.

In travels for any extended period I can be sure Dorothy will have along a personal medicine bundle that presumably is some kind of survival kit. It is not always large, nothing a stalwart burro could not handle. I have never seen its total contents, and it undoubtedly varies according to our prospects. It has been useful at times, and through long experience I confide that it will produce anything from a Band-Aid to a Stillson wrench. I dislike to speculate in such esoteric realms, but I suspect that Dorothy's penchant for gadgets is part of a providence-domesticity syndrome native to unliberated females.

All of which is to say that most of us, through lack of enough practice, are fighting the battle of too much wilderness finery. We will be more comfortable, even more secure, by depriving ourselves of many things we bought because the price was right.

Woodcraft and pioneer skills usually go with an appreciation of nature. But a common problem in today's world is that many people do not have the time to do things as they might like to do them. They fly where they can, take a boat from there, stay in a rustic lodge, get guide service, enjoy themselves, and get out again. It is a wholesome expenditure of leisure time and often makes possible a quickie communion with nature that would not be possible otherwise. We must acknowledge this demand and make provision for it.

Many of Isle Royale's visitors do it on about this basis, principally in the period mid-June to early September. They enjoy the facilities at Rock Harbor Lodge, do some offshore fishing, take a boat tour, and get out on local trails to see a moose.

This is a constructive kind of wild-country enjoyment. The time is ahead when I will do more of it myself. A problem develops when people devoted to these practices want the entire program to suit their needs, and it goes well beyond what I have described. The logic develops this way:

If Citizen A has a right to use the back country in the way he prefers, carrying a pack on the trail, then I, Citizen B, have a right to use it my way too. I will ride my trail bike, because that is the way I like it. A fallacy in this is being increasingly recognized, but most people have not yet caught up with it.

The critical dimension is carrying capacity. A hiker may cover 10 to 15 miles in a day. He has impinged on the privacy of people over that much trail. A trail-bike operator, with his noisy motor, covered three times as much ground and had an impact on people all the way. In effect, he has used at least three times the wilderness resources of the hiker.

A canoeist might cover 20 miles in a day comfortably. But a motorboat can outdo that by at least three times, so he too has overloaded the wilderness environment in terms of the experience of people who go there wanting to be alone.

A person on snowshoes might, we will say, cover 10 to 15 miles a day if he wanted to work at it. But a snowmobile can roar across the lakes and trails to a far greater distance (see Branson on snowmobiles).

The point is, of course, that the wilderness can accommodate many more people doing things the hard way than it can if individuals insist on the convenience of modern motorized equipment.

As the North American Wildlife Policy Committee pointed out, the "off-trail vehicle" probably has no legitimate status at all (12). Vehicles should be kept on trails where they will not tear up the land and cause damage that takes years of retirement to repair. I have enjoyed 4-wheel-drive tours of the back country myself, and I understand the charm of proper use in proper places. So I am not undercutting all such means of getting around.

The major issue is resolved by keeping the different kinds of interest well separated. We can have country where the latitude of public uses is broad. In other areas — the more remote and undeveloped kind of wilderness — restrictions are in order to protect the rights of the seeker after quiet and privacy. However, this does not solve all problems, because crowds of people get in their own way, whatever their interest.

At the present writing, 17,000 people visit Isle Royale in a 4-month season. The trails and campsites are functioning at about capacity. Any attempt to load them to a higher level will be counterproductive (pp. 410–411). What does it all mean? For a time in the 60s, Congress seemed to be appropriating funds on the basis of how many people were served in a given year.

Such an outlook fails to recognize that another Congress (1916) gave the Park Service a mandate to preserve the resources in their custody. This function requires a protective investment, and it implies that recreation for today is not the whole picture. In part we are preserving things because we are assured that people will always want a diversity of experience; which means we have not achieved a finality of wisdom that says, this we should keep and that we won't need.

When Congress passed the Wilderness Act of 1964, it appeared that to get the votes you had to represent wilderness as a place people could use. But alert scribes weasel-worded in the preservation idea, the future-generations concept, and the scientific-value afterthought. Perhaps it is the salvation of the democratic process that there usually is someone around with an impractical devotion to principle.

The 4th of February 1971 was a snowy day and we could not fly. We had decided to autopsy a young bull if possible, and I was making the rounds of the harbor. I crossed the western tip of Beaver Island and saw what I took to be a mink at the edge of the north shore. I stood and squeaked like a mouse to get a better look at the animal, but it did not reappear. Then, about 60 yards back in the brush, a bull stood up and looked at me. I could see only the top of its head above the thick growth. I took the plastic bag off the muzzle of the .30-06 and decided to take a shot, knowing I would either knock him down or miss completely.

I missed, and the bull simply walked ahead a few yards into the clear and turned to look at me. Its neck was now in full view, and my second shot did the job. As I approached the dead animal, I saw another bull about 20 yards away plodding deliberately through the deep snow. It did not appear to be alarmed and soon passed out of my view.

Why these moose were not spooked out of the area by the report of the rifle is a good question. Occasionally a moose is poached along the south shore in the fall, but most Isle Royale animals live and die without ever hearing a gun. Another example of their indifference was even more impressive.

In the winter of 1964, Pete brought his hunting rifle to camp, since at that time we had no collecting gun. He and I went part way down the hill and set up a cardboard box against the slope so he could sight-in the rifle. Our fox Sleepy had followed us and was sitting 10 feet from Pete looking off into the woods. I was watching the fox as Pete got off the first shot. Sleepy did not even turn his head.

I have wondered how to explain such nonreactions to what I thought was a crashing disturbance of the natural quiet. One explanation is that these animals have had no experience with any lethal kind of human activity. In areas where there is a hunting season I have not observed this passive aloofness on the part of wild creatures. As another angle on the question, the sound of the gun was not a great deal different from the thunder that animals hear during a good part of the year. Seldom is there any thunder in winter, but this is a fine point that need not be debated.

All other conditions aside, the wolves on Isle Royale are afraid of human beings. Does this mean they will always be afraid? Or is there a holdover of tradition, propagated by older individuals, that says these funny-smelling, two-legged, non-prey, socially anomalous creatures should be avoided? Will that tradition eventually wear out, and will we have half-tame wolves scavenging, like the foxes, around campgrounds?[4] It took a controversial closing of garbage dumps, a crackdown on roadside feeding, and severe penalties on foolhardy human familiarities to convert the bears of many parks back into wild animals.

In fact, it has not yet been totally successful with black bears, possibly because humans in the back country are so hard to reach and convince. On the other hand, Yellowstone and Glacier grizzlies have gone wild, but they seem to be turning back the clock to the time of Indian culture and Lewis and Clark. They are no longer *enough* afraid of people, with the result that tragic events are taking place.

There are complications in relating modern man to the wilderness and its creatures. The wild country is not just plants and scenery. It is not just magpies and marmots, whiskeyjacks and cute foxes. It is also big-game ungulates and the capable carnivores that weed them out and hold their numbers in check. Modern man is a nonadapted intruder, and it appears that he will need to meet wild nature on its own terms.

Our research on Isle Royale has helped to analyze the problem. We find that the principal concern must be to restrict the human impact so that, as provided in a far-sighted law, the wilderness will stay what it is. Those who use the wilderness will have to accept its native inhabitants. There is no way to foolproof it, nor is there a way to give total protection to many innocent users who are not fools.

It is unrealistic to assume that man can take over any area without changing it. Minor changes are taken for granted, although they jolt us suddenly at times from unexpected angles. Like the day when I stood at the ranger station and looked thoughtfully at a mountain ash tree that was totally without fruit, in contrast to the winter before. High in the tree I saw something; a siskin seemed to be swinging around in midair. With some difficulty I retrieved the dead and dried-up bird. Hooked through the lower jaw was a trout fly; the other end of the fine monofilament leader was tangled in an outer branch of the tree above the lake edge.

This made me reflect on the eagles and ospreys that nest along the west shore of Yellowstone Lake and feed on its abundant cutthroat trout. Many of these trout escape from fishermen carrying treble hooks, fish that may well thrash helplessly on the surface, easy prey for the watchful eagles or fish hawks. On Yellowstone Lake how much of a hazard is fishing to birds the park was intended to protect? We hear many accounts of such problems elsewhere; in Yellowstone we do not know the answer. However, several good reasons are evident for recent reductions of fishing pressure. One suspects this is all to the good in a park that supports birds, bears, and other fauna that must live, in part, on fish.

In our wilderness system the unconventional wisdom has finally acted. It has preserved something that steadily becomes more precious and unique as man exploits, infiltrates, subdivides, adulterates, and intoxicates everything else. As for the last term I use, the island national park and its surrounding waters — once conceived as the last word in wilderness purity — are not immune. About 1969 we began to see midwinter smog over the east end of the island, obviously drifting before a north wind from Thunder Bay, and several years later we could smell a papermill even on the southwest shore. The lake trout now being caught in Isle Royale waters are carrying such a load of PCBs that they are only marginally edible. The citizens of Duluth are restive over the finely divided asbestos found in the lake water they use for domestic purposes.

The newspapers and political solons treat these matters as technological problems. More realistically they are the expected and inevitable effects of overburdening the human habitat with too many humans — armed with a technology that does not know when or how to stop.

Irrespective of his specific interests, any ecologist meriting the name must inevitably project his science to the problems of our own species on earth. Which involves the fund and quality of resources, population, and living standards. In time his reflections will tell him that, amid wars and social injustice, man's greatest inhumanity to man is the biologically insensitive overproduction of his own kind. In this dimension we decide the equity that each individual is to have in the common environment, the standard of everyday life it will be his to enjoy or endure.

It is becoming evident that, as we use this earth, enough of it will need to remain in a natural or semi-natural state so that its energy exchange budget based on solar radiation, its atmosphere and water system, its weather and climate, can operate in the long-established manner. This entails a willingness to preserve space and other things lightly used. But we seem deadly intent on converting everything to some new design. Man has never yet created a self-operating ecosystem, and he can only botch up the ones we have.

Our most incredible policy commitment is the unqualified devotion to growth, growth that inevitably feeds on the property of people yet to come. We could learn from wild communities that too much of anything is at least as bad as too little. Most of the leaders we elect and tolerate have not discerned, or are afraid to recognize, the truth about economic expansion, which is spawning ills on every side and heading our entire operation into receivership. In resource affairs it might be practical guidance for a responsible electorate to adopt a critical assumption: that curiosity about the past and a visible concern for the future are the earmarks of a person who can be trusted today.

The ecologist looks genially at the mountain tops, the open waters, sandy lands, and "waste" areas. These, too, are space, and it will be a while before they are growing people. However, a doubling of numbers now appears to be built into the age structure of our population, and one wonders what the real future of wilderness is to be. Around the world, increasing numbers of humans living and reproducing at the subsistence level are forced to think of nothing but subsistence. Will wilderness, or simply the room necessary to achieve any grace in living, become a luxury we cannot afford? If that is our future, then it is good we studied the wolf when we did.

I have found that there is no place like a fire tower to do your thinking about Isle Royale. I enjoy believing it helps me to get the broad perspective. Also, it is easy to do a bit more cooking than over a campfire, especially in stormy weather, and it is good to employ one's hands. I do not say pass the time because time should not be held in so little esteem. The island has three towers: Ojibway, Ishpeming, and Feldtmann. The first and last have given me special inspiration because of the views they afford.

I felt like a Mayan priest in his stone observatory when I discovered that from Ojibway tower on the 4th of May the sun rises behind South Government Island. It will do that twice a year, and it will always be so until someone moves the tower. Of course that will happen; no one can ever be satisfied.

On that May morning in 1974 there still was ice on Sargent Lake and Moskey Basin. All the outer bays were clear. So was Ojibway Lake, which surprised me at first. Then I realized it had a flow of water through it or the beavers could not have flooded it years ago.

Earlier I had been awakened by two male sharptails dancing and hooting beneath the tower. The day before I had flushed a sharptail a quarter-mile east; it appeared to have been feeding on bearberries left over from last fall and preserved under the snow.

My finest recollection of a tower experience dates back to that first fall field trip, the sabbatical in 1972 when I hiked from Siskiwit camp to Feldtmann tower and made it my center of activity for three days. On the second day I went west on the ridge and then cut southward through the woods aiming for Long Point. I missed it by half a mile, which is easy to do. You try to keep a certain bearing on the sun — easier than forever looking at a compass — and you have to turn aside to avoid a low spot and a thick stand of cedars. Then you take a game trail that isn't *quite* in the right direction. The evident path eventually subdivides into obscure dendrites of travel that pose questions of which to follow, and finally they get lost in the ground cover. Then you try to take up where you left off.

At the shore I sat awhile, filled two canteens, and started back on a compass bearing. I recall a pause along the way.

On a low knoll there was a quarter-acre opening in the forest, which at that point was a close stand of tall aspens with little ground cover. As I lay back in the sun and rested I knew the primordial state. I did not have to be anywhere or go anywhere. No one had much idea where I was, and even I was not too sure.

It occurred to me that it would be appropriate for the moose to keep on browsing the small clump of white cedar and the foot-high sprouts of aspen and birch. They were preserving this spot of sunshine for me and other habitants of such places, though it was unlikely I could ever find it again. The strident call of a downy woodpecker roused me a bit. I dawdled, making a note that the ground cover here was Diervilla, bearberry, strawberry, and pearly everlasting. It was 12:29 P.M. I got into the harness of my nearly empty pack.

The wind was freshening, and by the time I reached the tower clouds were scudding across the island from the open lake to the southwest. A blow was on, and I knew it meant an early fall of leaves. Colors of the birches and maples were at their height.

The unheated tower grew steadily colder. I had a simple meal and gratefully slid into my down sleeping bag on a folding cot with a mattress, the *n*th degree of backwoods affluence. The wind rose steadily as the sun went down.

Latter-day fire towers are broad substantial structures assem-

Feldtmann tower.

bled from prefabricated members of stamped steel. Steps and other pieces have patterned openings and raw edges. When the wind blows, it is cut into a thousand strands of complaining sound. The gale built up to 50 knots that night, and the tower shook and strained, crying and caterwauling in melodic agony.

I went to sleep early but awakened at intervals listening in snug comfort to the screaming tempest in the framing under me. It rose in pitch, paused for breath, changed direction and force, wailing, moaning, howling, shrieking in the tortured fury of unblessed demons of the night.

Strangely, it was not especially disturbing; possibly I was that grateful to be under cover rather than out in the brush somewhere. I slept in spells and was roused again as rain pelted the tower and thunder rolled from the lake to the south. For a while the downpour streamed over the roof as from a waterfall. As it abated, a brilliant display of lightning tore the heavens apart. It was after midnight now; I slid out of the sack in longies, knitted cap, and heavy wool socks. Shivering with cold, I pulled on my down jacket and stood at the south windows.

The display was spectacular. Great lightning bolts seeming to originate high above me were stabbing the south shore of Lake Superior — not visible even from the ridge — half a dozen or more in quick succession. I wondered whether iron in the bedrocks was attracting the onslaught. It occurred to me that many of those giant arcs were well over a hundred miles long. As they lighted up the world momentarily, I glimpsed the white of breakers dashing against Long Point more than 2 miles away.

Through the rest of the night it blew, and all the next day. The maple forest was indeed shorn of its fall glory. Early on the 11th, when I headed off on a 9-mile trek to Windigo, the trail was carpeted with gold.

Water — glimpsed through trees at the end of a forest trail, or washing smoothed quartzites of the Canadian shield, or trapped in a glacial cirque on a western mountain — it gives character to country. It nurtures the life of hinterlands. On a June excursion across southern British Columbia I learned why that wondrous range is called the Cascades.

Let it be pellucid rivulets purling down from the highlands through clustered boulders. Let it be thwarted billows of our inland sea rearing in emerald translucence to strike again at a battered shore. Water in motion captures me in its ancient spell, waking to fantasy the primitive unctions of the inner man. Many have heard the singing of the Lorelei; if one should have to die with his boots on, it would at least be private in that green retreat — no doubt the most you could say for it. Fortunately, we have in common with creatures of the wild that certain decisions will be made for us, not by us. Meantime I may continue to glory in awesome sights, be comforted in my animal needs, and wander widely where the waters lure. Pursuing revelation, I shall seek the shore at nightfall where a brook debouches, and hide myself on the wash among the peavines, mayhap to observe the naiads bathing there.

The guardian waters of the great lake make Isle Royale what it is, providing almost unique isolation, subjecting its bays and inlets to endless change in beauty, charm, and challenge. On quiet evenings Dorothy and I have thrust our canoe among the craggy islands bordering Rock Harbor, where the lichen *Callopaca elegans* adorns the rocks like spatters of rich yellow pigment. On a scarcely perceptible swell the sensitive craft surges gracefully ahead, the paddler swaying at the hips. You ride a canoe, not *in* it. I salute a triumph of design, wrought by caring craftsmen in the old tradition. It is distinctly of this continent, yet kin to the elegant Viking ships that defied and survived the stormy northern seas. We prize our sixteen-footer as an heirloom of more frugal times.

A relished amenity of my first canoeing in the northland was the endless supply of pure water. Anywhere at all you dipped the ever-handy cup for instant refreshment. It personified the chastity of the forever wild.

We use the water with similar freedom around the island's open shores, and in harbors spring and fall, before and after the boating season. Immediately following the snow melt there is water in many pools and puddles, if you have time to boil it. Rolf carries with him one of the small purification filter pumps, and through the warm season they make use of any kind of water. In my own pack I have no weight leeway for such gear, and on occasion I have made do with whatever was available after a subjective site inspection. There is a spring near the foot of the ridge near Feldtmann tower, and I know of only one other, which may or may not be running in midsummer. I have drunk brown water out of a beaver-diked swamp that was surprisingly cool and refreshing. Its tannin content was not objectionable. Small brooks and trickles in the lowlands nearly always come from beaver works uphill somewhere, but on occasion one does not dwell on such things. With the hydatid tapeworm in mind, I do not recommend my casual practices in applying the park's water-boiling advisory.

In winter, hauling water in milk cans from the harbor was the main job of our snow machine. We did not like its noise and used it as little as possible. At any time there was always a chance of wolves on the harbor, and mornings we usually checked the view before anyone went down the hill. There were times, especially in earlier years, when we got our water by melting snow, or icicles off the roof, but that is the hard way.

On days when we were snowed-in, getting water was a regular chore, and usually I did not participate since I had cooking to do. However, I learned that quality control was my job, because no one else seemed to care.

For some reason the troops had a tendency to chop out the water hole, through 18 inches of ice, too near the dock. This caught up with us after the new dock was built as part of the NPS Mission '66 development program. In the winter of 1970 I got to feeling queasy at times, and the flying bothered me more than usual. I mentioned this, and John Keeler said he had noticed something similar. We traced the problem to the water cans and thence to the hole close beside some newly creosoted timbers under the dock.

Another winter, when similar symptoms developed, we detected a film of petroleum inside the water cans. We were baffled for a few days until I discovered that a new method of filling the cans was being used. Instead of dipping the water and ice out of the hole with an old 3-quart aluminum pitcher, someone had found it was easier to push the entire can down into the hole and let the water run in over the top. The cans were then hoisted into the trailer behind the snow machine and hauled up the trail.

Unfortunately, batteries, tools, and oily machine parts had also been carried in that trailer, leaving a coating of oil mixed with snow on the floor. The cans picked up this oil, which rose to the top and became a major pollutant when the cans were sunk in the water hole.

After every drawing of water, the hole in the ice was covered with snow to minimize freezing. The aluminum pitcher was turned over on the snow to mark the location. A winter or so after the engine-oil episode, we all noted that our water was getting a rich and musky odor — somewhat foxy, in fact. Several cans were so rank the water was unfit for anything but dish washing. Having had previous social-hour experience with yellowish icicles from the eaves, we quickly ran down the problem.

In traveling back and forth across the harbor, one or more of our foxes was detouring to mark the pitcher at the waterhole. We altered this program by sticking the ice spud a few feet from the hole and hanging the pitcher on top. The ice spud still got marked.

All of us agreed that water from the harbor was more satisfying and thirst quenching than melted snow. And in truth, aside from its presumed mineral content, it probably had nourishing properties somewhat beyond the level of an ordinary tap-water quaff. When you held a glass of it against the light, there came into view a host of minute copepods moving jerkily through their medium. To me it was somewhat unexpected to see these mainstays of the aquatic food chain thriving beneath the ice at a temperature close to freezing.

Like most of us, Don was moved to wonder as he inspected the contents of his drinking glass before the window and confronted, eyeball to eyeball, the translucent loveliness of a doomed and unsuspecting copepod. He was at some pains to discover, in all humanity, what admixture of bourbon might help to dull the trauma of cataclysm for our little friends of the deep. After all, the scale of one's world has no relationship to the experience therein, and a man with proper cosmic awareness might appropriately consider the poignancy of events within a drop of water.

Don and I derived some amusement from recalling occasions when, over the years, we revealed the biological activity in our drinking water to certain less impartial members of the team, who had betrayed strange suspicions of our Washington Harbor potions. Evidently, for some discriminating individuals there are important differences between a drink of water and a seafood cocktail. As they say on TV, to each their own.

CHAPTER SEVENTEEN

Wolves, People, and Parks

THE OUTLOOK FOR wolves, moose, and other creatures on Isle Royale depends much on management regulations and operations in the park. Our work since 1958 has contributed to the job of fact-finding on which management must be based.

In the present state of its vegetation, this 210 square miles has space and resources for a continuing population of about 1000 moose and between 1 and 2 thousand beavers. These, in turn, can support in comfort about 24 wolves. That is approximately five moose per square mile and a wolf per 10 square miles. Which equals a wolf per 50 moose as of midwinter.

There are important qualifications in the above statements: The vegetation does not hold still in its "present state," and average figures for a "continuing population" do not stay average. The period of our work showed constant change, and more of this is predictable.

The changes were tied together in ways we partially understand. Beavers increased through the 60s and into the 70s, and so did moose. Were the beavers a "buffer" for new generations of moose? In other words, was it easier for wolves to hunt around the ponds than to get by the hoofs of an irate cow to eat her calf? This relationship must have been involved and could be one reason why moose were building up. Also, 1968–70 was the period of lowest wolf numbers. If there were half a dozen less of them to feed, that should have been some saving of both moose and beavers.

As has appeared so vividly, trends in these animal populations tie back into the weather pattern. Weather provided water for beavers to prosper and for the annual regeneration of moose browse. It provided deep snows in at least five of eight years (beginning in '65), which made the plentiful moose increasingly vulnerable. Many beavers in summer and easy moose in winter equaled prosperity for the wolves. They brought through litters of thriving pups and formed new packs. Responding to an increased food supply, wolves multiplied to about twice what anyone thought a maximum population was likely to be. They did this at the expense of reducing the moose herd (as of 1976) by perhaps 20 percent. In terms of their basic estate, the wolves spent the annual earnings of interest and were digging into the principle.

That was not all bad, because it eased the pressure on winter foods of the moose. Which was not all good, because the forest was maturing (growing beyond the reach of moose) all the faster. We have no basis for summing up the long-term gain or loss.

These 18 years have been biologically productive, but happy times do not last forever. The wolves probably have reached the maximum density they can tolerate, and they will be reduced. This is in the cards because moose are declining and weather is sure to change. The wet years will inevitably give way to a dry phase of the climatic cycle. Then beavers will go down, moose browse will not grow so well, and compared with the good old days, the wolves will know privation. They will have to cut back on pup production, the overutilization of resources, the wastage of moose carcasses, and the device of killing one another to gain space. After they do this, they will find that everyone is better off, for a while.

What happens to Isle Royale's vegetation is the key to the future of its animal life. The forest is growing, shading out ground cover. Moose and beaver are gradually losing their food base, although we have no means of measuring accurately the decline in carrying capacity. All of us have overestimated the effects of heavy browsing. It kills some woody vegetation, but it preserves the openings so that more ground cover comes up. The canopy closes slowly, but over time it happens.

Dry years will bring something to cure all this: fire. Climatic extremes of drouth and fire on the one hand and deep snow and abundant rain on the other mediate a forest rotation from spruce-fir to aspen-birch and back again. Moose have a part in it too, since they bring about the nearly pure stands of spruce. These produce little but some clippings for hares, and without good ground coverts there will not be many hares.

Diversity in topography, drainage, and seasonal weather produces a diversity of successional stages in vegetation — habitats

In the winter of '72 we prepared for a visit by National Park Service director George B. Hartzog, Jr., but events in Washington intervened. The artwork was by Fred Montague.

for animals great and small. Fire plays a primary role in this diversity, and burnings also do something for the flow of nutrients. Various studies have shown that the mineral-rich ash helps to produce forage of high quality and palatability.[1]

This is no endorsement of uncontrolled conflagrations in our prize stands of old timber or on unstable watersheds. Patchy burnings are needed in the big areas of monotype, and the best way to achieve that is by setting fires in the right place at the right time.

However, except for remedial burning,[2] as now practiced among the great conifer forests of Yosemite and Sequoia-King's Canyon national parks and a few other places, I do not advocate deliberate burning in wilderness. This is an expression of my purism. I would prefer to let nature take its course, by way of lightning fires. Then we should control where we need to control for good reasons. That way any resulting devastation is an obvious act of God and not the malfeasance of a long-suffering superintendent.

The immediate effects of wildfire are unsightly (and in the wrong place unspeakably tragic), but after a few years the better side of this primordial scheme begins to show. Then a valid question arises: Is this the way to preserve our wilderness ecosystems? Counter question: Should the grouse, hare, goshawk, fox, deer, moose, and wolf be a part of wilderness?

For most thoughtful persons, there is only one way to come out on this issue: We must pay a price for reality and gear our policy to the long term. That is where the National Park Service comes out. Isle Royale has a "let burn" policy for natural fires. Such fires will be controlled only when necessary for the protection of property or for other worthy purposes. Fires started by people will be put out as soon as possible.

It is a responsible outlook. It will require administrative courage. It will produce desirable results for the future.

The personal relationship of wolves to men has long been a subject of interest to nearly anyone, as witnessed by the theme of many folk tales and an abundant literature on the subject in newspapers and magazines. There are sober accounts of people stalked, pursued, treed, and occasionally eaten by wolves.

Some of the stories were inherited from Europe, where they were so elaborately documented that one has to be a total cynic to brush them aside as fabrications. C. H. D. Clarke studied this matter and decided it was true that many people in the Old Country had been terrorized, attacked, and killed by wolves. He concluded that a part of the problem was the appearance of rabid wolves in various localities in earlier times (when there still were wolves). As another likelihood, he attributed some of the misdeeds to dog-wolf hybrids that did not have sufficient fear of man.[3]

In North America there are credible or documented records of people attacked by rabid wolves. Charles Larpenteur, a fur trader on the upper Missouri for 40 years, told of a "mad wolf" invading their camp at a rendezvous on the Green River in July 1833. The animal bit three men in the face as they slept. It also attacked a bull in their small herd of stock. The bull and at least one of the men died, obviously of rabies. Other accounts of rabies among wolves, as it affected Indians and frontiersmen, were compiled by Stanley Young (297). A well-authenticated attack by a wolf on a railway section foreman in Ontario was described in reports published by Randolph Peterson. The dead wolf was not examined for rabies, but the disease must be suspected.

After a lengthy review of accounts of wolf attacks on humans, Young said the Fish and Wildlife Service (meaning Young himself) had investigated such reports for 25 years and had not been able to substantiate any (297). He said an editor at Sault Sainte Marie, Ontario, had offered a reward for 14 years for proof of a wolf attack "and the prize is still in the editor's cash box." Lee Smits offered a similar award and found no takers. Shiras stated, "Nowhere in America have I ever been able to get an authentic account of a man being deliberately pursued or injured by a wolf . . ."

Both Shiras and Young prudently suggested that, in view of the multitude of such stories, it was quite possible some of them had a basis in fact. This seems an appropriate conclusion in that wild creatures are subject to the same variability as man himself. Given enough of them, in widely varying circumstances, it is not totally predictable what the individual will do.

Differences in the reactions of wolves to men are implicit in various accounts of wolf behavior in the wild. There is frequent mention in early journals of wolves digging up graves on the prairies and devouring the bodies.[4] Yet knowledgeable travelers had no fear of wolves and usually did not waste ammunition shooting them. The animals lurked about waiting to share the spoils of a hunt,[5] and at worst they were a nuisance. Ruxton mentioned their creeping into camp at night and gnawing ropes, saddles, and apishamores and sometimes even injuring the horses (223).

Seton said that while buffalo were on the plains wolves were relatively unsuspicious of man. After the passing of buffalo and the introduction of range livestock, the great drive to wipe out the wolf began. Large numbers were trapped and poisoned during the 80s, and the survivors became wary and difficult to take (250).

In areas where wolves are hunted, they obviously are afraid of the deadly primate who gives them no quarter. On Isle Royale they may be less afraid, but they take evasive action. It seems evident that man is neither prey to be eaten nor a competing predator to be attacked. It might seem that defense of the young would be a situation in which wolves would be most likely to exhibit aggression against man. Yet Murie entered a McKinley Park den and removed a cub while the distressed mother looked on.

The relationship seems to have social overtones. When I surprised a large (male?) wolf at a kill on 17 February 1961, the animal, though only 20 yards away, preserved its dignity, looked *past* me, rather than directly at me, and circled at a half-trot to a trail before moving away.

Such behavior is reminiscent of the affinity existing between people who keep confined wolves and the animals with whom they have daily contact. The human is accepted as a pack member, and dominance must be consistently on the human side if unfortunate incidents are to be avoided. Certain wolves, taken from the mother before their eyes open, can be socialized almost to the point of doglike docility. But these are rare cases. More commonly, wolves retain a strong individuality and engage in periodic testing to see who really is dominant. Klinghammer and others who know this situation well do not recommend the wolf as a "pet."

Wild wolves seem remarkably consistent and dependable in not attacking man. Yet they are not afraid to take on bears and other carnivores on occasion (cf. Haber). Their social system does not prohibit killing other wolves. Man does indeed have a special status, and whether he is to keep it may hinge on how he manages wolf ranges in the future (p. 389). Too much familiarity could defeat worthy preservation objectives, even in parks.

The fact that wolves ran from man and did not usually defend themselves when in traps gave rise to the frontier notion that the animals were "cowardly." While I dislike to use the word *shock* to describe something I do not understand, this term comes immediately to mind in dealing with the passive, disjoined condition in which a trapped wolf often is found.

An early record of this kind is from Audubon, who told of spending the night at a farmhouse in Kentucky and then accompanying its owner in the morning to visit his three wolf pits. The pit was a hole about 8 feet deep, wider at the bottom, fitted with a trap door on which a bait was fastened.

One of the pits contained three wolves, which, as Audubon said, "lay flat on the earth, their ears laid close over the head, their eyes indicating fear more than anger." The farmer lowered himself into the pit and, lifting each hind leg in turn, cut the hamstring tendon with his knife. He removed the animals by dropping a noose over the head of each and hoisting it to the surface. "We hauled it up motionless with fright, as if dead, its disabled legs swinging to and fro, its jaws wide open, and the gurgle in its throat alone indicating that it was still alive." The first and second wolves were dealt with in this manner and turned over to the dogs to be killed. The larger wolf, evidently an old female, was more active and defended herself successfully until shot.

George Shiras told of setting a large steel trap in the water of a creek on a deer runway near Whitefish Lake in upper Michigan. On his way to this trap he flushed a pair of blackducks and fired at them. He found the trap gone and a trail where a wolf was dragging the clog through the brush. He followed easily and soon came to a large wolf

lying on the ground with its head between its paws almost as if asleep. On approaching the animal it was found to be dead, its body still warm.

It had probably been held fast by the clog when it heard the shot fired at the ducks, which accounted for the bloody alders, as the animal frantically renewed its efforts to escape. On reaching the hilltop the accumulating terror of its position undoubtedly resulted in its death.

On another occasion, Shiras found that a wolf in a trap appeared near death from "an overpowering mental strain, producing a complete physical collapse, and that, too, of an animal weighing eighty pounds and in the best of physical condition."

These events recall the death of the fox on our back steps at Windigo. Whatever the proper term might be to describe the condition — whether shock, catalepsy, or something else — modern biologists trapping wolves (humanely) for examination and marking have had experience with it.

In case of need, a trapped wolf that must be handled can be immobilized with a drug injection. However, Kolenosky and Johnston found this unnecessary in their work at Algonquin Park. The wolf usually lay quietly as a forked stick was held over its neck while it was examined, tattooed, and tagged in the ear.[6] Mech and Frenzel mention handling without sedation a female wolf caught by means of a snare.

> Evidently she went into shock or some other psychophysiological state of unconsciousness, for after her release she remained on her side and did not move for 1.5 hours, despite our prodding during the first few minutes . . . Then suddenly she leaped up and ran off.

It appears that a wolf sometimes finds itself in a situation with which it is completely unprepared to deal, having no inherited reaction or experience to draw upon. As a result, the state of being "out of it" is induced. The stunning of mental faculties in this manner may be akin to what happens routinely to the opossum. The condition conceivably could date far back into the roots of mammalian evolution.

Many people unfamiliar with wild animals are inclined to view them as senseless automatons reacting to life in a day-to-day standardized existence. Contact with semi-tame creatures (e.g., our foxes) does much to dispel such notions. It is particularly true that those who work with kept wolves are impressed with their abilities to remember both persons and other wolves and to relate to them with keen perception over a period of years.

Probably no people of recent times have observed wild wolves more carefully or at greater length than the Nunamiut, caribou-hunting Eskimos of Anaktuvuk Pass in the Brooks Range of

Alaska. Stephenson and Ahgook, interpreting their attitudes and knowledge, said:

> The Nunamiut also stress the importance of recognizing the wolf's capability for high order behavior . . . and his great learning capacity. One unpleasant experience with a given phenomenon, such as an encounter with a hunter, is said to be sufficient for a wolf to avoid any similar situation in the future. Adults having one experience with traps are known to be very difficult to trap and will not permit their pups to visit, for instance a carcass that man has tampered with.

Significantly, also, they remarked:

> The Nunamiut do not begrudge the wolf its prey. Though they hunt the wolf and value his fur, they take his life without the hate, rancor, or guilt that has so often accompanied the killing of wolves by human societies.

In his *A journey . . . to the Northern Ocean,* Samuel Hearne commented on the amicable disposition of Indians toward wolves west of Hudson's Bay. Of the wolves he said:

> They always burrow underground to bring forth their young; and though it is natural to suppose them very fierce at those times, yet I have frequently seen the Indians go to their dens, and take out the young ones and play with them. I never knew a Northern Indian hurt one of them; on the contrary, they always put them carefully into the den again; and I have sometimes seen them paint the faces of the young Wolves with vermillion, or red ochre.

Nearly anywhere one can find people who feel it no indignity to share the earth with other kinds of life. But men called primitive and others called modern have subjected the wolf to every brutish artifice their ingenuity could devise. A gob of meat hiding a treble hook and suspended over a trail would hang up a wolf for leisurely collecting. A curl of sharpened whalebone frozen within a ball of fat to be swallowed by a passing wolf — the result a pierced stomach, peritonitis, and perhaps a pelt gained the easy way. The inhuman evils of thallium, long used as a poison on the western range, need not be described.

In the haunts of "civilized" man, the closer one gets to wolf range the more he encounters simplistic views toward nature. A holdover from the frontier is an assumption that man is the keeper of all creation, that his is the privilege and duty to organize the chaos of nature, to sort its creatures, some to be reserved for use, others to be destroyed as vagrants and outlaws. Often enough, the old-time backwoodsman had a certain appreciation for the beauty of a pelt, but not to the extreme of leaving it on the animal it was intended to protect.

It has been, and still is, an open question whether wolves in any appreciable portion of their original range, and in any majority of their remaining forms, can survive our pagan era of natural philosophy. The indigenous school of thought — more properly, attitude — has been a seminary of hostility wherever men of limited view have had the opportunity to react to the carnage of a wolf kill.

The image people have of the wolf is, in degree, a product of applying moralistic standards. In my lecturing on our findings I regularly encounter persons of compassion and good will who ask, sometimes in a spirit of reluctance, Are wolves cannibalistic?

I suppose the answer is that, given the right circumstances, almost anyone is. We think we know who killed Big Daddy, but we do not know who ate him. The fresh victims of three "executions" we recovered were not eaten. These were cases where the pack dealt with an alien who was careless enough to get caught. Evidently there were better things around to eat.

However, Mech told Peterson that he had a trapped wolf killed and eaten. It could have been a loner that was chased into the trap on a trail. Another wolf had only half a tail, bitten off while the animal was in a trap.

In making his collections of bountied wolves in Alaska (p. 127), Rausch learned of six occasions where a wolf caught in a trap or snare was eaten by other wolves. Aerial hunters told him of having skinned carcasses eaten. The Nunamiut said trapped wolves were sometimes eaten, even by members of their own pack.

No doubt much depends on the times.

The wolf is of the wilderness, and inseparable from it. But it can get by elsewhere, and it lives dangerously where the broad and scattered border lands of Minnesota's farm country merge into forest. Here the predator-livestock problem is aggravated. Traditionally the wolf's poor public relations in this respect have inflamed men to blind retribution. Today's need is for attitudes of reason, a search for truth, and a temperate control program where necessary.

Wolves in Minnesota are being monitored more intensively by good research than any other predators in history. Backed by quietly loyal supporters, this work is the most effective defense a species could have. Its continuation is the great hope for wolves in the future.

An appraisal of available evidence by the Eastern Timber Wolf Recovery Team indicated that Minnesota wolves increased in recent years (177). We have come far since those dark years of the early 60s when no one was listening. The wolf had a bounty on his head, and armed aircraft were patrolling the lakes. Snowmobilers penetrated the roadless interior to set hundreds of snares — many of them never checked again. These practices were all abolished by the state.

In a myopic kind of humanity, some active partisans now demand protection for all wolves, but this could love the species to destruction. Unreason in such things plays the game of that back-country underground who take the law into their own hands. Their snide attrition is sniping away our "protected" eagles, and even the condor. It is eliminating last remnants of the Northern Rocky Mountain wolf, a desperately endangered subspecies.[7] The few survivors may never have a chance to build up in those logical sanctuaries, Yellowstone and Glacier national parks.

More broadly, the outlook for wolf preservation on Isle Royale and elsewhere must be considered in the context of prospective social and economic trends in the next half century. A doubling of the human population and a vast increase in land-use intensity appear inevitable. We can expect a steady attenuation of inhabited wolf ranges to the point where last retreats of the species will be in designated wolf refuges and fully protected wilderness. The bulk of state and federal wildlife refuges and the National Wilderness Preservation System are open to hunting. Even where areas are ecologically suitable, any resident wolves would be competing with man for the annual yield of big game: a poor augury, even for legally protected animals. The gun often makes its own policy.

Properly controlled hunting is not destructive. Game regulations are calculated so that annual kill is compensatory with other causes of mortality. Thus, the animal population is not reduced from year to year. However, it will be evident that if Isle Royale had been hunted, the age structure of the moose herd would have been altered, and many moose would have been shot long before they were old enough to be killed by the wolf. If beavers had been subject to trapping, the summer food supply of wolves would have been pre-empted. If wolves had been exposed to control or illegal shooting, the social relationships of individuals and packs would have been disrupted. Under these conditions we would not have been able to learn what we did about the many-faceted relationship of predator, prey, and environmental conditions.

This is to say that for scientific purposes a few areas should be kept free of major disturbance, where we can find out how the natural system works. But this is also the preservation function of a national park. Safekeeping of the wolf is one of the things this island, dedicated to public service, is supposed to do.

In the sense of maximum security for wildlife, our "purest" wilderness is in the custody of the National Park Service, and this is likely to continue. It is on such areas that the wolf should be privileged to survive, pursue his system of free enterprise, entertain people with his howling, and husband the dependent herds. On this continent and in the world, Isle Royale is an almost unique repository of primitive conditions. Like a priceless antique, it will be even more valuable in times not far ahead.

The temper of the people and their Congress being what it is, no one is likely to convert this rocky island into a hunting ground or an amusement park. But there are more subtle threats to its integrity as a wolf range and a unit of the forever wild. These are questions of benevolent use and zealous management, of which administrators have been well aware.

Isle Royale's 174 miles of hiking trails are a major attraction of the park. The country is clean and uncluttered and surrounded by the magnificent lake spectacle on every side. The terrain varies so much there is little monotony in hiking the rocky miles. In addition, the construction of bridges over streams, and planked walkways across mucky drains, is an inestimable boon, especially when packs are heavy and weather is wet. The trails guide you to where you are going, and occasional rustic signs tell how far it is.

The trail system has more than doubled since 1958. Visitor use increased, wilderness campsites became overcrowded, and

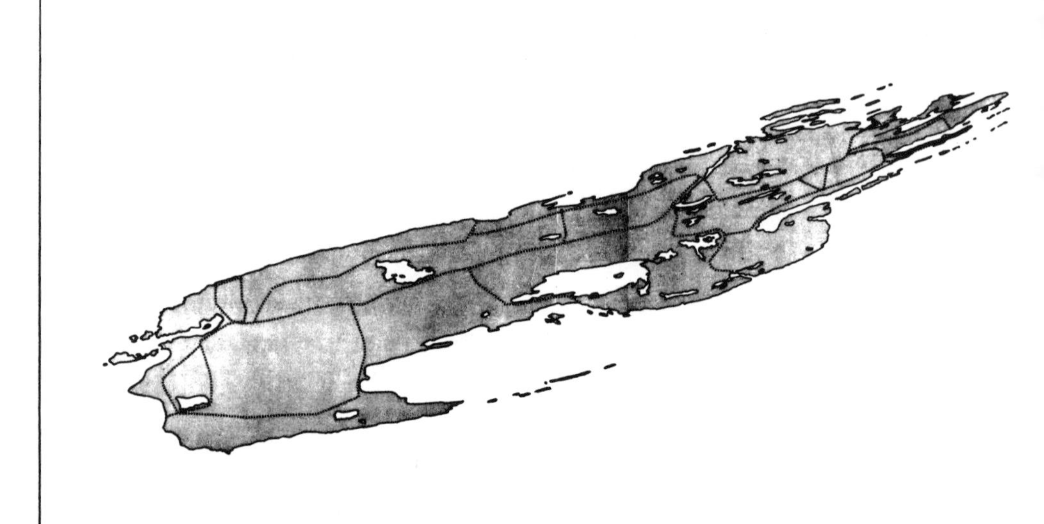

Figure 16. The 174 miles of trails.

segments of trails had to be relocated because of wear and erosion. In the early 70s there were plans for new trails to relieve pressure on the old ones.

At that time we were seeing a major increase in early visitation. As a result, where formerly scats had been plentiful on trails through mid-June, the wolves were now largely avoiding trails after the second week of May. At the request of Superintendent Hugh Beattie, Rolf summarized findings on scat collecting for the 1973 season: He found that up to mid-May — a scat deposition period of about 4 months — there were 5.6 scats per mile on 71 miles of trail. In the following 4-month period, to the middle of September, there were only 0.4 scats per mile.

Spring is the pupping season of the wolves; how much privacy do they need to rear their young? For once the park administration had a means of appraising this before a major development was carried out: Rolf's information on 16 dens and rendezvous sites showed that, with one exception, all of them were more than half a mile from any trail, and half were more than a mile. If this was a measure of requirements, then 75 percent of Isle Royale already was off limits for wolves in locating their homesites.

Superintendent Beattie had a problem. He needed to serve

the public to maximum degree consistent with a high-quality outdoor experience. Crowding vitiated all that most visitors came for. Isle Royale's trail development issue was a local model of the big, bifurcated, and somewhat self-defeating charge placed upon the National Park Service by the law of 1916: use with preservation. Overuse is not preservation. Oversolicitous preservation can void permissible uses.

In 1976 the park finally had its wilderness plan approved by Congressional act (p. 468). Some 98 percent of the island is now a part of the National Wilderness Preservation System. This places some limitation on major developments in that a rather lengthy public and Congressional review process must precede them. After construction of the west segment of the Feldtmann loop in the mid-70s, Hugh Beattie and his successor John Morehead made no further plans for an extension of the trail system. Our recommendations have been that remaining "remote" tracts of the park be left in isolation. These are not closed to people, but they are closed to camping, and penetrating them is something most people will not do.

In 1976, visitation during the May–September open season reached 17,000. Pressure on trails and campsites indicated that further increases would place in jeopardy the existing high standard of visitor service and park maintenance. Beattie and Morehead have regarded 20,000 as about the maximum that can be accommodated by present transportation facilities. However one looks at it, public use of the island is at or near carrying capacity.

A major amenity of Isle Royale as a wolf habitat is that it is deserted from November through April. This degree of isolation should help keep the wolves wild. They are unlikely to cultivate a close rapport with people during the summer season. However, winter closure is a matter of degree. For most of the 18 years, we have seen signs of plane landings and have had behavioral evidence that the wolves were being harassed.

Don and Rolf witnessed an example of this on 7 February 1971; it was Rolf's last flight before returning to the campus. Over Houghton Peninsula they saw a Piper Cherokee on wheels, circling at 200–400 feet. It moved up the island clear to Blake Point, pausing over Moskey Basin to buzz two wolves at low level; they ran south into the woods.

The plane came back along the north side to McCargoe Cove, then headed southwest to the end of the island. From here it worked back to Houghton Point and found the pack of 10. The wolves went down the slope and out onto Siskiwit Bay. Don and Rolf followed closely and saw the plane come down over the wolves repeatedly at 50 feet, scattering them to shore and into heavy cover. The Cherokee finally left the island, headed north toward Canada.

We need no explanation why it has sometimes been necessary to circle the packs at a distance for long periods to accustom them again to the research aircraft.

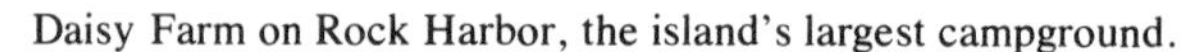

Daisy Farm on Rock Harbor, the island's largest campground.

Each year the park gets requests for winter camping. These have had to be denied. Any change in this policy would greatly degrade the island as a wolf range. Activity on the lakes and bays — in periods when this was possible — would interfere vitally with the travel, killing, and feeding routine of the wolves.[8]

The island obviously needs more winter patrol (early and late), more cooperation between the National Park Service and the Federal Aeronautics Administration, more control of air traffic and illegal activities. The Isle Royale wolves are cherished by a host of people as living memorabilia of an earlier period in our history. In a sense they are biological Americana. We the citizens have surmounted barriers and gone to great expense to provide for them in this one-of-a-kind sanctuary. We have given custody to the National Park Service, and we expect our interest to be served.

In its history as we know it, and in its destiny to come, Isle Royale witnesses a lofty thought in human enterprise. The setting aside of wilderness is more than the wisdom of holding space for recreation and a check area for scientific learning. It is also a matter of principle, of refinement in human character. Man's mastery of the earth pays him greatest tribute when he achieves the humility to leave some of it alone.

In this sense, "preservation" is not an idea to sweep our hard-scrabbling world. But it offers options, and what it does today will have social resonance far beyond our time. A culture universally at want for subsistence or gain will improve its lot in tending its estate.

Beyond the technological junkyard, provident men can ease their tensions and find, for the moment, some promise of security. Their sense of weal is not alone in scenic beauty or uninvaded quiet; it is part of being well off. For that wilderness, that protected remnant of a stability that was, is now a resource bank account, a fund of hoarded assets. It makes no one rich at once; it draws interest reliably, year by year.

For some, an untimely thought. But for many more a comfort we feel amid the luxury of assets unrealized, the opulence of unused space, idle waters, resting lands — and, as the law says, the wildlife therein.

The moose and the wolf need no one to lead them . . .

but only a place to be left alone.

APPENDICES NOTES LITERATURE CITED

APPENDIX I-A ADMINISTRATIVE STAFF, ISLE ROYALE NATIONAL PARK, 1958–76

Superintendents

John G. Lewis	Sept. 1956 - Oct. 1959
George W. Fry	Oct. 1959 - Aug. 1961
Henry G. Schmidt	Aug. 1961 - Jan. 1965
Carlock E. Johnson	Jan. 1965 - Sept. 1967
Bruce J. Miller	Sept. 1967 - March 1968
Hugh P. Beattie	March 1968 - March 1975
John M. Morehead	March 1975 - March 1979

Chief Rangers

Benjamin J. Zerbey	1958 - 1961
John C. Raftery	1961 - 1966
Robert W. Rogers	1966 - 1976

Chief Naturalists

Robert M. Linn	1958 - 1963
William W. Dunmire	1963 - 1966

APPENDIX I-B STUDENT FIELD ASSISTANTS, 1962-76

Bell, Ronald L.
Purdue student, Wildlife Science
Summer 1969: 2 June - 24 August
1970: 9 June - 7 July
1972: 13 June - 8 September
(Ohio State University)

Coble, John A.
Purdue student, Wildlife Science
Summer 1966: 7 June - 31 August
1967: 6 June - 30 August
1968: 11-30 June

Dietz, James M.
Purdue graduate student, Wildlife Science
Winter 1971: 26 January - 11 March
Summer 1971: 18 May - 20 August

Doskocil, Michael J.
Purdue student, Wildlife Science
Summer 1968: 11 June - 5 September

Irmiger, Robert A.
Michigan Technological University, Wildlife Science
Summer 1976: 1 June - 1 September

Keeler, John C.
Purdue student, Wildlife Science
Summer 1967: 6 June - 30 August
1968: 11 June - 28 August
1969: Park seasonal ranger
Winter 1970: Assisted in winter study
29 January - 7 March

Knauer, William A.
Purdue student, Wildlife Science
Summer 1966: 7 June - 14 August

Kochert, Michael N.
Purdue student, Wildlife Science
Summer 1969: 20 May - 23 July

Lawrence, Timothy C.
  Purdue student, Engineering Science
  Summer 1973: 15 May - 16 August
Long, Michael T.
  Purdue student, Wildlife Science
  Summer 1962: 5 June - 23 August
    Red squirrel study
Roop, Larry J.
  Purdue student, Wildlife Science
  Summer 1962: 5 June - 23 August
    Snowshoe hare study
  1963: 4 June - 25 August
    Mammal and bird studies
  1964: Park seasonal naturalist
    Assisted on moose study September - October
  1968: Park seasonal naturalist
    Assisted on squirrel studies September - October
  1969: Park seasonal naturalist
    Assisted on moose-wolf work in September - October
Ruckel, Steven W.
  Purdue student, Wildlife Science
  Summer 1969: 2 June - 24 August
Scheidler, Joseph M.
  Purdue student, Wildlife Science
  Summer 1974: 5 May - 14 August
  1975: 13 May - 14 August
  1976: 26 May - 29 October
Seitz, William K.
  Iowa State University student, Wildlife Science
  Summer 1964: 2 June - 26 August
    Beaver study
  1965: 6 June - 26 August
Simpson, Philip W.
  Purdue student, Wildlife Science
  Summer 1972: 13 June - 18 August
  1973: 15 May - 25 July
Stauber, Erik
  Purdue student, Veterinary Medicine
  Summer 1963: 4 June - 3 September
    Birds of prey study
  1964: 2 June - 5 August
  1965: 6 June - 14 August
Stephens, Philip W.
  Michigan Technological University, Biological Science
  Summer 1976: 1 June - 1 September
Vanada, John D.
  Purdue student, Wildlife Science
  Summer 1970: 9 June - 4 September

Woolington, James D.
Purdue student, Wildlife Science
Summer 1974: 5 May - 14 August
1975: 13 May - 1 September

Wrighthouse, Michael W.
Purdue student, Wildlife Science
Summer 1973: 26 July - 16 August

APPENDIX I-C FIELD SCHEDULE, PERSONNEL, AND COOPERATORS, 1958-76*

1958 — Summer: L. David Mech, 28 June - 30 August.
Meeting, ad hoc advisors, 28 June - 1 July.
Ontario Department of Lands and Forest: Douglas H. Pimlott, Roger Stanfield.
Michigan Department of Conservation: Raymond D. Schofield.
Minnesota Department of Conservation: Milton H. Stenlund.
U.S. Fish and Wildlife Service: Laurits W. Krefting.
National Park Service: C. Gordon Fredine, John G. Lewis, Robert M. Linn, Robert H. Rose.
Purdue University: Durward L. Allen.

1959 — Winter: L. David Mech, 3 February - 14 March.
Pilots: Jack Burgess, 3-11 February; Donald E. Murray, 11 February - 3 March; Lee Schwartz, 3-14 March.
IRNP: Robert M. Linn, 3 February - 3 March; David G. Stimson, 3-14 March.
Summer: L. David Mech, 7 May - 19 August; 27 October - 1 November (fall pilot Jack Burgess).

1960 — Winter: L. David Mech, 3 February - 22 March.
Pilot: Donald E. Murray
IRNP: David G. Stimson, 3-17 February; Robert M. Linn, 17 February - 8 March; Benjamin J. Zerbey, 8-22 March.
Summer: L. David Mech, 9 May - 1 September; Philip C. Shelton, 18 June - 2 September, 11 - 17 October (fall pilot Jack Burgess).

1961 — Winter: L. David Mech, 30 January - 21 March; Durward L. Allen, 30 January - 25 February.
Pilot: Donald E. Murray
IRNP: Roy W. Stamey, 30 January - 17 February; Peter L. Parry, 17 February - 8 March; Benjamin J. Zerbey, 8-25 March.

* My own irregular trips to the island, varying from 3 days to 2 weeks, in spring, summer, or fall are not included — DLA

Summer: Philip C. Shelton, 15 June - 8 November; L. David Mech, 10 May - 15 June (fall pilot Jack Burgess).

1962 — Winter: Philip C. Shelton, 30 January - 20 March; Durward L. Allen, 30 January - 1 March.

Pilot: Donald E. Murray

IRNP: Richard W. Igo, 30 January - 5 February; Robert M. Linn, 5-23 February; Henry G. Schmidt, 23 February - 1 March; John C. Raftery, 1-7 March; Peter L. Parry, 7-20 March.

Summer: Philip C. Shelton, 8 May - 12 December (fall pilot William Caza).

Summer students and assistants: Michael T. Long, 5 June - 23 August; Larry J. Roop, 5 June - 23 August.

1963 — Winter: Philip C. Shelton, 28 January - 21 March; Durward L. Allen, 28 January - 26 February.

Pilot: Donald E. Murray

IRNP: Richard W. Igo, 28 January - 8 February; John C. Raftery, 8-21 February; Peter L. Parry, 21 February - 5 March; Robert L. Peterson, 5-21 March.

Summer: Philip C. Shelton, 13 May - 19 September (employed by IRNP); Peter A. Jordan, 4 August - 27 October (fall pilot William J. Martila).

Summer students and assistants: Larry J. Roop, 4 June - 25 August; Erik Stauber, 4 June - 3 September.

1964 — Winter: Peter A. Jordan, 31 January - 25 March; Durward L. Allen, 31 January - 26 February.

Pilot: Donald E. Murray

IRNP: Richard W. Igo, 31 January - 13 February; William W. Dunmire, 13-26 February; David R. Kangas, 26 February - 11 March; William Bromberg, 11-25 March.

Summer: Peter A. Jordan, 4 May - 29 October (fall pilot William J. Martila).

Summer students and assistants: William K. Seitz, 2 June - 26 August; Erik Stauber, 2 June - 5 August; Larry J. Roop (summer employee of park), 1 September - 1 November.

1965 — Winter: Peter A. Jordan, 2 February - 23 March; Durward L. Allen, 2-26 February.

Pilot: Donald E. Murray

IRNP: Jon B. Abrams, 2-15 February; William W. Dunmire, 15 February-3 March; Richard W. Igo, 3-12 March; Jon B. Abrams, 12-21 March.

Summer: Peter A. Jordan, 16 May - 27 October (fall pilot William J. Martila).

Summer students and assistants: William K. Seitz, 6 June - 14 August; Erik Stauber, 6 June - 14 August.

1966 — Winter: Peter A. Jordan, 2 February - 24 March; Durward L. Allen, 2 February - 1 March.

Pilot: Donald E. Murray

IRNP: Richard W. Igo, 2-7 February; David R. Kangas, 7-18 February; William W. Dunmire, 1-12 March; Jon B. Abrams, 12-20 March.

Summer: Peter A. Jordan, 4 May - 12 June; Wendel J. Johnson, 7 June - 30 September.

Summer students and assistants: John A. Coble, 7 June - 31 August; William A. Knauer, 7 June - 14 August.

1967 - Winter: Wendel J. Johnson, 2 February - 21 March; Michael L. Wolfe, 11 February - 21 March; Durward L. Allen, 2 February - 2 March.

Pilot: Donald E. Murray

IRNP: Richard W. Igo, 2-11 February; Zeb V. McKinney, 22 February - 2 March; Robert W. Rogers, 2-11 March; Warner M. Forsell, 11-21 March.

Grand Portage NM: C. Newton Sikes, 11-22 February.

Summer: Wendel J. Johnson, 9 May - 31 October; Michael L. Wolfe, 9 May - 31 October (fall pilot William J. Martila).

Summer students and assistants: John A. Coble, 6 June - 30 August; John C. Keeler, 6 June - 30 August.

1968 — Winter: Michael L. Wolfe, 30 January - 17 March; Wendel J. Johnson, 30 January - 17 March; Durward L. Allen, 30 January - 29 February.

Pilot: Donald E. Murray

IRNP: Alvin E. Olson, 30 January - 2 February; Zeb V. McKinney, 14-24 February; Bruce J. Miller, 24-29 February; Warner M. Forsell, 29 February - 10 March; Richard W. Igo, 10-17 March.

Grand Portage NM: C. Newton Sikes, 2-14 February.

Summer: Wendel J. Johnson, 6 May - 31 October; Michael L. Wolfe, 6-25 May, 3-17 September, 8-31 October (fall pilot William J. Martila).

Summer students and assistants: John A. Coble, 11-30 June; Michael J. Doskocil, 11 June - 5 September; John C. Keeler, 11 June - 28 August; Larry J. Roop (summer employee of park) 5-24 October.

1969 - Winter: Michael L. Wolfe, 31 January - 14 March; Durward L. Allen, 31 January - 22 February. Note: MGM - Canadian Film Board group in camp 6-18 February.

Pilots: Donald E. Murray; Donald E. Glaser

IRNP: Richard W. Igo, 31 January - 3 February; Donald S. Anderson, 3-22 February; Warner M. Forsell, 22 February - 14 March.

Summer: Michael L. Wolfe, 20 May - 5 August, 29 September - 26 October; Philip C. Shelton 3-23 October (fall pilot William J. Martila).

Summer students and assistants: Michael N. Kochert, 20 May - 23 July; Ronald L. Bell, 2 June - 24 August; Steven W. Ruckel, 2 June - 24 August; Larry J. Roop, (summer employee of park) 18-26 October.

1970 — Winter: Michael L. Wolfe, 29 January - 17 February; Durward L. Allen, 14 February - 14 March; John C. Keeler, 29 January - 7 March.

Pilot: Donald E. Murray

IRNP: William E. Dohrn, 29 January - 2 February; Donald S. Anderson, 2-17 February; Alan D. Eliason, 17 February - 4 March; Zeb V. McKinney, 4-14 March.

Summer: Rolf O. Peterson, 14 June - 27 August.

Summer students and assistants: Ronald L. Bell, 9 June - 8 July; John D. Vanada, 9 June - 4 September.

1971 — Winter: Durward L. Allen, 26 January - 11 March; James M. Dietz, 26 January - 11 March; Rolf O. Peterson, 26 January - 7 February.

Pilot: Donald E. Murray

IRNP: William E. Dohrn, 26 January - 7 February; Charles F. Atwood, Jr., 7-15 February; Alan D. Eliason, 15-24 February; Hugh P. Beattie, 24 February - 3 March; Frank J. Deckert, 3-11 March; Irving L. Dunton, 3-11 March.

Summer: Rolf O. Peterson, 8 June - 7 September.

Summer students and assistants: James M. Dietz, 18 May - 20 August.

1972 — Winter: Rolf O. Peterson, 23 January - 10 March; Durward L. Allen, 23 January - 29 February; Fredrick H. Montague, 23 January - 5 February.

Pilot: Donald E. Murray

IRNP: William E. Dohrn, 23-28 January; Frank J. Deckert, 28 January - 5 February; Alan D. Eliason, 5-19 February; Hugh P. Beattie, 19-22 February; Richard E. Hoffman, 29 February - 10 March.

Grand Portage N. Monument: Arnold J. Long, 22 February - 2 March.

Summer: Rolf O. Peterson, 9 May - 11 November (fall pilot Robert R. Mohr).

Summer students and assistants: Ronald L. Bell, 13 June - 8 September; Philip W. Simpson, 13 June - 18 August.

1973 — Winter: Rolf O. Peterson, 24 January - 16 March; Durward L. Allen, 24 January - 3 March; Fredrick H. Montague, 24 January - 2 February.

Pilot: Donald E. Murray

IRNP: William E. Dohrn, 24 January - 2 February; Alan D. Eliason, 12-21 February; Richard E. Hoffman, 21 February - 3 March; Irving L. Dunton, 3-16 March.

Pictured Rocks N. Lakeshore: Fred H. Young, 2-12 February.

Summer: Rolf O. Peterson, 4 May - 19 November; Philip C. Shelton, 14 - 20 October (fall pilot Robert R. Mohr).

Summer students and assistants: Timothy C. Lawrence, 15 May - 17 August; Philip W. Simpson, 15 May - 25 July; Michael W. Wrighthouse, 26 July - 17 August.

1974 — Winter: Rolf O. Peterson, 23 January - 17 March; Durward L. Allen, 23 January - 1 March.

Pilot: Donald E. Murray

IRNP: William E. Dohrn, 23-31 January; Carl M. Fleming, 31 January - 9 February; Ivan R. Tolley, 9 February - 1 March; Dale Peterson, 1-17 March.

Summer: Rolf O. Peterson, 29 April - 31 October; Philip C. Shelton, 10 September - 24 October (fall pilot Robert R. Mohr).

Summer students and assistants: Carolyn C. Peterson, 29 April - 31 October; Joseph M. Scheidler, 5 May - 14 August; James D. Woolington, 5 May - 14 August.

1975 — Winter: Rolf O. Peterson, 26 January - 8 March; Durward L. Allen, 26 January - 28 February.

Pilot: Donald E. Murray

IRNP: Ivan R. Tolley, 26-30 January, 7-19 February; Carl M. Fleming, 30 January - 7 February, 28 February - 8 March.

Grand Portage N. Monument: W. Michael Quick, 19-28 February.

Summer: Rolf O. Peterson, 29 April - 1 September, 21 September - 30 October (fall pilot John Brandrup).

Summer students and assistants: Carolyn C. Peterson, 29 April - 1 September, 21 September - 30 October; Joseph M. Scheidler, 13 May - 13 August; James D. Woolington, 13 May - 1 September.

1976 — Winter: Rolf O. Peterson, 26 January - 13 March.

Pilot: Donald E. Murray

IRNP: Warren L. Rigby, 26 January - 5 February; John M. Morehead, 5-19 February; Donald S. Anderson, 19-25 February; Delbert C. Galloway, 25 February - 6 March; Noel R. Poe, 6-13 March.

Summer: Joseph M. Scheidler, 26 May - 29 September (fall pilot John Brandrup).

Summer students and assistants: Lee Scheidler, 26 May - 29 September; Robert A. Irmiger, 1 June - 1 September; Philip W. Stephens, 1 June - 1 September.

APPENDIX II WOODY PLANTS AS MOOSE BROWSE

Trees and shrubs of Isle Royale are here roughly categorized as food for moose, based on unsystematic observations of preferences and occurrence in various habitats.

Class 1. Highly Preferred

Abundant

Balsam fir	*Abies balsamea*
Quaking aspen	*Populus tremuloides*
Paper birch	*Betula papyrifera*

Less plentiful

Mountain ash	*Sorbus americana*
Juneberry	*Amelanchier sp.*
Red-osier dogwood	*Cornus stolonifera*
Red maple	*Acer rubrum*
Fire cherry	*Prunus pennsylvanica*
Squashberry	*Viburnum edule*
Willow	*Salix spp.*
Black ash	*Fraxinus nigra*

Scarce

American yew	*Taxus canadensis*

Note: It is possible that other less plentiful or less available (because of age) species should be listed here, such as balsam poplar, highbush cranberry, and yellow birch. My own observations, on which this is based, were not intensive enough to decide the position of these. This applies also to the plentiful bush honeysuckle, which is a summer food, although it is a low shrub and provides little in winter. Chokecherry is much less common than fire cherry, but the stands on Feldtmann Ridge seem to be equal in acceptability to the latter. Practically all of the Class 1 list are used both summer and winter, with the exception of balsam fir, exclusively a winter food. The highly palatable red-osier dogwood is browsed in summer and fall, but the season's crop probably is entirely gone by winter. Aspen is notable in furnishing important food year round: leaves in summer, twigs and buds in winter, and bark (from green blowdowns) in spring.

Class 2. Acceptable but Less Preferred

White cedar	*Thuja occidentalis*
White pine	*Pinus strobus*
Mountain maple	*Acer spicatum*
Sugar maple	*Acer saccharum*
Mountain alder	*Alnus crispa*
Red elderberry	*Sambucus pubens*
Beaked hazel	*Corylus cornuta*

Class 3. Seldom Taken

White spruce	*Picea glauca*
Black spruce	*Picea mariana*
Ground juniper	*Juniperus communis*
Ninebark	*Physocarpus opulifolius*
Speckled alder	*Alnus rugosa*
Thimbleberry	*Rubus parviflorus*
Sweet gale	*Myrica gale*

Note: The above lists include most of the more common trees and shrubs of the island, with certain exceptions. I have made no personal appraisal of the extent of browsing on many small shrubs, such as blueberry and rose, which are available only in summer. I have no notes on jack pine, which grows on dry ridges little frequented by moose in winter.

By far the most thorough studies of moose food habits on the island have been made by Krefting over a period of nearly 30 years. He has summarized this information in his bulletin on the ecology of Isle Royale moose. Included also is available information on the extensive feeding by moose on herbaceous and aquatic plants in summer and a review of the literature in this field. More recently, Snyder and Janke made comparisons of forest composition and growth on browsed and unbrowsed areas of Isle Royale and outlying islands.

APPENDIX III AGE AND SEX OF KNOWN AND PROBABLE WOLF KILLS (WINTER), 1959–1976 (MALE, FEMALE, UNKNOWN)*

	1959			1960			1961			1962			1963			1964		
Age	M	F	U	M	F	U	M	F	U	M	F	U	M	F	U	M	F	U
Calf			4	1	1	8			6			2		2	2		3	5
1																2		
2																		
3										1								
4		1																
5																		
6	1				2			4			1							
7		1	1	2	2		1			1	1							
8			1		1		1			1	2			1				
9					1		1	1		1	3		1	2				
10		2					1	1		1	1		1	2	1	2	2	
11								1		1	1		2	1		1	1	
12							1	3		3			1	1			5	
13					1		1	1						2				
14																		
15																		
16																		
17																		
18																		
19																		
20																		
Unk			2				1		3									1
Total	1	4	8	3	8	8	7	11	9	9	9	2	5	11	3	5	11	6

* Includes winter-spotted kills and random pick-ups.

	1965			1966			1967			1968			1969			1970		
Age	M	F	U	M	F	U	M	F	U	M	F	U	M	F	U	M	F	U
Calf		1	2	1	1	2			1		1	1	1	3	13	2	3	3
1	2		1		2					1			1	1		1	2	
2										1								
3				1	2				1								1	
4				1														
5								1										
6	1				1					1								
7							1			1			1	1				
8								1		1			2	2		1		
9		1																
10					1						1		1	3				
11	2	3				1	1			1				1				
12		2		1	2			2										
13	1	1			1		1	1		1	1			1		1		
14	1	1			1		1	2						1		1		
15	1	1										1						
16																		
17														1				
18																		
19																		
20																		
Unk			3		1	3					1			1				
Total	8	10	6	4	12	6	4	7	2	7	4	2	6	15	13	6	6	3

Age	1971			1972			1973			1974			1975			1976		
	M	F	U	M	F	U	M	F	U	M	F	U	M	F	U	M	F	U
Calf	6	12	10	5	7	16	4	2	6		8	17	2	3	7		2	16
1				3	2		4	1	1		1							
2	1	1		5			1	4		1	1		1	2	2	1	1	2
3	2	1		1	3		1	2					1		1			
4	1						1	1		1				2				
5				1						2						1	1	
6		1	1		3		1			1	1		1					
7	1	1			2		1			1				4		2	1	
8	1	2		1	1		1			2			1	3		1	3	
9	2				3		1			1	1	1	1	2		2	3	
10	4						1			1	1		2	1		1	4	1
11		1		2				3			1	1		3		1	1	
12							1	1			1	1		1		2	1	
13	1							2		2			1	1		2	2	
14														3	1	1	1	
15				1							1			1			2	
16					1						2			1			1	
17											1		1					
18										1								
19																		
20																		
Unk	1					1		1	2						2			3
Total	20	19	11	19	22	17	17	17	9	13	19	20	11	27	13	14	23	22

APPENDIX IV NOTES ON WINTER WEATHER, SNOW AND ICE CONDITIONS, FRUIT CROPS, AND WINTER BIRDS

1959 Snow depth 16-24"; ice to Canada intermittently solid; plenty of ice in bays; good flying weather.

1960 Snow depth 12-16"; intermittent ice to Canada; warm; harbors and bays not frozen until March. A few redpolls seen.

1961 Snow on ground 20-26"; ice bridge solid mid-Feb.-21 Mar.; good shelf ice; cold; bays and harbors solidly frozen. Heavy crop of birch seed; many redpolls, some siskins.

1962 Snow depth 17-22" increasing to 25" by mid-Feb.; ice to Canada solid in Feb.-Mar.; much shelf ice. A year of much conifer blowdown. Birch seed scarce, a few siskins.

1963 Snow on ground 16-24", 19 Mar. new snow to 36". Ice to Canada solid for most of period. Open leads 3 Feb. Snow with 10" base and crust. Cold winter; good shelf ice. Good crop birch seed, abundant mt. ash. Many small fringillids, mostly redpolls, a few siskins; many pine grosbeaks on Siskiwit Islands.

1964 Snow depth 20-22", soft and powdery to ground, bay ice open, warm winter. Nearly total failure of birch and mt. ash. No small fringillids or grosbeaks.

1965 Snow depth Windigo 2 Feb. 25-30"; 3" fell 5 Feb.; 9 Feb. thaw to 2 feet; 10 Feb. a foot new snow, tot. 40"; 12 Feb. 3 ft. on level down to 30" on 17th; 3-4" new snow 20 Feb.; 1 Mar. 27" crusty and melting, sleet. Good crop birch seed, excellent mt. ash, conifer cones. Good bird year; mixed flocks siskins, redpolls, a few goldfinches; some wh. winged crossbills, a flock wintering robins.

1966 Snow over 3 feet; 9 Feb. big thaw, ice storm. Shifting floating ice and open leads to Canada. No mt. ash, little birch seed, a few spruce cones. No small fringillids.

1967 Snow depth 23-33", over 30" much of Feb., 25" 21 Mar. After 1st week the most extensive and level ice field we have seen to Canada. Coldest winter in 9 years. Snow fluffy and difficult. Quiet water and much ice everywhere. No mt. ash, little birch seed, few cones, fair crop green alder. Rarely a few redpolls seen.

1968 Snow depth 9.3-15″, declining to 13″; ice Washington H. 17.5″. Ice to Canada intact to 14 Feb., then open leads; froze over again 3d week; much wind; little bay or shelf ice. Excellent fruit and mast year: birch, mt. ash; spruce, balsam, cedar cones. Many siskins (no redpolls), a few goldfinches, small flocks of purple finches, red cro sbills, pine and evening grosbeaks.

1969 Depth greatest yet, 44″ declining to 32″ by Mar.; floes and skim ice to Canada; lakes slushy, no shore ice, bays open. Jan. ice storm stripped birches of seed; mt. ash and cone crops a failure. Seed-eating birds rare. A few pine grosbeaks.

1970 Snow depth ''normal,'' 23″ on 31 Jan., 22″ on 23 Feb., ice bridge intact to mid-month and then open leads; much slush on harbors and lakes; much wind. Failure of birch, mt. ash, and conifer cones. Some seeds on green alder. A few redpolls. No other small fringillids.

1971 New snow 4 Feb. to 36″ depth; down to 29″ on 15 Feb.; ice bridge intact from 2 Feb.; windy, with much slush. High production of mt. ash; small crop birch seeds. Many goldfinches, a few siskins, purple finches, evening grosbeaks.

1972 On 23 Jan. 35″ (3d winter of deep snow in last 4), mid-Feb. 28″, to 40″ on 10 Mar.; ice to Canada formed in late Feb., high winds, slush. Little white birch seed, some yellow; no mt. ash fruit. A few redpolls, siskins, pine grosbeaks.

1973 To 22 Feb. average snow about 24″, to 34″ in late Feb., packed and crusty. Warm winter; N. channel frozen 8 Feb. but partly broken up a week later; ice out of Siskiwit Bay 13 Feb.; rain and thawing in March. Nearly impossible working conditions. No mt. ash; birch largely a failure. No small fringillids. A few (budding) pine grosbeaks.

1974 Snow depth 23″ on 23 Jan., soft and fluffy, later ranging to 30″; N. channel freezing 4 Feb.; good shelf ice; bays well frozen, breaking up in thaw after 26 Feb. and into March; then crusty snow. Fair crop birch seed, good crop of alder; mt. ash gone in Windigo area; locally plentiful on S. shore. Small flocks siskins and a few redpolls (dozen or less); a few pine and evening grosbeaks.

1975 Snow depth 26″ on 28 Jan., 36″ on 12 Feb. to 25″ last week of Feb.; little ice to Canada, gone by 21 Feb.; 10″ ice on Washington H., least ice and worst conditions we have seen. Little birch seed, no mt. ash, few cones. Occasional flocks redpolls.

1976 Range of snow depth 22-39″, average about 28″. On 26 Jan. packed snow base 20″ plus 6″ new snow. On 11 Mar. 9″ new snow, 39″ accum. Lakes slushy. Channel mostly frozen, with open leads by 10 Mar. Fair crop birch seeds; some redpolls.

APPENDIX V TECHNICAL NAMES OF PLANTS AND ANIMALS

In this listing, vernacular names are arranged alphabetically and identified to family, genus, species, or subspecies, as may be appropriate for present purposes. Forms of life for which no common name is known are identified sufficiently in the text and are not included here.

Alder, green, mountain *Alnus crispa*
speckled *A. rugosa*
Antelope, pronghorn *Antilocapra americana*
Ash, black *Fraxinus nigra*
mountain, see mountain ash
Aspen, quaking *Populus tremuloides*
Aster, bigleaf *Aster macrophyllus*
Balsam fir, balsam *Abies balsamea*
Balsam poplar, balm of Gilead *Populus balsamifera*
Baneberry, red *Actaea rubra*
white *A. alba*
Barasingha *Cervus duvauceli*
Bat, red *Lasiurus borealis*
Beadlily, blue *Clintonia borealis*
Bear, black *Ursus americana*
grizzly *U. horribilis*
Bearberry *Arctostaphylos uva-ursi*
Beaver *Castor canadensis*
Birch, paper, white *Betula papyrifera*
yellow *B. lutea*
Bittern, American *Botaurus lentiginosus*
Blueberry *Vaccinium*
Bluebird *Sialia sialis*
Bluejoint, Canada blue joint *Calamagrostis canadensis*
Bobcat *Lynx rufus*
Bracken, bracken fern *Pteridium aquilinum*
Budworm, spruce *Archips fumiferana*
Buffalo *Bison bison*
Buffaloberry *Shepherdia canadensis*
Bunchberry, Canada dogwood *Cornus canadensis*
Bunting, indigo *Passerina cyanea*
snow *Plectrophenax nivalis*
Bur reed *Sparganium angustifolium*
Caribou *Rangifer tarandus*
Cedar, northern white *Thuja occidentalis*
Cheetah *Acinonyx jubatus*
Cherry, choke *Prunus virginiana*
fire, pin *P. pennsylvanica*
Chickadee, black-capped *Parus atricapillus*
Clover, white *Trifolium repens*
Commandra, northern *Commandra livida*
Cormorant, double-crested *Phalacrocorax auritus*
Cougar, mountain lion *Felis concolor*
Coyote *Canis latrans*
Cranberry, highbush *Viburnum opulus*
large *Vaccinium macrocarpum*
small *V. oxycoccus*
Crossbill, red *Loxia curvirostra*
white-winged *L. leucoptera*
Crow *Corvus brachyrhynchos*
Crowberry *Empetrum nigrum*
Currant *Ribes*
Dandelion *Taraxacum officinale*
Deer, red *Cervus elaphus*
roe *Capreolus capreolus*
white-tailed *Dama virginiana*
Deermouse, see woodmouse
Devil's club *Oplopanax horridum*
Dog, African wild, cape hunting *Lycaon pictus*
Dogwood, Canada, bunchberry *Cornus canadensis*
red-osier *C. stolonifera*
Duck, black, blackduck *Anas rubripes*
goldeneye *Bucephala clangula*

mallard *Anas platyrhynchos*
oldsquaw *Clangula hyemalis*
ring-necked *Aythya collaris*
wood *Aix sponsa*
Eagle, bald *Haliaetus leucocephalus*
golden *Aquila chrysaetos*
Eelgrass, wild celery *Valisneria americana*
Elder, red-berried *Sambucus pubens*
Ermine, see short-tailed weasel
Everlasting, pearly *Anaphalis margaritacea*
Falcon, peregrine, duck hawk *Falco peregrinus*
Finch, purple *Carpodacus purpureus*
see goldfinch
Fir, balsam *Abies balsamea*
Fireweed *Epilobium angustifolium*
Flicker, yellow-shafted *Colaptes auratus*
Fly, black *Simulium venustum*
moose *Lyperiosiops alcis*
Flycatcher, olive-sided *Nuttalornis borealis*
yellow-bellied *Empidonax flaviventris*
Fox, red *Vulpes fulva*
Frog, green *Rana clamitans*
mink *R. septentrionalis*
western chorus *Pseudacris nigrita*
wood *Rana sylvestris*
Goldfinch *Spinus tristis*
Goose, Canada *Branta canadensis*
blue and snow *Chen caerulescens*
Gooseberry *Ribes*
Goshawk *Accipiter gentilis*
Grackle, common *Quiscalus quiscula*
Grebe, horned *Podiceps auritus*
Grosbeak, evening *Hesperiphona vespertina*
pine *Pinicola enucleator*
Ground hemlock, American yew *Taxus canadensis*
Ground pine, club moss *Lycopodium* or *Selaginella*
Grouse, ruffed *Bonasa umbellus*
sharp-tailed *Pedioecetes phasianellus*
spruce *Canachites canadensis*
Gull, herring *Larus argentatus*
Hare, snowshoe *Lepus americanus*
Harebell *Campanula rotundifolia*
Hawk, broad-winged *Buteo platypterus*
Cooper's *Accipiter cooperi*
duck, see falcon, peregrine
marsh *Circus cyaneus*
pigeon, see merlin
red-tailed *Buteo jamaicensis*
sharp-shinned *Accipiter striatus*
sparrow, see kestrel
Hawthorne *Crataegus*
Hazel, beaked *Corylus rostrata*
Hepatica *Hepatica americana*
Heron, great blue *Ardea herodias*
Honeysuckle, dwarf bush *Diervilla lonicera*
Horsetail *Equisetum*
Hyena, spotted *Crocuta crocuta*
striped *Hyena vulgaris*
Indian pipe *Monotropa uniflora*
Iris, blue flag *Iris versicolor*
Ivy, poison *Rhus radicans*
Jay, blue *Cyanocitta cristata*
Canada, gray, see whiskeyjack
Junco, slate-colored *Junco hyemalis*
Juneberry, serviceberry *Amelanchier*
Juniper, creeping *Juniperus horizontalis*
ground *J. communis*
Kestrel, sparrow hawk *Falco sparverius*
Kinglet, ruby-crowned *Regulus calendula*
Kob, Uganda *Adenota kob thomasi*
Komodo dragon, monitor *Varanus komodoensis*
Lark, horned *Eremophila alpestris*
Leatherleaf *Chamaedaphne calyculata*
Lemming *Lemmus, Dicrostonyx*
Leopard *Panthera pardus*
Lily, yellow pond *Nuphar variegatum*
white water *Nymphaea odorata*
Lily-of-the-valley, wild *Maianthemum canadensis*
Lion, African *Panthera leo*
mountain, puma, see cougar
Longspur, Lapland *Calcarius lapponicus*
Loon *Gavia immer*
Macaque, gray *Macaca irus*
Magpie *Pica pica*
Maple, mountain *Acer spicatum*
red *A. rubrum*
sugar, hard *A. saccharum*
Marsh marigold *Caltha palustris*
Marten *Martes americana*
Meningeal worm *Paraelaphostrongylus tenuis*
Merganser, American *Mergus merganser*
hooded *Lophodytes cucullatus*
red-breasted *Mergus serrator*
Merlin, pigeon hawk *Falco columbarius*
Mink *Mustela vison*
Moose *Alces alces*
Morel mushroom *Morchella*
Mosquito *Culicidae*
Mountain ash *Sorbus americana*
Mouse, see woodmouse
Meadow *Microtus pennsylvanicus*
White-footed *Peromyscus leucopus*
Muskrat *Ondatra zibethica*
Nighthawk *Chordeiles minor*
Ninebark *Physocarpus opulifolius*
No-see-um, midge, sandfly, punky *Culicoides*

Nuthatch, red-breasted *Sitta canadensis*
Oak, red *Quercus rubra*
Orchid, calypso *Calypso bulbosa*
Osprey *Pandion haliaetus*
Otter *Lutra canadensis*
Owl, great horned *Bubo virginianus*
  snowy *Nyctea scandiaca*
Parsnip, cow *Heracleum lanatum*
Peeper, spring *Hyla crucifer*
Pine, jack *Pinus banksiana*
  red *P. resinosa*
  white *P. strobus*
Pipit, water *Anthus spinoletta*
Pipsissewa *Chemaphila umbellata*
Plover, black-bellied *Squatarola squatarola*
  golden *Pluvialis dominica*
Polygala, fringed *Polygala paucifolia*
Pondweed *Potamogeton*
Poplar, balsam, balm of Gilead *Populus balsamifera*
  see aspen
Porcupine *Erethizon dorsatum*
Pronghorn, see antelope
Pyrola *Pyrola*
Raccoon *Procyon lotor*
Raspberry, red *Rubus idaeus*
Raven *Corvus corax*
Redpoll *Acanthus flammea*
Redwood, coastal *Sequoia sempervirens*
  giant sequoia *S. washingtonia*
Robin *Turdus migratorius*
Rose, prickly *Rosa acicularis*
Rush *Juncaceae*
Sarsaparilla *Aralia nudicaulis*
Sawfly, larch *Pristiphora erichsonii*
Scoter, white-winged *Melanitta deglandi*
Sedge *Cyperaceae*
Sequoia, giant *Sequoia washingtonia*
  coastal, see redwood
Serviceberry, see Juneberry
Sheep, Dall *Ovis dalli*
Siskin, pine *Spinus pinus*
Skunk, striped *Mephitis nigra*
Skunk cabbage *Symplocarpus foetidus*
Smelt, American *Osmerus mordax*
Snake, garter *Thamnophis sirtalis*
  red-bellied *Storeria occipitomaculata*
Snipe, Wilson *Capella gallinago*
Snowberry *Symphoricarpos albus*
  creeping *Gaultheria hispidula*
Solomon's seal *Smilacina trifolia*
Sparrow, chipping *Spizella passerina*
  field *S. pusilla*
  fox *Passerella iliaca*
  song *Melospiza melodia*
  tree *Spizella arborea*
  vesper *Pooecetes gramineus*
  white-crowned *Zonotrichia leucophrys*
  white-throated *Z. albicollis*
Spruce, black *Picea mariana*
  white *P. glauca*
Squashberry *Viburnum edule*
Squirrel, red, pine *Tamiasciurus hudsonicus*
  Isle Royale red *T. h. regalis*
Starling *Sturnus vulgaris*
Strawberry, barren *Waldsteinia fragarioides*
  wild *Fragaria virginiana*
Sturgeon, lake *Acipenser*
Sucker, white *Catostomus comersonnii*
Swallow, tree *Iridoprocne bicolor*
Swan, whistling *Olor columbianus*
Sweet gale *Myrica gale*
Tamarack *Larix laricina*
Tapeworm, hydatid *Echinococcus granulosus*
Tea, Labrador *Ledum groenlandicum*
Teal, blue-winged *Anas discors*
  green-winged *A. carolinensis*
Thimbleberry, salmon berry *Rubus parviflorus*
Thrush, Swainson's *Hylocichla ustulata*
  wood *H. mustelina*
Tick, winter *Dermacentor albipictus*
Tiger *Panthera tigris*
Timothy *Phleum pratense*
Toad, American *Bufo terrestris*
Turtle, western painted *Chrysemys picta*
Twinflower *Linnaea borealis*
Twistedstalk, rosy *Streptopus roseus*
Veery *Hylocichla fuscescens*
Vireo, red-eyed *Vireo olivaceus*
Warbler, black-throated green *Dendroica virens*
Water shield *Brasenia schreberi*
Weasel, long-tailed *Mustela frenata*
  short-tailed, ermine *M. erminea*
Whale, blue *Sibbaldus musculus*
  killer *Orcinus orca*
Whiskeyjack, gray or Canada jay *Perisoreus canadensis*
Wildebeest *Connochaetes taurinus*
Willow *Salix*
Wintergreen *Gaultheria procumbens*
Wolf, gray *Canis lupus*
  eastern timber *C. l. lycaon*
  northern Rocky Mountain *C. l. irremotus*
Woodcock *Philohela minor*
Woodmouse, deermouse *Peromyscus maniculatus*
Woodpecker, downy *Dendrocopos pubescens*
  hairy *D. villosus*
  pileated *Dryocopus pileatus*
Wren, winter *Troglodytes troglodytes*
Yew, American *Taxus canadensis*

Notes

Prelude How It All Got Started

1. This cast was made by ranger D. Robert Hakala on Washington Creek, 7 May 1952. He saw the first two wolves the following October.

Chapter 1 The Founders

1. Fisherman Milford Johnson, of Crystal Cove, told David Mech that he spent the winters of 1924, 1925, and 1931 on Isle Royale. Each winter he saw the tracks of a single timber wolf.

2. The details of political maneuvering by exploitive interests and the great campaign for preservation of Yellowstone as public property have been ably documented by Reiger. Much credit belongs to George Bird Grinnell, editor of *Forest and Stream,* a founder of the Boone and Crockett Club. Protective legislation for Yellowstone National Park was one of the first projects of the club (1888) and its dynamic president, Theodore Roosevelt (see Trefethen).

3. This correspondence was in connection with Dr. McCleery's efforts to get from Biological Survey trappers young wolves from the northern plains before the "buffalo wolves" of that region became extinct. The animals he obtained were mostly from Montana and were propagated at his farm in Pennsylvania. The work was later taken over by Jack and Marjorie Lynch, who moved the operation to Gardiner, Washington. I am indebted to Russell A. Fink, of the McCleery foundation for this reference from their files.

4. Evidently this policy of protecting predators was not without its exceptions. In reviewing the history of wolf control in and around Mount McKinley National Park, Haber stated that from 1917, when the park was established, until 1952 it was policy to control wolves in the park, which he interpreted to mean killing as many as possible. Two predator control agents of the Fish and Wildlife Service were assigned to the park and adjacent areas. After 1952 there was evidence of continued wolf killing by poachers (97).

5. A more recent attempt to restore wolves to Michigan's upper peninsula failed through human antipathy and intervention (285). This will inevitably be a major obstacle to the reintroduction of wolves in any area that might become ecologically suitable. The homing tendencies, even of captive-reared animals, must also be taken into account, as was evident in an experiment at Barrow, Alaska (104).

Chapter 2 Moose Haven

1. The browse line that is so conspicuous on old balsam trees and on lake-edge cedars may be well above 8 feet because the animals were standing on winter snow. In Aldous's browse survey he considered the upper limit of available forage to be 12 feet (5).

2. A climax vegetation type is the final association of plants that will come to occupy an area under given climatic conditions. The process of succession through temporary stages may require long time periods during which sites and soils evolve from extremes of dryness or wetness to intermediate moisture conditions more favorable to the final stage of stability. The association preceding the climax, like the birch-aspen on Isle Royale, is termed a "subclimax." Today on the island we have every stage in plant succession from bare rock on the upland and open water in the lowlands to the two climaxes. These forests denote two climatic zones: the high, dry, and warm hard maple-yellow birch and the low, moist, and cool spruce-fir, as detailed in the studies of Robert M. Linn.

3. Henry Schoolcraft's wife, Jane, was half Ojibwa. It was on legends of the Ojibwas collected by Schoolcraft that Longfellow based his epic poem of Hiawatha.

4. A counterpart of these changes took place in British Columbia after the mid-20s. Forest fires destroyed large areas of mature lowland forest required by the mountain caribou for winter range. The caribou underwent a decline, while major increases occurred in moose and deer (71).

5. According to Hickie, the Michigan legislature gave complete protection to moose in the state in 1889, and hunting has not been legal since then. In frontier communities such as Isle Royale, subsistence hunting was commonly practiced and there was little concern with controlling it. By a series of closing orders beginning in 1925, the Conservation Commission prohibited all hunting and trapping on Isle Royale until the island became a national park in 1940. This policy was inconsistent with the fact that the state had its own coyote trapper on the island in the late 20s (145).

6. Such effects also apply to our own kind. It is interesting to speculate on the physical and cultural vigor that must have attended the first colonization of the Americas by primitive men, who spread across the Bering Plain and invaded the continent to exploit the previously undisturbed megafauna of late Pleistocene times.

7. Among ungulate animals, the Uganda kob is one of the most specialized in its breeding behavior. Rutting males collect on restricted areas strongly reminiscent of the booming grounds of prairie grouse. Each male defends his own small territory against other males and attempts to hold any female that crosses his "property." Helmut K. Buechner, who first described this behavior, said that only 13 known breeding grounds were being used in a range of about 100 square miles. In certain other species of antelope the claiming of breeding territories by males tends to space out the population over the range largely in accordance with available forage. This social and economic system is not so well developed in any of our North American big-game species.

My use of the term *territory* is broad. It refers to an area of habitat claimed for exclusive use by individuals or groups. Necessarily it includes the resources needed for rearing the young and other essential purposes. Aliens are repelled through various social devices that may involve open combat, aggressive displays, covert signals, and scent marking.

8. In his review of coyote history, Krefting indicated that there is no record of the build-up in the period 1918-25. However, since the state placed a

trapper (Bill Lively) on the island to control the species from 1926 to 1929, Krefting logically concludes that they must have been numerous.

9. "Irruption" is a useful and highly descriptive term meaning a great outpouring or increase of animals. The peak of a population cycle is typically irruptive in nature. It is common practice for editors, who think biologists can't spell, to change the *i* to an *e,* thus shifting the scene of action from biology to geology.

10. This valuable bulletin, undated like some others of the period, was published in 1937 by the National Park Service. The work was a follow-up on the significant earlier studies by Holt (with the Adams party) in 1905 and Cooper in 1909-10.

11. For more than 10 years various organizations and individuals in Michigan had worked to make Isle Royale a national park. Among leaders in the movement was Albert Stoll, Jr., of the *Detroit News.* In recognition of these services, the trail now leading from Rock Harbor Lodge northeast to Scoville Point is known as Albert Stoll trail. The park was dedicated in 1940.

12. The hiker through stands of old aspens finds himself tripping over footworn roots that crisscross the trail. The network of sucker roots surrounding male or female trees an inch or so below the soil surface provides sprouts whenever disturbance of any kind — including the wind-throwing of old trees — lets sunlight in to the forest floor. Roots may produce suckers 80 feet from the parent tree and give rise to a clone of genetically similar trees that may cover as much as 35 acres (91).

13. Clifford C. Presnall, Laurits W. Krefting, and Shaler E. Aldous, of the Fish and Wildlife Service, were in the park together on this occasion. It will be evident that the historic record of the vegetation and animal life on Isle Royale from the late 20s to the mid-50s is heavily dependent on the work of Murie and Hickie in Michigan and Aldous and Krefting from the federal bureau. All who do further work in these fields on the island will be indebted for the painstaking and sometimes inadequately supported efforts of these men.

14. There is a voluminous literature on the so-called 10-year cycle of many creatures in the north. I have put together a brief review of the situation elsewhere (7).

15. Krefting's terminology was "yearlings and adults." He undoubtedly was using *yearling* in the sense of "short yearling," or an animal approaching one year of age. According to our standardized usage, employed in this book, a young moose is a calf until it is a year old, which means about 1 June. A yearling is an animal between one and two years old. On its second birthday the yearling becomes an adult.

Chapter 3 Island Laboratory

1. This quotation from historian Reuben G. Thwaites's *The French Regime in Wisconsin-II, 1634-1727,* Wisconsin Historical Collection, vol. 15, is taken from James B. Griffen, *Lake Superior Copper and the Indians,* p. 41.

2. This translation by Louise P. Kellogg is from a passage in *Découvertes et Établissements des Français dans l'Amérique Septentrionale,* by Pierre Margry, Paris, 1896 (129).

3. In classifying the "biotic provinces" of North America, Lee R. Dice designated the broad belt of spruce-fir forests that stretches across the continent from Newfoundland to Alaska as the Hudsonian. From western Minne-

sota eastward to New England and the maritime provinces, is the Canadian Biotic Province, taking in the northern halves of Wisconsin and lower Michigan. As noted previously (Chapter 2) it is a region of northern hardwoods on the better, more moist, upland soils and pines in dry sandy areas. The Canadian extends southward in the Appalachians at elevations having climatic conditions more common farther north. Isle Royale lies in the transition zone between the two biotic provinces and supports plants and animals common to both.

Chapter 4 Our Winters North

1. Sourdough and its uses: This information is furnished since there seems to be interest in it. My sourdough culture is kept warm under a loose lid and is fed daily with nothing but unbleached white flour. The day before some of it is used, the separate batch should be made up in another bowl in which various other ingredients (such as whole wheat or rye flour) can be added. On the day of bread making, if you are in a hurry, mix in some commercial dry yeast (a compromising expedient) when you add the salt, sugar, and oil — or honey, raisins, or whatever.

When I close camp for the season, I spray a sheet of heavy foil with lecithin and paint on a thin coat of the sourdough. When it is dry I shake the flakes loose and store them in envelopes. On occasion these dry starters can be sent to deserving and grateful people. The dough can be activated by putting some dry culture in the crock, and mixing in warm water and unbleached flour. I do this patiently for about five days, at which time it will take off and perform in its usual magnificent fashion. The action of sourdough can be stopped and it can be stored wet in a refrigerator jar for a week or so.

Recipe for sourdough pancakes: The night before, put 3 cups of sourdough in separate bowl and add enough whole wheat flour, wheat germ, and warm water to make it like thick gravy.

Next morning do this:

Beat 2 eggs to a fluff in a separate dish.

Put 3 teaspoons baking powder in a separate cup, dry.

Add to the sourdough (without mixing) some cooking oil, a little powdered milk, a teaspoon of salt, and 2 tablespoons sugar (white or brown).

When griddles are very hot, add water to the baking powder, stir, and dump into the sourdough, along with the beaten eggs.

With a large spoon work the mixture over gently — get it well mixed but don't beat. It will fizz, puff up, and become incredibly light.

Spoon immediately onto the griddle and, *violà — crêpes de la maison*!

Uneaten pancakes can be dried and used for dressing.

Unused pancake batter can be kept cold and worked into the next batch of bread or biscuits.

2. After the winter field work in 1960 and 1961, we visited Benson Ginsburg and his students Jerome Woolpy and Devra Kleiman at the University of Chicago, who were studying the problems of "socializing" adult wolves with people. They collaborated also in the work of George Rabb of the Chicago Zoological Park, who had established a socially organized pack of wolves for observation in an enclosure. Under the direction of Erich Klinghammer of Purdue and helped by Dr. Rabb, Rolf Peterson made systematic observations of these captive animals in the spring of 1971. All of the above were helpful to us in the interpretation of our field observations, as were their publications (90, 136, 138, 228, 293, 294).

Dr. Klinghammer joined the Purdue staff as animal behaviorist in the Department of Psychological Sciences in 1968 and immediately became our close collaborator and consultant in the wolf work. He was with us in the winter camp from 21 February to 1 March in 1972. His establishment of The Wolf Park at Battleground and the studies carried on there provided our students with continuing benefits of his counsel and facilities in the field of ethology. The Wolf Park is part of the program of the North American Wildlife Park Foundation, of which Klinghammer is founder and president.

Chapter 5 Wolves and Their Hunting: Prey Selection

1. The word *ecosystem* is becoming familiar, but its meaning can be confused by diverse usages. Some employ it to describe any aggregation of living things through which energy is transferred from one (trophic) level of utilization to another. Thus a city or a cornfield could be an ecosystem. I regard this as too broad a context, in which the word loses its value. It is used here in the sense of the original Tansley definition: the natural association of living organisms plus the physical environment in which they live. Under conditions in a given area (climate, soil, water levels, etc.) the ecosystem develops to a state of "dynamic equilibrium," in other words, the climax. I consider the term ecosystem most useful in describing natural systems that are large enough, old enough, and complex enough to be self-perpetuating. For present purposes I use such terminology as life community, plant community, or animal community casually as a nontechnical designation of components of the ecosystem. For the most part, they do not need explaining (10).

My statement that every living thing in the ecosystem plays some part that has survival value for the whole would be difficult to prove as universally true. However, over time it probably is true, because in a state of nature structures and organisms that are nonadaptive tend to disappear. In a mature (seasoned by time) community of life there probably are no free riders.

2. It may appear strange that the wolf, a dominant carnivore, had a higher effective reproductive rate than the deer, a typical prey animal. However, this is not entirely illogical because it is a widespread phenomenon that where an animal population is drastically thinned out, it responds with an increase in fecundity and survival. This is the biologist's "inversity" principle, and it could apply to the Minnesota wolves under a regime of heavy trapping, snaring, and shooting for the bounty. The annual take of deer was not high in terms of the reproductive capabilities of the species. Under optimum range conditions, and with heavy annual losses, deer can produce up to 200 fawns per 100 adult does. Depending on sex ratio, a hunting kill of at least 40 percent of the herd can be replaced annually. However, Stenlund indicated that the range in the Superior National Forest was deteriorating because of forest maturation and overbrowsing in the "starvation" winters that came six times in twenty years. Only 61 fawns were being produced per 100 does.

3. This subject was dealt with extensively in the various works of Paul L. Errington, especially relative to his studies of the muskrat and mink. High population density, food exhaustion, flooding, drouth, and disease were factors determining the toll of muskrats by their most important predator (73, 74, 75, 76, 77, 78).

In their studies of raptorial birds in Michigan, Frank and John Craighead had impressive evidence of changes in the vulnerability of meadow mice with changing habitat conditions.

A late-winter thaw followed by freezing filled burrows and tunnels with ice and exposed the mice to heavy inroads by rough-legged hawks. "Only on this site were partially eaten meadow mice observed, good evidence that the hawks actually were catching more than they could consume."

A striking example of man-induced vulnerability in prey was cited by Carl B. Koford in his monograph on the prairie dog. The observation was by Joseph Keyes during rodent-control work in Colorado: "He saw about 20 hawks, mostly rough-legged hawks and prairie falcons, feeding on dead and dying dogs which had been poisoned the previous day. The hawks did not kill healthy rodents but rather seemed to follow the poison crew as scavengers."

In Sweden predators are looked upon as a serious threat to the welfare of game and the domesticated reindeer. In a survey of the situation the carcasses of 1031 roe deer (a small species in which a mature buck weighs about 60 pounds) that were found dead in various parts of the country were sent to the State Veterinary Medical Institute for autopsy. The report revealed that about 15 percent (156 individuals) had been killed by such predators as dogs, foxes, wolverines (3 only), and eagles (2 only).

Of the deer killed by predators, 71 were considered to be "normal" in health, and 85 (54.5 percent) were found to be "diseased." About half the predator kills were ascribed to dogs. It appeared that 60 percent of all the losses were caused by winter starvation. The 156 deer killed certainly are not a random sample, since "it seems incredible that the roe deer population should consist of about as many diseased individuals as healthy ones . . . Thus, the predation has certainly been a selection of weak animals, so that in the long run what seems so harmful may be beneficial to the species" (34).

4. In 1963, at my suggestion, the Outdoor Writers Association of America invited Mrs. Crisler to show a wolf film on their program at the 28th North American Wildlife and Natural Resources Conference. On this occasion I discussed with her the reported hamstringing of the caribou by her dog. In her field notes, now at the Denver Conservation Library, the statement is like that in the book. She told me that she had seen no evidence of hamstringing by wolves in the Arctic.

5. The fecal deposits of most animals, including birds, are commonly referred to as droppings. However, field biologists have widely adopted the term *scat* in dealing with the droppings of carnivores and omnivores. Its etymology is obscure. Ancient fossilized scats, including those of man, are designated more elegantly as *coprolites*.

6. Northeast Airways, our flight contractor, furnished three successive field pilots that winter, of which Don Murray was one. Don became so interested in the study and performed so well that we were delighted with his offer to do all the field flying in subsequent winters. Don's wife, Helen, carried on alone during 7 weeks each winter for 18 years. Of their seven children the youngest never had his father at home for a birthday — even the one we shall have to number 0.

7. This was a method developed by Passmore, Peterson, and Cringen and was based on dividing moose mandibles into wear classes that roughly corresponded to periods of years. It was a useful method, but we replaced it profitably by another method after Michael Wolfe joined the project in 1966. Mike had been exposed to Mitchell's method of sectioning the teeth of Scottish red deer and counting cementum layers. He applied this successfully to our moose collections and it has been used ever since. Our usual procedure is to slice the upper first molar longitudinally, polish the cut surface of the root, and count the layers of cementum that are laid down annually. Usually this method

is accurate to plus or minus one year. (See bibliographical titles by above authors.)

8. I have raised this point with Mech, relative to the extensive radio-telemetry work in Minnesota. He has not found both sexes in a group of nonbreeding adults in the breeding season. All aggregations of males and females contained a breeding pair, although, as on Isle Royale, pups were not always produced.

9. In his paper of 1977 Mike Wolfe discusses age and sex characteristics of wolf-killed moose through 1969 (290).

Chapter 6 Wolves and Their Hunting: Killing Methods

1. In addition to his thesis of 1966, published as a National Parks Fauna, the Isle Royale work was the firm basis of personal research that enabled David Mech to write his more general book. This is the most definitive and authoritative treatise available on *Canis lupus*. Except for several years in the mid-60s, Dave has devoted nearly full time to field studies and writing on the timber wolf since he first entered graduate work in 1958. Using radio telemetry, extensive flying, and other modern techniques, he and his coworkers have sifted nearly every aspect of the biology and relationships of Minnesota wolves. Mech is on the wildlife research staff of the U.S. Fish and Wildlife Service, although funds for the research have been obtained from many sources. The wolf in Minnesota is the subject of controversy between those who would wipe it out and those who would give it total protection. With the work in progress, there should be a basis of reliable facts to aid necessary management decisions by the state and the federal government. This work is under the surveillance of a "wolf recovery team" of technical consultants who confine their efforts to the eastern timber wolf, *Canis lupus lycaon,* of which the only viable populations south of Canada are in Minnesota and Isle Royale (177). However, these are contiguous with a much larger and presumably secure population in Canada. This subspecies is not at present endangered, but it can be properly classed as threatened.

2. In his book on the wolves of North America, Stanley Young told of keeping 10 young wolves until they were 3 years old and the pelts were prime. They averaged 90 pounds each. After 2 days without food, and before being killed, they were fed as much as they would consume. It was found that the average weight of the stomachs — filled with fresh meat and fat — was 18 pounds.

3. On 5 February 1975 Rolf and Don watched the east pack (of 15) chase a cow for 1.5 miles near Lake Whittlesey. Rolf said that "Three times I saw her lash out with a rear foot as she ran, each time apparently hitting a wolf and stopping them in their tracks for a second or two." Later Don said it was the first time he had seen a moose kick like that without looking back. This cow evidently exhausted the wolves in the long chase — 10 minutes through the snow at top speed — and as they gave up and filed away Rolf said she actually appeared to join them "as one of the pack." She moved along with them on the opposite side of a stand of spruce as the wolves walked slowly away.

4. Another example of this phenomenon was described by Hoogerwerf, who witnessed the killing of a macaque monkey by a Komodo "dragon," or monitor, the largest of all living lizards, on the island of Rincha, in Indonesia. The author and a companion heard the screaming of a large group of gray macaques, and on approaching saw one of the excited monkeys on the ground.

All the monkeys disappeared, but farther on the observers came upon a large lizard holding an injured monkey by the head. The monkey was in such a position that it could look around, and it showed no sign of alarm or fear. As the monitor moved, its victim stepped along with its feet, making no effort to escape. "Apparently the macaque was in some sort of shock condition that suppressed any efforts toward escape. When three quarters of the monkey's body had disappeared into the lizard's mouth, breathing was still to be noted. Finally, the hind legs and tail disappeared: the process of ingestion had lasted only about twenty minutes."

It is not surprising that, with his extensive experience in watching predatory animals, Schaller has made similar observations. He told of an uninjured buffalo lying on its side, while a lioness chewed on its tail. "Animals in such situations seem to be in a state of shock."

5. The occasional mention of our using an ax to open a moose carcass may need some explanation. After a moose dies, its appendages freeze quickly in cold weather. Under these conditions even the jaw cannot be removed, except in fragments, and usually we leave it for further reference after the wolves have done their work. However, the interior of a moose may still be unfrozen several days after death. In such a bulky animal this appears to result from a continuation of the heat-producing fermentation of food in the chambers of the stomach (p. 202) and decomposition of the viscera. Sometimes nearly intact carcasses are opened and the viscera pulled out by only one or two wolves. We do not know how long the internal organs remain unfrozen, but the failure to freeze immediately might keep the larvae of hydatid tapeworms alive and contribute to the infection of a wolf.

6. It will be noted that, for possible purposes of reference, I have recorded the autopsy number of specimens where some condition of interest is mentioned. Our autopsy records are on consecutively numbered cards (4 × 6, printed both sides) that were used for any species, although the great majority represent moose. Since no specimen received an autopsy number until it was examined, we had no systematic way to record kills seen from the air until 1967. It was in that winter that we started the new "kill numbering series" (e.g., 72-32) that henceforth was set up for each year at the beginning of the winter study (p. 79). Since then, the kill maps and card files for each year's kills have been a valuable crosscheck for many purposes.

Most particularly, it sometimes happens that the wolves make a kill on offshore ice that we are unable to examine before it disappears. In these cases the kill number is the only identification we have of a moose taken by the wolves at that time in that locality. Had I been more provident in my thinking, there would have been great value in having such maps for the earlier years.

During Wolfe's tenure on the project we began working out a computer sorting system to analyze the data on our autopsy cards. This was continually refined and reached useful efficiency when Peterson was extracting information on 16 years of records for purposes of his thesis in 1974.

Chapter 7 Summers in the Field

1. Krefting and a crew from the Fish and Wildlife Service have taken pellet-group counts, the last in 1970, that were used as a basis for their moose population estimates (146).

2. See Sigurd Olson, "The way of a canoe," in *The singing wilderness,* p. 77.

3. There is always a tendency to carry too much. Foods must be dry, and they are packaged in plastic refrigerator bags. If I am to stay over in a trail cabin (with a stove), I take along a small bag of self-rising cornmeal, powdered eggs, and a miniature plastic bottle with a couple of ounces of oil. I always have powdered milk, and with these ingredients it is easy to make pancakes for breakfast. A pancake can be spread with peanut butter and folded into a sandwich for lunch. Often I take along a 3-pound coffee can punctured under the top rim for a thin wire bail. This is for boiling water over campfires. It gets black and has its own heavy plastic bag in the pack, where it is filled with food bags. At the end of a trip the can goes into the trash. Cheese is an excellent — but heavy — trail food. I sometimes have a piece (and even an apple) to use the first day.

4. Most amazingly, I updated it on 15 October 1976 when a magpie flew up in front of me near the Windigo Ranger Station.

5. Reports, and also the evidence of old nests, indicate that bald eagles coming to Isle Royale were more productive in the 40s and 50s. The failure to breed in the 60s suggests that these birds were tied into the widespread decline of bald eagles and other fish-eating birds that resulted, in large part, from the build-up of organochlorine pesticides and other contaminants in their environments. It appeared that a pair of eagles attempted to nest on Malone Bay in 1962, but Shelton said nothing came of it. We have a few records of the golden eagle in fall and winter.

6. I recorded my last duck hawk, seen over Birch Island, when Dorothy and I were camped there 22 July 1964. Larry Roop saw one the year before at Lake Ojibway on 7 August. See papers in Hickey for information on the peregrine (105).

7. Several bulletins published by the Isle Royale Natural History Association are outstandingly useful to everyone interested in the flora and fauna. Each of us has frequently carried along the handy-sized "Wildflowers of Isle Royale," written by Robert A. Janke, our long-term authority on plants of the island. Janke has worked as a summer naturalist on the park staff and in winter is Professor of Botany at Michigan Technological University. His bulletin on the common flowering plants was ably illustrated by his wife, Nadine.

Two other useful bulletins in this series are "Forests and trees of Isle Royale National Park," by Robert M. Linn, and "The vertebrates of Isle Royale National Park," by Robert G. Johnsson and Philip C. Shelton. In all national parks, literature of this kind for the aid of the visiting public is sold at nominal prices in the concession stand. On Isle Royale this is located at the Rock Harbor Lodge.

Our most up-to-date listing of Isle Royale birds is the bulletin by Krefting, Lee, Shelton, and Gilbert, "Birds of Isle Royale in Lake Superior," published by the Fish and Wildlife Service.

8. Relative to baneberries, which may be either red or white, Martin, Zim, and Nelson said, "The berries of these woodland plants are somewhat poisonous to man." However, they had records of use by the ruffed grouse and white-footed mouse. It can be deduced with some certainty that such a fruit represents an adaptation to seed distribution by birds; otherwise there would be no reason for the bright color, which birds perceive. Where fruits are eaten by mammals, scent and taste may be involved. These senses are practically nil in the bird.

Chapter 8 New Generations

1. The most recent glacial advance reached its southern limit about 18,000 years ago. At such times the ocean level is diminished (in this case by about 150 feet) by the tying up of much of the earth's water in continental and alpine ice fields. Great areas of offshore shallows are laid bare exposing such "land bridges" as the Bering Plain. Animal populations spread into unoccupied habitat wherever it was available and in this manner made the "crossing" to North America. Horses and camels, which originated on this continent, got to Asia in the same way. Then they died out here in the great and somewhat mysterious disappearance of megafauna that took place during some thousands of years following retreat of the glacial ice.

Randolph Peterson's book, *North American moose,* is the one comprehensive work available on this species in North America. A good literature on moose is now being compiled in transactions of the annual North American Moose Workshop, the 12th of which was held in 1976. These have been hosted by various Canadian provincial agencies.

2. My interpretation of population counts does not necessarily agree with all earlier calculations. I expect that others involved in this work will come out differently on some of the figures even though my conclusions are influenced by their judgment. In any case we probably have revealed the approximate size and trends of the moose population. If it were possible to quantify adequately the complex interaction of weather variables affecting moose activity and visibility, we probably would have an explanation for the improbably low count in 1967 and the high one in 1969. Adding 20 percent to each year's count brings most of them closer to the truth; but it might well stretch an error in such a year as 1969. That was my assumption in reducing the calculated population figure for that winter.

3. Relative to the minimum number of moose on Isle Royale since Murie's time, Aldous and Krefting said in 1946: "How many moose were left on the island by 1935 is not known but the number perhaps did not exceed 200. By 1936 the carrying capacity for moose on the island had probably reached its lowest point. In addition to overbrowsing, a great reduction of balsam fir browse was inflicted by the spruce bud worm."

4. Yearlings usually have spikes or forked antlers with only minor suggestion of palmation. Animals a year older commonly show three tines on a side branching from two flattened forks. After that, palmation increases to the large "shovelhorns" of old animals. The most definitive study of antlers was made by A. T. Cringen, published as an appendix of Peterson's book on the moose.

As Bubenik interprets the evolution of mooselike cervids, the tined antlers of the young animal reflect a primitive condition. In other words, in some early ancestor antlers were more like those of a deer, and that ancestor probably lived in thick woods or brushlands where this kind of headgear could be maneuvered to advantage. The broadly palmate conformation probably developed with the progressive use of more open habitats where threat display at a distance could be more effective in rut competition. The shovelhorn antler is a sexual characteristic presumably functioning to intimidate other bulls and perhaps to attract the cows.

We are slowly collecting an antler series composed of specimens from which teeth are available for aging. Ultimately, it is hoped that this can be on exhibit in a new Rock Harbor visitor center, where it will illustrate graphically how antler development is associated with age. Some bulls that reach advanced

Table 4. Moose Counts and Population Estimates

Year	Investigator	Percent /island sampled	Moose count	Extended total	20% added	Percent calves fall herd*	Adjusted population estimate	Remarks
1960	Mech	100	529	529 (±222)	661	21.6	700	For strip counting 20% may be low adjustment
1966	Jordan	12.1	143	705 (±184)	881	8.8	881	Stratified sampling began here. No basis for adjustment.
1967	Wolfe	12.2	96	531 (±230)	664	—	—	Count considered low from undet. causes. No basis for est.
1968	Wolfe	12.6	128	1015 (±266)	1269	17.5	1000	
1969	Wolfe	12.6	185	1150 (±242)	1438	18.5	1100	Appeared good count. 20% "correction" probably high.
1970	Wolfe	8.1	101	945 (±234)	1181	16.3	1000	
1972	Peterson	8.8	117	818 (±234)	1023	—	1000	
1973	Peterson	—	—	—	—	10.5	—	
1974	Peterson	6.9	113	875 (±260)	1094	15.6	900	Estimate reduced on basis of high recent mortality rate.
1975	Peterson	—	—	—	—	13.7	—	
1976	Peterson	—	—	—	—	10.2	800	Estimate based on continued low calf production and high mortality rate.

* Refers to previous fall.

age (possibly 12+ years) appear to lose their great palmations and go back to tined antlers that may be heavy and malformed.

5. The recent breeding history of many female mammals can be read with reasonable accuracy by studying the succession of pigmented follicles that develop in the ovaries following ovulation, conception, and birth. We obviously had no such collection on Isle Royale, and it could be obtained only in a population where cows are hunted. There is no guarantee that birth statistics obtained from an exploited population apply to Isle Royale's moose, which are hunted by wolves rather than people. Different age classes are being killed. Simkin compared his results with those of workers in other regions and found fair agreement (72, 219, 202).

6. On several occasions our field men have escaped embarrassment by going up a tree, and every year or so an episode of this kind is reported by hikers. We have direly predicted that "someone will get it yet." However, it has not happened, thanks in part, no doubt, to timely warnings given by the park staff to all visitors not to trifle with an alert cow in May or June.

7. On 3 June 1974, Don and B. Kaye Moll saw a cow and three small, similar-sized calves swim from the shore of Tobin Harbor to a wooded island, a distance of about 50 yards. This seems a valid record of triplets.

8. Two days after the episode of the twin calves, in the same general area, Phil made another unusual observation: two sets of what appeared to be twin yearlings. One pair was with the cow and the other two were accompanying a large and a small bull.

9. The bulls do not always welcome this association by younger animals. On 3 July 1971 Jim Dietz saw two moose feeding on the shoreline of Sumner Lake. One of them was a large bull with a massive rack (in velvet). The other was a small bull that tried to accompany the other. Jim said the large bull would not tolerate the small one in his immediate feeding area and chased him away twice.

10. In central Newfoundland the introduced moose multiplied so abundantly that the population attained a density of more than 12 per square mile. There were no wolves to support, and hunting was the principal controlling factor on the herd. To mitigate heavy damage to birch and fir reproduction, regulations were liberalized in 1960 making it legal to kill three moose of any age or sex. Two years of heavy hunting were effective; the herd was reduced, and some recovery of the browse was evident. It was estimated that the cutovers of this region could support about 6 moose per square mile (probably slightly over our maximum density on Isle Royale in 1969–70). The outlook was that the use of heavy hunting would need to be continued as a check on the herd (29, 30).

11. Conditions for work and living in our winter camp illustrated impressively that the complexity of human relationships grows on an exponential scale with increases in "population" density. Our space and equipment were excellent for three or four men. We could accommodate five for short periods. Beyond that our efficiency rapidly dropped. During 8 days of the MGM filming, with seven or eight people in camp, our regular work was at a standstill. We expected this, but it was evident that the National Park Service had a stake in the forthcoming TV show, as did the wolf as a species. So, for the only period during our 18 years of research, we devoted ourselves to this public relations effort. William C. Mason, a Canadian, did a highly professional job on the photography, and Irwin Rosten and Nick Noxon directed and assembled from many sources a documentary that jarred the outdoor world. It was aired by NBC on 18 November 1969. The response was such that we can regard this date as the turning point in public awareness of what has

happened to the continent's most influential carnivore of primitive times and its need for rational management and protection.

Because our plane could accommodate only the pilot and one passenger, and because our research time was so precious, we have had to decline many offers by writers and newsmen (sometimes long-term friends) to visit us in winter.

12. At 9:30 Don and I had checked this kill and found a single wolf feeding on it, with the usual ravens sitting around. Tracks of the pack led out toward Senter Point. This moose (73-23) had been wounded by the west pack on 8 February, and the wolves hung around for a couple of days before abandoning it. It was alive for 8 or 9 days and then died or was killed while we were grounded by weather. The kill was plainly visible in an area with a large tree as landmark. The following 2 June Rolf, Phil Simpson, and Tim Lawrence went to the tree where the moose had been wounded and found half a dozen pellet groups and some wolf scats. Then they took a compass bearing that should have spotted the kill. The bearing led instead to the bones of an old kill (aut. 872) not previously recorded. The 12-year-old bull had died 2 to 5 years before without benefit of wolf chewing. Kill 73-23 disappeared and never was found, despite 16 man-hours of searching. A year or so later Joe Scheidler remarked that it was one of three that Rolf had missed in 4 years.

Chapter 9 Moose Ways

1. As has been widely verified, the growing tips of woody browse (including, most significantly, aspen) and other plants represent the seasonal stage highest in protein and nutrients in general. This is especially characteristic of the first stages after a fire. Such plants are correspondingly palatable to most herbivorous animals (56, 66, 272). The prevalence of a wide variety of tender greens early in the year is spoken of as the "nutrient flush," a time of recuperation for the browsers, grazers, and such herb-feeding omnivores as bears. J. L. Oldemeyer has compiled a review of available information on the nutritive value of moose foods. He noted that this animal does quite well in areas where the main browse species are relatively low in protein. It is of particular interest that red-osier dogwood, a highly preferred food on Isle Royale and elsewhere, is low in protein. We are reminded that, while our present crude generalizations have value, the profitable studies of nutrition relative to habitat in wild creatures have only just begun.

2. The work referred to is the vegetation-type map of Isle Royale (1970) by Krefting, Hansen, and Meyer; the bulletin on the forest, fire history, and wildlife (1973) by Hansen, Krefting, and Kurmis; and Krefting's bulletin on the ecology of Isle Royale moose (1974). His article on the history of the coyote (1969) is another valuable item in the series.

3. Koelz carried out these studies for the Michigan Institute for Fisheries Research, Ann Arbor. His manuscript, written about 1937–38, was entitled "A survey of the lakes of Isle Royale, with an account of the fishes occurring in them." It was examined and quoted by Cain, and is here cited secondhand.

4. Ford Kellum carried on these studies of the captive moose at Cusino some years before they were taken over by Louis Verme.

5. Randolph Peterson (1953) found this habit to be "quite common" in Algonquin Provincial Park, Ontario, although he added that few instances were noted in the St. Ignace area. Elsewhere in the literature straddling or riding down is mentioned as taken for granted (e.g., Hatter), or there is a

denial that it occurs at all (anonymous editorial in *British Columbia Wildlife Review,* 19). Such disagreements raise the valid question as to whether there are "traditions" in certain types of behavior among animals in different areas. This one needs to be looked into.

6. This kind of feeding was particularly evident in the winter of 1968, when we had only a foot of snow. Often there was little browse at levels above two feet from the ground because of heavy use of the young balsams in previous years. But that winter moose were nosing aside the snow and eating the bottom branches that ordinarily are protected by the snow pack. Moose do not paw for food, but sometimes they do for water.

The reddish color of urine probably results from the excretion of carotene, derived from the green balsam needles, that has not been converted to the colorless vitamin A. Some breeds of cattle are known to be less efficient at this conversion than others, notably the Guernsey. The milk has a "rich" appearance because of its carotene content.

Studies by Cowan et al. of the nutritive value of various stages of forest growth demonstrated that the older trees were highest in carotene content (though lower in other nutrients), which helps to explain the change in urine color when the blown-out tops became available as forage.

7. Relative to the use of licks in Minnesota, see Surber.

8. The *Ruminantia* represent a suborder of the even-toed ungulates (Order *Artiodactyla*) that includes the deer, giraffes, antelope, cattle, bison, goats, and sheep. This large group of browsing and grazing animals is characterized by the cud-chewing (rumination) type of digestion featuring the four-chambered stomach, the compartments being the rumen, reticulum, omasum, and abomasum. The odd-toed ungulates (*Perissodactyla*), horse, tapir, rhinoceros, and allies, are non-cud-chewing. They accomplish the same digestive functions by means of large storage caeca (appendices) that branch off the intestine.

9. Our park permit allowed us six animals, but this was mainly to cover some unforeseen emergency in which several moose might be injured by wolves or accidents and become good subjects for autopsy. This did not happen to more than one animal per winter, and a total of three was the most we collected during a single study period. A "complete" autopsy was a demanding and time-consuming business that left us all exhausted.

10. When an animal goes into poor condition from any cause, the fat in its bone marrow is last to be utilized. Thus the examination of bone marrow has long been used by biologists in judging the nutritional status of deer in hard winters. The fat content of marrow in moose bones left by the wolves has been one criterion by which we could sometimes determine that something was physically wrong with the victim. "Normal" bone marrow looks to be solid with fat, as in a cut of round steak, although it may be pinkish or somewhat bloodshot. Totally fat-depleted marrow is like clear red jelly, and of course there are many intermediate conditions. Although we thought our subjective records were satisfactory for most purposes, we wanted to know how much fat actually was present. Most measurements of this kind have been in terms of percent of fat in a fresh marrow specimen. It bothered me, however, to see that bones lying around in the field for weeks or months obviously were losing water, and the weight of fat as a percentage was biased. What we needed was a standardized method of getting the weight of fat per cubic centimeter of bone marrow cavity.

With this in view, after Jim Dietz and I came back from the island in the winter of 1971, he developed a laboratory technique and ran our backlog of frozen bone sections. Then Rolf compared the results with our own field descriptions and published records from other areas of fat-content percentages in fresh specimens (mostly deer). We decided that we could standardize field observations by using three subjective classes: (1) first class, evidently normal; (2) class two, those appearing partially depleted (as analysis showed they were); (3) extremely or totally depleted. The lab work showed a maximum of 0.84 and a mean of 0.58 grams of fat per cc of bone cavity in undepleted marrow. The intermediate class had a mean of 0.43 grams. The depleted class averaged about 0.02 grams of fat, or practically none. In terms of animal health, class III bone marrow may well indicate an individual past the point of recovery, but this is speculative for the present.

11. An outstanding exception to this statement was a 4.5-year-old bull (aut. 172) collected by Phil on Washington Creek 11 February 1963. It had more fat in the coelum than we have seen in any other male, and even some fat under the skin. It was obviously a healthy animal in spite of the fact that we collected it because of extensive bare patches on the skin: an indication of the heavy tick load, which we estimated to average 1-2 ticks per square inch over most of the body. This bull had nine hydatid cysts in the lungs and one in a kidney. In the liver were four cysticerci of another tapeworm in which the wolf is the primary host, *Taenia hydatigena.* While it probably is unusual to find fat animals in winter, our sample of these young moose is too small to judge their average condition.

12. As a case in point, Jedediah Smith was conducting a group of trappers north along California's Klamath River in May 1828. Dale Morgan (p. 262) said the party made only 3 miles on the 18th, which taxed their strength to the utmost. "The men were almost as weak as the horses, for the poor [spring] venison of this country contained little nourishment."

Chapter 10 More about Moose

1. Both Rolf and I have recorded seeing a moose scratch the back of its head with the hoof of a hind foot. Evidently, for a limited area, this is another way in which the moult might be helped along or tick irritation alleviated.

2. In plants and animals the rapid spread of disease and parasites is commonly observed in times of stress, overpopulation, and environmental decline. Specific to ticks, Lee and Martha Talbot said of the Wildebeeste, "Sick or otherwise weakened animals carry more ticks than healthy ones." Relative to moose, elk, and deer in Alberta, Stelfox said that the young are most subject to attack. "The stronger animals recover and the weaker die." We have not been able to associate tick infestation with malnutrition or other physical ills on Isle Royale, possibly owing to the rapid weeding out of inferior individuals by the wolves. This, of course, is speculative.

3. We have not collected any of the several species of flies that obviously are a plague to the moose. Murie collected moose flies, a relative of the house fly, which Randolph Peterson suggested was the principal cause of hock sores. They attack the head and posterior region of the moose, entering the rectum and laying eggs that are expelled with the fecal material.

4. Phillips, Berg, and Siniff studied the movements of 36 marked moose on the Agassiz National Wildlife Refuge in northwestern Minnesota. They found that in summer and fall the home ranges occupied by cows averaged 6.9 square miles, and for bulls the figure was 5.6 square miles. In winter the ranges were much restricted, averaging 1.4 and 1.2 square miles, respectively.

In studies of the Shiras moose in Yellowstone National Park, J. F. McMillan recognized and made location records on 38 individual moose in the Willow Park-Swan Lake Flat area. He found summer populations of 33 and 29, respectively, in 1948 and 1949 in a 9-square-mile area including the willow-grown bottomland of Obsidian Creek. His conclusions, based on animals with the most records, were that many moose pass the summer within a home range of less than a square mile and that it is rare for an animal to use more than 2 square miles.

5. Scent glands of various members of the deer family were first described by John Dean Caton, a lawyer, in a book published in 1877. Caton was a meticulous observer, who studied every detail of the structure and characteristics of the many species in his deer park. Unfortunately, he had no moose to provide firsthand information.

6. The marking of the bull was experimental, and no extensive work of this kind was contemplated as part of our studies.

7. In his studies of pre-rut behavior in Newfoundland, Dodds recorded some sparring by bulls in August, when antlers were still in velvet. He said it appeared that the animals were treating their antlers delicately. On Isle Royale we see loose aggregations of bulls and a continuance of dominance interactions after mid-October. Moose tend to be in the open at this time, which is an aid in our fall sex-and-age counts from the air. On a single flight, 21 October 1967, Mike Wolfe counted 43 bulls, 24 cows, and 12 calves. Evidently the situation is the same elsewhere. Peek et al. reported, "Largest group sizes occurred when moose were primarily on the most open parts of their habitat: alpine tundra on the Kenai; recent cutover areas in northeastern Minnesota; and willow bottoms in Montana" (204).

8. In her paper on group dynamics, Altmann described the antler display behavior as a "swaying gait with the heavy antlers dipping from side to side, executed in stiff, long strides, usually in formal circles around the rival bull." Rolf is the only member of our group who saw the full routine from beginning to end. He said it took about an hour.

9. In his thesis of December 1974, Rolf Peterson included a more informative and detailed discussion (pp. 270–291) of bone abnormalities, especially jaw necrosis and arthritis. For technical guidance in this work we have depended heavily on the generous help of George W. Neher and other staff members in Veterinary Medicine at Purdue.

10. A discussion of lumpy-jaw in various species is given by Cass. O. J. Murie found this condition — also referred to as necrotic stomatitis — to be the most important cause of elk mortality at the National Elk Refuge in Jackson Hole.

11. In the present work no general account of parasites and pathology not important on Isle Royale is considered necessary. Reviews of this kind are found in Fenstermacher and Olsen, in Cowan (1951), and in Anderson and Lankester. The meningeal worm, carried by deer, and an important moose disease in Minnesota and elsewhere (but not on Isle Royale) is described by Anderson (1972). Our most important parasitic disease of Isle Royale moose is caused by the hydatid tapeworm, as discussed. More information is given by Cowan (1948) and Robert L. Rausch.

12. In September 1972 antibody tests were given to island workers and our crew in connection with a survey by the Atlanta Center for Communicable Disease of the U.S. Public Health Service. Superintendent John M. Morehead informed me in 1976 that they had received no official report. However, he learned by telephone inquiry that the tests were negative.

Chapter 11 Heavy Industry

1. The spring course was taken over by my colleague Charles M. Kirkpatrick. This made it possible for me to concentrate on teaching in the fall and research in the spring. Thus I was able to spend parts of 15 consecutive winters on Isle Royale, 1961-75.

It was my original plan to overlap successive graduate students by one year in order to propagate their knowledge of the island, boat operation, and other matters. It worked out that way for Mech and Shelton, but later our main dependence for this function was on summer helpers.

2. The aerial photos we have regularly used in the field were made in 1957 by the U.S. Geological Survey. They are out of date, and a new flight is needed for all park purposes.

3. This was also the mean number of beavers per colony (5.1) for a total of nine studies, in various parts of the country, which Shelton reviewed in the literature. Later, he found an average of 6.4 beavers per colony on Isle Royale, necessitating some recalculation of early estimates.

4. There was logic in Krefting's estimate, for he based it on Bradt's average of 4 beavers in colonies with only one lodge. On Isle Royale he regarded that as usual, and it is also true that in many areas conditions for bank dens are not favorable. Shelton's results, showing 6.4 animals per colony, came closer to Morgan's figure of 7, based on early fur-trapping reports.

5. Epizoötic refers to an outbreak of disease among wild animals. For humans the corresponding term is epidemic.

6. As with other animals, however, the beaver sustains a marked seasonal depletion of its fat reserves. The tail is its most important fat repository and 50-60 percent of tail weight is fat at the beginning of winter. In a 2-year-old (and, no doubt, in other adults) this fat was mobilized rapidly from January to March and declined to a minimum in May, although it still constituted 10-20 percent of the weight of the tail (6).

7. In 57 colonies Bradt found three with 12 beavers present. The most common number was six (10 colonies).

8. Morgan said of the three heaviest Lake Superior beavers of which he had reliable knowledge, two weighed 58 pounds and the other 60.

9. The beaver's genital organs are in its cloacal chamber, into which the intestine and urinary and scent ducts open. The large castor glands secrete a yellowish strong-scented musk that the animal evidently produces at will to mark the mud pies. This scent is attractive to other beavers and has been much used by trappers in lieu of bait.

10. In his notes for 23 June 1974, Rolf wrote that "the dam below Benson Lake broke this A.M. washing out the bridges on the trails (crossing Benson Creek) above Daisy Farm."

11. Brenner, who studied the food consumption of beavers in Pennsylvania, calculated that a 10-year-old aspen would be 2 inches in diameter and would contain nearly 10 pounds of beaver food. He said that an acre of aspen would not mature rapidly enough to maintain itself, but on the average it would produce 5840 pounds of food "and would support 10 beavers for one year provided a high degree of waste does not occur and herbaceous material is utilized during spring and summer months."

At many feeding areas we have observed that beavers eat twigs up to about half an inch in diameter. From larger sizes they chisel off the bark and cambium layer. Lodges and dams are littered with the bare sticks and segments of log up to about 4 inches in thickness.

12. The largest beaver-cut tree measured was a birch 21.6 inches in diameter just above the cut. When the Lewis and Clark expedition was on the Yellowstone River in April 1805, Lewis mentioned in his journal that beavers had cut great quantities of timber, including a tree (no doubt cottonwood) nearly 3 feet in diameter (44).

13. In our terms beavers are indeed "wasteful." In 1938 Shaler Aldous made good use of available CCC personnel to measure the use and waste of aspen bark by beavers at a site near Ely, Minnesota. At five active ponds the animals cut 456 trees up to 11 inches in diameter. The cuttings made nearly 10 tons (19,420 pounds) of bark available as food, of which the beavers ate just over 3.5 tons (6987 pounds). The rate of utilization was 36 percent, and 64 percent went unused.

14. U.S. Weather Bureau records for Grand Marais, Minnesota, show that for each of the 3 consecutive years, 1969-71, total precipitation was higher than in any other years of this 18-year study.

Chapter 12 Wolf Society

1. The collection by Robert A. Rausch of 1262 wolf carcasses taken under Alaska's bounty system from 1959 to 1966 is the most representative sample from which information on breeding incidence and litter size has been obtained. In 175 adult females (age 3 years or older) the number of fetuses averaged 6.5, with a range of 3-11. In 69 2-year-olds the average was 5.3. Litters of viable pups must average slightly less than these figures. It is likely also that the social disruption and thinning of numbers brought about by bounty hunting was a stimulation to the production of young. In fully protected populations a smaller number of young would be expected.

As for incidence of breeding and conception, the exploitation factor probably is involved also in Rausch's conclusion, "The information obtained during this study shows that a high proportion of all females two years old and older did ovulate, conceive, and probably give birth to pups annually . . ." Relative to observations on the protected and socially well-organized Brookfield wolves, Woolpy remarked, "that little reciprocal courtship by either sex has been observed before three years of age suggests that this is the age at which wolves become socially integrated and serve as effective members of the mating population" (293).

2. We cleave to that great democratic ideal that all men are created equal — in terms of whatever opportunities it is within the power of societies and governments to offer. However, in the realm of biology this principle of original equality does not work. All wolves were not created equal nor were all men. In a state of nature there is no such thing as an unstratified society, nor will there be among humankind.

3. In dealing with wolves we follow the same practice as with moose: A young wolf is a pup until its first birthday, after which it is a yearling until it is 2 years old.

4. Details of the early lives of pups are taken mainly from the literature on captive-reared animals, including those at Brookfield (Rabb, Woolpy, Peterson's notes), the Wolf Park (Klinghammer, personal communication), and the Bavarian Forest (Zimen), as well as others kept incidentally by various people such as Crisler, Fentress, Kuyt, and Mech. The observations at dens in the wild by Murie and Haber add valuable correlates, as do their records on rendezvous sites. Our Isle Royale information on wolves in summer was gathered almost entirely in the 70s by Peterson's work with packs and their

young at rendezvous and the examination of dens after abandonment. We had no open ground where observations could be made at a distance, and Rolf scrupulously avoided any close approach and disturbance of an occupied den, although he knew approximately where they were.

5. These behaviorisms that have become fixed in the genetic determinants of the wolf must be of ancient origin. Schenkel regards scraping after urination and the habit of circling before lying down (seen sometimes in both wolf and dog) as "rudiments," left over from times when they probably were more functionally significant than now. The common habit of rolling in carrion, feces, or anything else that is strong-smelling probably is currently useful, although only a wolf or dog could explain it adequately, and neither has done so.

6. Klinghammer has watched this "play" activity carefully at the Wolf Park. He feels certain that in the wolf (not the dog) the riotous gamboling that appears to be all fun is actually a part of the rank-order testing process. If an individual momentarily betrays signs of inferiority or weakness, the mock combat can instantly become the real thing.

7. It speaks significantly for the durability of favorable habitat conditions, as well as pack adaptations and possible traditions, that Haber found in 1966-67 and later that packs of the Denali region were using essentially the same home ranges, territories, and denning locations that Murie found them using more than 25 years before (95).

8. In 1961 I had purchased a portable record player and obtained some records of the howling of captive wolves. In June, the first time Dave and I tried the equipment — on the ridge behind Daisy Farm about 9:00 P.M. — we had a distant response from the north. It began with the low moaning call that we have heard frequently through the years (it could not always have been the same wolf) and became a full-blown chorus howl. Dave carried this gear over the island all summer and had little success. We used it many times after that. Rolf decided it needed more amplification, and the bull horn proved to be much more portable. The human voice seemed to do better than our scratchy records.

9. Rutter and Pimlott cited two examples — one concerned Lois Crisler and the other Pimlott himself — of a pair of immature yearling wolves taking over a litter of young pups and rearing them as though they were their own. This behavior is obviously programmed in the wolf's genetic code.

10. It is true, however, that some of the nonbreeding duos and trios reject a loner that has all the earmarks of a regular trailing wolf. In 1974 our McGinty Cove duo had a subordinate satellite that was sometimes accepted but more often kept its distance. On 20 February Rolf and Don found the duo bedded about 20 feet apart at the harbor entrance, and 100 yards away was the loner. He evidently had followed them to kill 74-18, on which all had fed.

11. If my total figures for a given year do not match exactly the totals we have published elsewhere, it usually is because I have subtracted any known losses from our maximum figure. In other words, the total population figure here is the minimum for the year, essentially the breeding stock.

12. The piece of light-colored pelage measuring 54 inches long is now tanned and in the Purdue mammal collection. The bones were examined in the Veterinary Diagnostic Laboratory, and some evidence of an arthritic condition was found in the joints. It might have accounted for the animal's problem. With wolves, foxes, and ravens in the area, we could not speculate constructively on the last rites of Big Daddy.

13. Details of this summary of events in 1967 and 1968 are given in Wolfe and Allen.

14. The following summer (1968) Dorothy and I took a trip around the north side of Lake Superior to investigate this situation. Allen Elsey, of the Ontario Department of Lands and Forests at Port Arthur (later a part of Thunder Bay), told me that black wolves are fairly common in the area north of Isle Royale. James L. Nuttall of Hurcutt, Ontario, said that of about 100 per year that his brother killed (by aerial hunting) for the bounty, one in 15 or 20 would be black, and others would be darker than average. Kolenosky and Standfield had records on 76 wolves in the area west of Thunder Bay, and about 7 percent of them were classed as black.

Chapter 13 The Followers

1. A year before, Erik had seen something similar in front of Pete Edisen's fish house — it could have been the same pigeon hawk that nested farther east. In "flying upside down," as Stauber expressed it, the raven extended its feet to ward off the falcon.

2. When Prince Maximilian was near the mouth of the Little Missouri in 1833, he observed that both wolves and ravens would immediately gather at the report of a gun. Indian hunters told him that wolves followed these birds to learn the direction of the prey. We may extrapolate from this that wolves and ravens had the same relationship on the northern prairies that we have seen on Isle Royale.

3. Holger Johnson of Chippewa Harbor told Clifford Presnall (p. 28) that in 1923 snowshoe hares were so thick the children could pet them. A year later they were scarce. Pete Edisen spent the winter of 1935-36 on the island, and hares were abundant everywhere. The following spring they were being found dead in the woods, and by summer not a hare could be seen. Based on nearly continentwide increases, we can infer that on Isle Royale, too, there was an abundance of hares in 1942. Then the spring of 1943 initiated a series of years that saw the most disastrous and widespread crash of northern cyclic species in this century (7). However, as expected, the hares recovered, and ranger Dave Stimson told Mech that in 1953 they had reached another major high, after which they declined to 1957. In '60, Stimson's last year on the island, tracks were again more plentiful than he had seen since 1953. We watched that increase top off in 1963 before the next reduction set in. As Wendel's work was terminating in the late 60s, another gain in both hares and foxes was under way, and these species were plentiful in 1970-72. After that there were fewer tracks in marginal areas, but we continued to see many tracks in the best habitat. The population was spotty, but there was no general decline of hares in the mid-70s, as might have been expected.

4. Recently Rolf has looked more carefully into the alleged early presence of martens on Isle Royale, and he believes that all reports go back to a statement by a single trapper. The Adams party mentioned martens several times in their report and assumed them to be present. However, there is no evidence that they verified this. My several references to martens on Isle Royale must be open to question until someone reveals more specific information on the animal.

5. We have four copulation dates for the Isle Royale squirrels:

23 May 1962	2 May 1974
19 May 1973	17 May 1974

On the early date, I witnessed at 15 feet what I believe to have been the full act of coition and timed it at approximately 3 minutes, 20 seconds. The gestation period of this squirrel is 40 days.

6. On being consulted about "vacant niches," I was inclined to favor this — until I discussed it with superintendent Carlock E. Johnson. He knew about martens from personal experience (which I did not), including their camp-robbing and general destructive behavior. I now think that if martens are to rejoin the soft-living squirrels on Isle Royale they should be allowed to work it out on their own.

7. Over most of the island, foxes seemed to hold up better than hares in the mid-70s, possibly because of the excellent supply of moose carrion that was available during these years. In the annual report for 1976 Rolf compiled his own records (the most reliable way to standardize this) on foxes seen at least half a mile from any known kill. Fox observations per 100 hours of flying for 5 years were as follows:

1972	*1973*	*1974*	*1975*	*1976*
25	24	21	16	22

8. On 8 February 1967 Wendel rigged a padded trap and (while all watched) caught Fido and the gray female. Both were ear-tagged and weighed. On 1 February 1973 we did the same for two more females. It should be noted that these animals were weighed in essentially their wild condition, and we knew something of their ages. While the weights may be fairly representative of Isle Royale foxes, it is possible that all of these individuals had a good start in life through the handouts received in their natal summer and also in winters that some enjoyed at our camp. In the days before being caught they had not been fed enough to affect their weights significantly.

Adult male: 14 lb. 4 oz. (Fido)
Adult female: 10 lb. 4 oz. (Gray female)
Adult female: 14 lb. 8 oz. (BK, age 2 yr. 10 mo.)
Subadult (?) female: 11 lb. (Little One, age 10 mo.)

9. They do not even socialize overtly with other foxes. Only in obvious courting would our animals approach and solicit one another. Even at such times there was no grooming or close physical contact, as occurs in other species. As Kleiman observed, red foxes usually move and act independently.

10. This could need some qualification. As breeding activity intensified in late February, there were brief episodes when certain females had unusual feeding privileges. As an example, I recall an occasion when Fido was eating out of a pie pan and Angie, an old and decrepit female, walked deliberately in behind him and crowded him to one side to get the food. He suddenly saw something of interest off in the woods and walked away. This was highly irregular.

11. A summary of copulation dates for wild (except one) foxes follows. Those designated "tracks" were from the characteristic tracks made by a copulating pair (per Murray, a "two-headed fox"). These all appeared fresh in the snow, and most probably are accurate within a day.

16 February 1963 (tracks)
24 February 1963 (tracks)
24 February 1965 (observed)
24 February 1973 (tracks)
29 February 1972 (BK, 2-minute tie seen at Windigo)
2 March 1972 (observed)
3 March 1971 (tracks, with 2 foxes still present)

5 March 1971 (observed)
8 March 1972 (complete copulation period, 31 minutes, observed by Peterson and Murray)
9 March 1971 (observed)

12. The foxes readily ate any kind of small bone, such as chicken or small pork-chop bones. These were efficiently chopped up by the shearing teeth (carnassials) and were fully digested. This was in contrast to some vegetable material. I found something in a fall scat collected near Windigo that had me puzzled for a time. Then I saw that chunks of raw carrot (from the campground) had gone through undigested.

13. In caching food, the fox digs a hole with its forefeet, deposits the item, then scrapes snow back into the hole with its nose.

14. The building to which I frequently refer was the new summer base for the trail crew. It was constructed in 1970 of materials from the old building, farther down the hill, which had collapsed under the weight of snow before we got to the island in 1969.

Chapter 14 The Whole of Nature

1. I have avoided the word *instinct* because it has been damaged by misuse. Other degraded terms are *conservation, balance of nature,* and even *ecology.* When people talk about *the* ecology, it usually means they do not know how to explain something.

2. It could be that the scent of a wolf has a different effect on beavers. Kolenosky and Johnston said that in Algonquin Park when one of their instrumented wolves approached a pond, beavers could be heard slapping their tails on the water.

3. Birch seeds are not all he may have. In March 1961, chief ranger Ben Zerbey was with the research team, and he was doing some burning of deteriorating shacks at the site of the old CCC camp on Siskiwit Bay. One of these old buildings has been kept up, after a fashion, as an overnight trail cabin.

Don went over to pick up Ben, and they made a pot of coffee from snow scraped up outside the door. Before leaving, Ben insisted that Don finish off the pot, after which the grounds were thrown out on the snow. Among the grounds were half a dozen still-intact hare pellets. Ben and Don were greatly amused and raised a question as to whether the Indians had made full use of their resources.

4. I had evidence that foxes are well aware of their limitations. On 1 February 1968 I saw eight siskins come down on the snow within 15 feet of our fox Fido, who looked around at them then paid no further attention. He knew they would be gone in an instant if he made a move toward them. These birds are "nervous" and alert, almost constantly moving on the ground. At the slightest cause for alarm (or when none is evident) the flock rises into the trees. They dribble down a few at a time when the coast is clear.

5. Moose can move about at night under any conditions, but bright moonlit nights may be especially favorable in terms of visibility. This seems evident in the case of the snowshoe hare.

6. Scavenger niches often include specialized jobs, such as those of maggots, which empty leg bones of marrow after weather warms in spring, and the squirrels, mice, and hares that chisel away and recycle the calcium of cast antlers and other bones.

7. I have been impressed repeatedly with evidence that the heavy-bodied,

ground-feeding flicker is vulnerable to attack by pigeon hawks, sharpshins, Coopers, and other birds of prey. We find their feathers on the ground, and their remains regularly show up at nests and feeding perches.

8. The broadwing has a particular fondness for snakes, and our numerous field records indicate it to be the main predator of the island's one large species, the garter snake. I saw one carrying a snake as early as 4 May 1974. In southern Michigan I found a broadwing nest lined with a couple of shed snake skins. Habits of the broadwing have been nicely detailed by Paul Matray.

Chapter 15 New Rules

1. How Rolf's results contributed to our interpretation of changes in moose-wolf relationships is evident in two papers, of 1974 and 1976, respectively (215, 213). In this work we had the advantage of earlier investigations by Formosov, Des Meules, and Kelsall et al. (84, 62, 131, 132, 133).

2. Kelsall found that moose became restricted in their movements when snow depth was approximately 27.5 to 35.5 inches (70–90 cm). Chest height of the adult moose is about 40 inches. Deer were considerably impeded in their movements when snow reached a depth of about 16 inches. Chest height of the deer averaged 40 percent less than in the moose (131). Moose are better adapted to northern ranges of greater snowfall.

3. Peek's findings in Minnesota during the winter of 1969 were similar to ours. Snow averaged 45 inches in depth, and moose movements were severely restricted. Moose concentrated in the conifers, where snow depth was about 20 percent less. He mentioned hard late-winter crusts in open areas (203, 204).

In my notes for 5 February 1969 I wrote: "This morning there was a cow moose just behind the shack. When it went off through the deep snow it was wallowing badly and having a difficult time making progress. It was breaking through a hard layer a foot or two above ground, and this made real work of it."

Michael Wolfe found that water content varied from 17 to 35 percent by volume from top to bottom of the snow profile. In early February snow depth in the burn above Siskiwit Lake was 4 inches less than at Windigo, and it was 12–15 inches less at the east end of the island. Less than 6 inches of new snow fell during February and early March, and by the end of the study period on 14 March, snow cover at the camp had settled to 32 inches. The heavy snow caused slush to form on lakes and bays, precluding landings in many areas.

4. It will be noted that in some ways the winter of 1970 appeared unfavorable to the moose, despite snow on the ground of only about two feet: Of 17 kills examined, 8 were calves and 4 were in the 1-to-5 age group. Moose were on lake edges and largely absent from the big burn. We saw moose aggregations, and killing and feeding habits of the wolves were a continuation of the pattern established in 1969. Presumably in 1970 moose were still feeling effects of the rigors they had endured a year before. We have tended to think of the above indications (and others) as a "bioassay" revealing hard times in 1970, even though our weather information does not show this clearly.

5. Beginning with 1969, deer underwent a drastic decline in the Superior National Forest of Minnesota. Mech and Karns found that in a core area of the interior deer were extirpated. Wolves increased in the 1969 breeding season but after that were reduced by about 40 percent in the absence of an adequate food supply. In deep snow they indulged in surplus killing and incomplete utilization of carcasses (178, 180).

In the Pakesley area (Canadian shield country) of east central Ontario, Kolenosky found excessive killing and under-use of deer carcasses during the winter of 1969. He calculated a hunting success of 63 percent, as compared with 25 percent in 1968.

Stenlund said that hard winters in Minnesota came six times in 20 years. In Algonquin Park (Ontario) the winter of 1959 was one of severe conditions for deer. Pimlott, Shannon, and Kolenosky estimated losses from weather and starvation to be 22 per square mile in the Bonnechere deer yard. Wolves had easy hunting but did not kill to excess. Initial consumption of carcasses was reduced, and certain deer were hardly fed upon at all. It was noticed that the flesh of these starved animals had a strong odor.

In most cases of incomplete utilization of kills, it must be assumed that wolves and other scavengers will return later and consume more of the remains.

6. In Minnesota, Peek et al. found a "decrease in survival of calves following a particularly severe winter [1968–69] indicated by a drop in the percentage of yearlings observed from the air in fall of the following year . . ." (205).

In work on the Superior National Forest these authors found highest mortality rates among moose less than five years old. Vulnerability of the early age group to wolves appeared correlated with the presence of cerebrospinal nematodiasis among the moose. The causal pathogen, a meningeal worm (*Paraelaphostrongylus tenuis*), is carried by the deer (128). The absence of deer on Isle Royale can be presumed a benefit in this, as in other, ways (i.e., browse competition).

7. In discussing breeding productivity of the wolf and dog, Scott observed that no figures are available on prenatal mortality, "but in human populations this is at least 20% of conceptions, and figures from various domestic and laboratory animals indicate that the true figure is in the neighborhood of 40% to 60%." Some of this he ascribed to lethal genes, but he suggested that environmental factors also are involved.

8. In case this situation is not clear, the best browse next to trails had been taken, and this calf could not negotiate the deep snow away from trails. The animal probably did not survive the winter and should have been collected for autopsy. However, autopsying a moose you *know* ("Baby Moose," per Murray) is something different.

9. Autopsy 333 on 11 February 1966 was a young cow (3.5 years) that Don and I found by virtue of many wolf tracks 100 yards north of Sargent Lake. The bones were scattered, and "all legs were folded as though the animal died on its hunkers." The situation was so novel I brought two of the legs in for Pete to see.

In the mid-60s, when the big pack was cleaning up all carcasses, deaths from malnutrition might have gone undetected. We had no previous exposure to this. Depleted bone marrow can result from various kinds of pathology, and we may not have recognized the most obvious one. Calves and old debilitated moose are most susceptible to food shortage. Deep snow, overbrowsing, and wolf predation serve to hasten the end.

10. Our "biological year" for purposes of annual reporting ended with termination of the winter study. The report usually was distributed by the end of April. Since moose dying in winter often were not examined until May or June, this information appears in reports for the year following.

11. Another year, Rolf and Don saw this happen to a different female on the outer harbor from our camp. The estrous subordinate was pursued by pack members for a mile and a half. Several days later Rolf watched from the ranger station as she furtively approached the pack on our autopsy bait. When

the well-fed wolves were scattered, she tied with a subordinate male. There was a general scramble, and while the coupling lasted the female was severely bitten up. It showed there *are* ways to get around the rules.

12. Wolfe's original composite survivorship curve appeared in our annual report for 1968-69. In a later publication (290), he expanded the sample to 439 and constructed two curves, based on known winter wolf kills and random collections, respectively. This, which he regarded as a preliminary treatment, allowed for the possibility that a combined curve might contain biases according to the method of collection. In further studies, including specimens through 1974, Peterson and I concluded that the least bias occurs in a consolidated sample, and that was the viewpoint taken by Rolf in constructing the survivorship curves here reproduced from his thesis.

Chapter 16 Contemplations on Wilderness

1. I have heard this low characteristic call a number of times on Isle Royale, most often at night. Usually daytime howling has not started this way, nor has the howling of captive wolves I have heard. An exception was on 22 February 1970, when six wolves of the west pack were on Washington Harbor. They had fed on an autopsy carcass on the north side of Beaver Island and then appeared headed toward Washington Creek. John Keeler and I quickly occupied a couple of the shelters in the campground to watch them go by. We were disappointed, but between 3:00 and 3:30 P.M., when they were still out of sight on the harbor, they howled for about a minute. In my notes I speculated: "The short 'chorus' was begun by a single wolf that was over in the woods north of the creek mouth. It was the low, moaning call that I have heard several times over the years (as far back as 1958). I have the impression that an adult male may call this way to initiate the howling of the pack."

The first chorus howl that Dave and I heard was most impressively of this type (p. 461). It has been mentioned by several of our field workers at various times in their notes. On 28 May 1962 Shelton said: "About dark (2120 hours), with occasional very light rain falling from low clouds, at least two wolves howled from the direction of the southwest end of Sargent Lake. One began, very low pitched and long sustained, then another joined in at a higher pitch. The howls then broke into a coyote-like yammering, the whole performance lasting no more than a minute, probably not that long."

This kind of howling also occurs on other areas, as indicated by Rutter's quote of Abbott Conway: "It started with a long, low howl, almost a moaning sound, and then began to rise higher and higher. Other wolves joined in with that curious off-key effect that we know so well, until the sound seemed to fill the whole bush."

2. Elsewhere (9) I have explored this complexity phenomenon, which is all about us. It means that with every increase in human density and every step in the growth of technology the effects are at a new point on some kind of exponential scale. Problems of this kind do not grow in linear fashion (as, in degree, Malthus realized); they are geometric in their dimensions. Increasing numbers of world leaders understand this, but for good reasons few of them undertake to explain it.

To pursue the matter a bit further, compartmentalized science — some social, some biological, some physical, and the twain so seldom meet — is in large part responsible for our general confusion. Such statements as "We can feed our world," perennially a fresh new discovery by bright new optimists, are insipidly misleading (in a world eager to be misled) unless qualified by the answers to questions that should be obvious: (1) How many people are we feeding and for how long? (2) At what cost to the environment? (3) At what standard of living? A world population doubling in 41 years and adding (1978) 76 million more people a year, can be fed for only so long in a finite world, regardless of the standard. As a disturbing afterthought, some human beings demand more from life than just being fed.

A readily available source of population information is the annual *World Population Data Sheet* of the Population Reference Bureau, Washington, D.C.

3. The wilderness plan for Isle Royale National Park, as provided for in the Wilderness Act of 1964, was finally approved by the Congress in 1976. It was a subject of controversy for nearly 10 years. As is typical of such issues, there were those who wanted more trails, a marina, even roads on the island — development that would benefit the local economy and business in general. There were others who battled to keep development at a minimum and place the major part of the island under the protection of the wilderness designation. A part of their aim was to maintain scarce and interesting fauna: the wolf. In the supposedly final act, Isle Royale received a high level of wilderness protection.

4. In their studies of wolves along the Lake Superior shore of Minnesota, Van Ballenberghe et al. found them making frequent use of garbage dumps and refuse piles. They fed on road-killed deer and interacted with dogs. In four known cases wolves were shot near buildings.

Chapter 17 Wolves, People, and Parks

1. Burning experiments on four species of deer browse in Maryland by DeWitt and Derby produced increases in the protein of sprout growth for at least two years. In Texas, Lay analyzed 25 species of browse in burned and unburned areas. He found an immediate increase in protein by as much as 42.8 percent and in phosphoric acid up to 77.8 percent. These benefits disappeared in from one to two years. It appears that after the initial nutritional advantages from certain types of burning, the longer-term favorable effects will be in the availability of low-growth plants.

2. The great conifer forests of the Pacific Coast, including the coastal redwoods and sequoias, are adapted to regular ground fires that clear away trash accumulation and provide favorable seedbed conditions. Long-continued fire protection in parks and other areas has allowed the build-up of a highly inflammable understory and accumulation of pine thatch and dead wood. Under drouth conditions this is a hazard that could result in a devastating crown fire in these treasured and irreplaceable forests. A program of controlled burning is gradually removing the accumulation and restoring natural conditions by way of a regime of regular nondestructive ground fires. This is essential wilderness maintenance (32, 33, 134, 135).

3. In his survey of the mammals of North Dakota, Vernon Bailey reported that hybridization has also occurred on this continent. He said that in 1856 Lt. G. K. Warren collected a large series of wolf skulls at Fort Union and sent them to the National Museum. Examination showed that some of the specimens were not full-blooded wolves. "Many stories have been current of the ferocity of these hybrid wolf-dogs, and it is not improbable that their tameness and lack of fear of man . . . was due in part to their mixture with domestic animals."

4. By way of example, when Lewis H. Garrard was with a wagon train on the Arkansas River in 1847 he participated in two unceremonious burials. The first, he said, was in a grave "not quite three feet deep" and was "mere form, for the wolves scratch up the bodies again, and often before we were out of sight the prairie ghouls were at their horrid work." The following morning two more bodies were disposed of in similar fashion.

5. During his crossing of the Kansas prairie on the way west in 1859, Horace Greeley wrote (p. 93): "It is impossible for a stranger to realize the impudence of these prairie-lawyers. Of some twenty of them that I have seen within the last two days, I think not six have really run from us. One that we saw just before us, kept on his way across the prairie stopping occasionally to take a good look at us, but not hurrying himself in the least on our account, though for some minutes within good rifle shot . . . It is very common for the wolves to follow at night a man traveling the road on a mule, not making any belligerent demonstrations, but waiting for whatever will turn up."

Under some circumstances it is apparent that the scent of man was enough to send the wolves on their way. When Edward Harris was on the upper Missouri with Audubon in 1843, he wrote in his notes of a common stratagem in preserving meat from the wolves. He said one of the hunters had left the carcass of an antelope to be picked up by the cart (166). "Alexis had put his jacket over the head of the antelope to keep off the Wolves and we found all safe. A few hundred yards out in the prairie we found the other Kabri as they are here called with some other article of clothing attached to it, with like good effect."

6. Wolves handled by humans do not appear to have any social stigma attached to them. As found by Mech et al. (181), and also Kolenosky and Johnston, animals that had been trapped and radio-collared were quickly reassociated with their packs.

7. Since the late 60s there have been scattered sightings of wolves in the vicinity of Yellowstone and Glacier and the country between. Logically, these are survivors of the nearly extinct subspecies *irremotus,* which formerly ranged from central Wyoming to southern Alberta (100). With abundant elk and other ungulates available in the two parks and adjacent national forests, there is no evident reason why the remaining wolves could not organize packs and build up in numbers. The Rocky Mountain Wolf Recovery Team has had little support from either the federal government or the states. Here is a cause for any who need one.

8. In these discussions I have given little attention to the matter of public safety, for the reason that this is no argument that impresses anyone. People are willing to take their chances, even though any disaster will be a public problem and a public expense. Annual expenditures of the National Park Service for mountain-climbing rescue teams, helicopters, and the like are substantial, and they serve few people. Around such an area as Isle Royale, a naive public often are willing to accept marginal flying practices. The exploits of pilots they knew about had Murray and Martila shaking their heads. Bill died in a crash on a Canadian lake while riding as a passenger. Our crew felt they had witnessed an ultimate folly in February 1976, when the tracks of two snow machines were seen on Thompsonite Beach on the north shore. They had come across from Thunder Bay.

Literature Cited

Works of authors mentioned in the text can be identified in this alphabetical listing. Where no reference is made, or where uncertainty might exist among several citations, the number at left is used in parentheses. With a few exceptions, unpublished material is described in the text but not included in this list of references.

1. Adams, Charles C. 1909. An ecological survey of Isle Royale, Lake Superior. Michigan Bd. Geol. Survey Rept., 1908, State Biol. Survey, 468pp.
2. Albright, Horace M. 1931. The National Park Service's policy on predatory mammals. J. Mamm. 12(2):185-186.
3. Aldous, Shaler E. 1938. Beaver food utilization studies. J. Wildl. Mgt. 2(4):215-222.
4. ———. 1944. A deer browse survey method. J. Mamm., 25(2):130-136.
5. ———, and Laurits W. Krefting. 1946. The present status of moose on Isle Royale. 11th N. Amer. Wildl. Conf. Trans., 296-308.
6. Aleksiuk, Michael. 1970. The function of the tail as a fat storage depot in the beaver (*Castor canadensis*). J. Mamm. 51(1):145-148.
7. Allen, Durward L. 1962. Our wildlife legacy (rev.). Funk & Wagnalls, N.Y., 422pp.
8. ———. 1963. The costly and needless war on predators. Audubon Mag. 65(2):82-89, 120-121.
9. ———. 1969. Population, resources, and the great complexity. 34th N. Am. Wildl. and Nat. Res. Conf. Trans., 450-461.
10. ———. 1976. The worth of wilderness: with interpretations from a study of wolves and moose on Isle Royale. *In* Research in the Parks, Robert M. Linn, ed., USDI, Nat. Park Service, Symposium Ser. 1, 169-181.
11. ———. 1977. Wolf research on Isle Royale. *In* North American Big Game, 7th ed., William H. Nesbitt and Jack S. Parker, eds., Boone and Crockett Club and Nat. Rifle Assoc. Amer., 48-52.
12. ———, et al. 1973. Report of the Committee on North American Wildlife Policy. 38th N. Amer. Wildl. and Nat. Resources Conf. Trans., 149-181.
13. Altmann, Margaret. 1956. Patterns of social behavior in big game. 21st. N. Amer. Wildl. Conf. Trans., 538-544.
14. ———. 1958. Social integration of the moose calf. Anim. Behavior VI(3-4):155-159.

15. ———. 1959. Group dynamics in Wyoming moose during the rutting season. J. Mamm. 40(3):420–424.
16. ———. 1960. The role of juvenile elk and moose in the social dynamics of their species. Zoologica 45(1):35–39.
17. Anderson, Roy C. 1972. The ecological relationships of meningeal worm and native cervids in North America. J. Wildl. Diseases 8:304–310.
18. ———, and M. W. Lankester. 1974. Infectious and parasitic diseases and arthropod pests of moose in North America. Naturaliste Canadien 101(1–2):23–50.
19. Anon. 1967. Moose in British Columbia, Canada. British Columbia Wildl. Rev. 4(6)15–18.
20. Audubon, Maria R. 1897. Audubon and his journals. Charles Scribner's Sons, N.Y., 2 vol.
21. Bailey, Vernon. 1907. Wolves in relation to stock, game, and the National Forest Reserves. USDA, Forest Service, Bul. 72, 31pp.
22. ———. 1926. A biological survey of North Dakota. USDA, Bur. Biol. Survey, N. Amer. Fauna 49, 226pp.
23. ———. 1930. Animal life of Yellowstone National Park. Charles C Thomas, Springfield, Ill., 241pp.
24. ———. 1936. The mammals and life zones of Oregon. USDA, Bur. Biol. Survey, N. Amer. Fauna 55, 416pp.
25. Banfield, A. W. F. 1951. The barren-ground caribou. Canada Dept. Resources and Develop., 52pp.
26. Barnett, S. A. 1967. Instinct and intelligence. Prentice-Hall, Englewood Cliffs, N.J., 224pp.
27. Benson, Adolph B., ed. 1937. Peter Kalm's travels in North America. Wilson-Erickson, N.Y., 2 vol.
28. Berger, A. K. 1934. The Isle Royale moose herd. Nat. Hist., 34(7):678.
29. Bergerud, Arthur T., and Frank Manuel. 1968. Moose damage to balsam fir–white birch forests in central Newfoundland. J. Wildl. Mgt. 32(4):729–746.
30. ———, Frank Manuel, and Heman Whalen. 1968. The harvest reduction of a moose population in Newfoundland. J. Wildl. Mgt. 32(4):722–728.
31. Bird, Ralph D. 1961. Ecology of the aspen parkland of western Canada in relation to land use. Canada Dept. Agr., Research Br., 155pp.
32. Biswell, Harold H., and R. P. Gibbens. 1968. Fuel conditions and fire hazard reduction costs in a giant sequoia forest. Nat. Parks Mag. 42(251):16.
33. ———, Harry R. Kallander, Roy Komarek, Richard J. Bogl, and Harold Weaver. 1973. Ponderosa fire management. Tall Timbers Research Sta., Misc. Pub. 2, 49pp.
34. Borg, Karl. 1962. Predation on roe deer in Sweden. J. Wildl. Mgt. 26(2):133–136.
35. Bradt, Glenn W. 1938. A study of beaver colonies in Michigan. J. Mamm. 19(2):139–162.
36. ———. 1939. Breeding habits of beaver. J. Mamm. 20(4):486–489.
37. ———. 1947. Michigan beaver management. Michigan Dept. Cons., Game Div., 56pp.
38. Branson, Norman B. 1975. A rational rebuttal to snowmobile cynics. Manitoba Nature, Winter, 34–37.
39. Brenner, Fred J. 1962. Foods consumed by beavers in Crawford County, Pennsylvania. J. Wildl. Mgt. 26(1):104–107.
40. Brown, Clair A. n.d. [1937]. Ferns and flowering plants of Isle Royale, Michigan. USDI, Nat. Park Service, 90pp.

41. Bubenik, A. B. 1973. Hypothesis concerning the morphogenesis in moose antlers. 9th N. Amer. Moose Conf. and Workshop, Proc., 195-231.
42. Buechner, Helmut K. 1961. Territorial behavior in Uganda kob. Sci. 133(3454):698-699.
43. Burkholder, Bob L. 1959. Movements and behavior of a wolf pack in Alaska. J. Wildl. Mgt. 23(1):1-9.
44. Burroughs, Raymond D. 1961. The natural history of the Lewis and Clark Expedition. Michigan State Univ. Press, 340pp.
45. Cahalane, Victor H. 1939. The evolution of predator control policy in the national parks. J. Wildl. Mgt. 3(3):229-237.
46. ———. 1947. A deer-coyote episode. J. Mamm. 28(1):36-39.
47. Cass, Jules S. 1947. Buccal food impaction in whitetailed deer and *Actinomyces necrophorus* in big game. J. Wildl. Mgt. 11(1):91-94.
48. Caton, John Dean. 1877. The antelope and deer of North America. Forest and Stream Publ. Co., N.Y., 426pp.
49. Chittenden, Hiram M. 1917. The Yellowstone National Park. Stewart and Kidd Co., Cincinnati, 350pp.
50. Clarke, C. H. D. 1971. The Beast of Gévaudan. Nat. Hist. 80(4):44-51.
51. Cole, James E. 1957. Isle Royale wildlife investigations, winter of 1956-57. IRNP files, Spec. Rept., 42pp. (unpub. ms.).
52. Cooper, W. S. 1913. The climax forest of Isle Royale, Lake Superior, and its development. Bot. Gaz. 55:1-44, 115-140, 189-235.
53. Cowan, Ian McT. 1947. The timber wolf in the Rocky Mountain national parks of Canada. Canadian J. Research, Sec. D 25(5):139-174.
54. ———. 1948. The occurrence of the granular tape-worm *Echinococcus granulosus* in wild game in North America. J. Wildl. Mgt. 12(1):105-106.
55. ———. 1951. The diseases and parasites of big game mammals of western Canada. British Columbia Game Conf. Proc., 5:37-64.
56. ———, W. S. Hoar, and James Hatter. 1950. The effect of forest succession upon the quantity and upon the nutritive values of woody vegetation. Canadian J. Res., 28, Sec. D (5):249-271.
57. Craighead, John J., and Frank C. Craighead, Jr. 1956. Hawks, owls and wildlife. Stackpole Co. and Wild. Mgt. Inst., 443pp.
58. Cringen, A. T. 1955. Studies of moose antler development in relation to age. *In* Randolph L. Peterson, North American Moose. Univ. Toronto Press, 239-246.
59. Crisler, Lois. 1956. Arctic wild. Harper and Bros., N.Y., 301pp.
60. Cross, E. C. 1941. Colour phases of the red fox (*Vulpes fulva*) in Ontario. J. Mamm. 22(1):25-39.
61. Denniston, Rollin H., II. 1956. Ecology, behavior and population dynamics of the Wyoming or Rocky Mountain moose, *Alces alces shirasi*. Zoologica 41(3):105-118.
62. Des Meules, Pierre. 1965. Hyemal food and shelter of moose (*Alces alces americana* Cl.) in Laurentide Park, Quebec. Univ. Guelph, M.S. Thesis (unpubl.), 138pp.
63. de Vos, Antoon. 1950. Timber wolf movements on Sibley Peninsula, Ontario. J. Mamm. 31(2):169-175.
64. ———. 1958. Summer observations on moose behavior in Ontario. J. Mamm. 39(1):128-139.
65. ———, and Randolph L. Peterson. 1951. A review of the status of woodland caribou (*Rangifer caribou*) in Ontario. J. Mamm. 32(3):329-337.
66. DeWitt, James B., and James V. Derby, Jr. 1955. Change in nutritive val-

ue of browse plants following forest fires. J. Wildl. Mgt. 19(1):65-70.
67. Dice, Lee R. 1943. The biotic provinces of North America. Univ. Michigan Press, 78pp.
68. Dodds, Donald G. 1958. Observations of pre-rutting behavior in Newfoundland moose. J. Mamm. 39(3):412-416.
69. ———. 1960. Food competition and range relationships of moose and showshoe hare in Newfoundland. J. Wildl. Mgt. 24(1):52-60.
70. Eaton, Randall L. 1974. The cheetah. Van Nostrand Reinhold Co., N.Y., 178pp.
71. Edwards, R. Y. 1954. Fire and the decline of a mountain caribou herd. J. Wildl. Mgt. 18(4):521-526.
72. ———, and R. W. Ritcey. 1958. Reproduction in a moose population. J. Wildl. Mgt. 22(3):261-268.
73. Errington, Paul L. 1943. An analysis of mink predation upon muskrats in northcentral United States. Iowa Agr. Exp. Sta., Research Bul. 320:797-924.
74. ———. 1946. Predation and vertebrate populations. Quart. Rev. Biol. 21:144-177, 221-245.
75. ———. 1954. The special responsiveness of minks to epizoötics in muskrat populations. Ecol. Mon. 24:377-393.
76. ———. 1963. The phenomenon of predation. Amer. Sci. 51(2):180-192.
77. ———. 1963. Muskrat populations. Iowa State Univ. Press, 665pp.
78. ———. 1967. Of predation and life. Iowa State Univ. Press., 277pp.
79. Estes, Richard D., and John Goddard. 1967. Prey selection and hunting behavior of the African wild dog. J. Wildl. Mgt. 31(1):52-70.
80. Farrand, William R. 1969. The Quaternary history of Lake Superior. Internat. Assoc. Great Lakes Research, 12th Conf. Proc., 181-197.
81. Fay, L. D. 1970. Skin Tumors of the Cervidae. *In* Infectious diseases of wild mammals, John W. Davis, Lars H. Karstad and Daniel O. Trainer, eds., Iowa State Univ. Press, 385-392.
82. Fenstermacher, R., and O. Wilford Olsen. 1941. Further studies of diseases affecting moose III. Cornell Vet. 32(3):241-256.
83. Fentress, John C. 1967. Observations on the behavioral development of a hand-reared male timber wolf. Amer. Zool. 7:339-351.
84. Formosov, A. N. 1946. Snow cover as an integral factor of the environment and its importance in the ecology of mammals and birds. Univ. Alberta Boreal Inst., Occ. Publ., 141pp.
85. Fox, Michael W. 1971. Behavior of wolves, dogs and related canines. Harper and Row, N.Y., 220pp.
86. ———. 1972. The social significance of genital licking in the wolf, *Canis lupus*. J. Mamm. 53(3):637-640.
87. Fuller, William A. 1962. The biology and management of the bison of Wood Buffalo National Park. Canadian Wildl. Service, Wildl. Mgt. Bul. Ser. 1, No. 16, 52pp.
88. Garrard, Lewis H. 1955. Wah-to-yah and the Taos Trail. Univ. Oklahoma Press, 298pp.
89. Geist, Valerius. 1963. On the behaviour of the North American moose. Behaviour 20:377-416.
90. Ginsburg, Benson E. 1965. Coaction of genetical and nongenetical factors influencing sexual behavior. *In* Sex and Behavior, F. A. Beach, ed., Wiley and Sons, N.Y., 35-75.
91. Graham, Samuel A., Robert P. Harrison, Jr., and Casey E. Westell, Jr. 1963. Aspens: Phoenix trees of the Great Lakes region. Univ. Michigan Press, 272pp.

92. Greeley, Horace. 1860. An overland journey. C. M. Saxton, Barker and Co., N.Y., 386pp.
93. Griffin, James B., ed. 1961. Lake Superior copper and the Indians: Miscellaneous studies of Great Lakes prehistory. Univ. Michigan Mus. Anthrop., Pap. 17, 189pp.
94. Guggisberg, C. A. W. 1975. Wild cats of the world. Taplinger Pub. Co., N.Y., 328pp.
95. Haber, Gordon C. 1968. The social structure and behavior of an Alaskan wolf population. N. Michigan Univ., M.A. Thesis (unpub.), 198pp.
96. ———. 1973. Eight years of wolf research at McKinley Park. Alaska 39(4):7-11, 55-58.
97. ———. 1977. Socio-ecological dynamics of wolves and prey in a subarctic ecosystem. Univ. British Columbia, Ph.D. Thesis (unpub.), xxx+786pp.
98. Hakala, D. Robert. 1954. Wolf on Isle Royale! Nature Mag. 47(1):35-37.
99. Hall, Alex M. 1971. Ecology of beaver and selection of prey by wolves in central Ontario. Univ. Toronto, Dept. Zool., M.S. Thesis (unpub.), 116pp.
100. Hall, E. Raymond, and Keith R. Kelson. 1959. The mammals of North America. Ronald Press Co., N.Y., 2 vol.
101. Hansen, Henry L., Laurits W. Krefting, and Vilis Kurmis. 1973. The forest of Isle Royale in relation to fire history and wildlife. Univ. Minnesota Agr. Expt. Sta., Forestry Ser. 13, Tech. Bul. 294, 43pp.
102. Hatter, James. 1950. The moose of central British Columbia. State Col. Washington, Ph.D. Thesis (unpub.), 359pp.
103. Hearne, Samuel. 1958. A journey from Prince of Wales's Fort in Hudson's Bay to the Northern Ocean, 1769-1772. [Richard Glover, ed.] Macmillan Co. Canada, Toronto, 301pp.
104. Henshaw, Robert E., and Robert O. Stephenson. 1974. Homing in the gray wolf (*Canis lupus*). J. Mamm. 55(1):234-237.
105. Hickey, Joseph J., ed. 1969. Peregrine falcon populations. Univ. Wisconsin Press, 596pp.
106. Hickie, Paul F. 1936. Isle Royale moose studies. 1st N. Amer. Wildl. Conf. Trans., 396-398.
107. ———. n.d. [1943]. Michigan moose. Michigan Dept. Cons., Game Div., 57pp.
108. Holt, W. P. 1909. Notes on the vegetation of Isle Royale, Michigan. *In* Adams, Charles C. 1909. An Ecological Survey of Isle Royale, Lake Superior. Michigan Geol. Survey Rept. 1908, 217-248.
109. Hoogerwerf, A. 1958. The Indonesian giant. Nat. Hist. 67(3):136-141.
110. Hornocker, Maurice G. 1970. An analysis of mountain lion predation upon mule deer and elk in the Idaho primitive area. Wildl. Mon. 21, 39pp.
111. Hoskinson, Reed L., and L. David Mech. 1976. White-tailed deer migration and its role in wolf predation. J. Wildl. Mgt. 40(3):429-441.
112. Houston, Douglas B. 1968. The Shiras moose in Jackson Hole, Wyoming. Grand Teton Nat. Hist. Assoc., Tech. Bul. 1, 110pp.
113. Hubbs, Carl L., and Karl F. Lagler. 1949. Fishes of Isle Royale, Lake Superior, Michigan. Michigan Acad. Sci. Pap. 33:73-133.
114. Huber, N. King. 1973. Glacial and postglacial geological history of Isle Royale National Park, Michigan. U.S. Geol. Survey, Prof. Pap. 754-A, 15pp.
115. ———. 1973. Geologic map of Isle Royale National Park, Keweenaw

County, Michigan. U.S. Geol. Survey, Misc. Geol. Invest., Map I-796.
116. ———. 1975. The geologic story of Isle Royale National Park. U.S. Geol. Survey, Bul. 1309, 66pp.
117. Irving, Washington. 1956. A tour on the prairies. Univ. Oklahoma Press, 214pp.
118. James, Edwin, ed. 1956. A narrative of the captivity and adventures of John Tanner, during thirty years among the Indians. Ross and Haines, Minneapolis, 427pp.
119. Janke, Robert A. 1962. The wildflowers of Isle Royale National Park. Isle Royale Nat. Hist. Assoc., 85pp.
120. Johnson, Wendel J. 1969. Food habits of the Isle Royale red fox and population aspects of three of its principal prey species. Purdue Univ., Agr. Exp. Sta., Ph.D. Thesis (unpub.), 268pp.
121. ———. 1970. Food habits of the red fox in Isle Royale National Park, Lake Superior. Amer. Midl. Nat. 84(2):568-572.
122. ———, and John A. Coble. 1967. Notes on the food habits of pigeon hawks. Jack-pine Warbler 45(3):97-98.
123. Johnsson, Robert G., and Philip C. Shelton. 1965. The vertebrates of Isle Royale National Park. Isle Royale Nat. Hist. Assoc., 20pp.
124. Jordan, Peter A., Daniel B. Botkin, and Michael L. Wolfe. 1971. Biomass dynamics in a moose population. Ecol. 52(1):147-152.
125. ———, Philip C. Shelton, and Durward L. Allen. 1967. Numbers, turnover, and social structure of the Isle Royale wolf population. Amer. Zool. 7:233-252.
126. Joslin, Paul W. B. 1966. Summer activities of two timber wolf (*Canis lupus*) packs in Algonquin Park. Univ. Toronto, Dept. Zoology, M.S. Thesis (unpub.), 99pp.
127. ———. 1967. Movements and home sites of timber wolves in Algonquin Park. Amer. Zool. 7:279-288.
128. Karns, Patrick D. 1967. *Pneumostrongylus tenuis* in deer in Minnesota and implications for moose. J. Wildl. Mgt. 31(2):299-303.
129. Kellogg, Louise P. 1925. The French regime in Wisconsin and the Northwest. State Hist. Soc. Wisconsin, Madison, 474pp.
130. Kellum, Ford. 1941. Cusino's captive moose. Michigan Cons. 10(7):4-5.
131. Kelsall, John P. 1969. Structural adaptations of moose and deer for snow. J. Mamm. 50(2):302-310.
132. ———, and William H. Prescott. 1971. Moose and deer behavior in snow. Canadian Wildl. Service, Reprint Ser. 15, 27pp.
133. ———, and Edmund S. Telfer. 1971. Studies of the physical adaptation of big game for snow. *In* Snow and ice in Relation to Wildlife and Recreation Symposium, Proc., Arnold O. Haugen, ed. Iowa State Univ., 134-147.
134. Kilgore, Bruce M. 1973. The ecological role of fire in Sierran conifer forests. J. Quat. Research 3(3):496-513.
135. ———. 1975. Restoring fire to national park wilderness. Amer. For. 81(3):16-19, 57-59.
136. Kleiman, Devra G. 1967. Some aspects of social behavior in the Canidae. Amer. Zool. 7:365-372.
137. Klein, David R., and Sigurd T. Olson. 1960. Natural mortality patterns of deer in southeast Alaska. J. Wildl. Mgt. 24(1):80-88.
138. Klinghammer, Erich, and Eugene P. Brantley, Jr. 1977. Social rank and mating behavior in two captive wolf packs (*Canis lupus*). Anim. Behav. Soc., Ann. Meet. Trans. Boulder, Col.

139. Koford, Carl B. 1958. Prairie dogs, whitefaces, and blue grama. Wildl. Mon. 3, 78pp.
140. Kolenosky, George B. 1972. Wolf predation on wintering deer in east-central Ontario. J. Wildl. Mgt. 36(2):357-369.
141. ———, and David H. Johnston. 1967. Radio-tracking timber wolves in Ontario. Amer. Zool. 7:289-303.
142. ———, and Roger O. Standfield. 1975. Morphological and ecological variation among gray wolves (*Canis lupus*) of Ontario, Canada. *In* The Wild Canids, Michael W. Fox, ed. Van Nostrand Reinhold Co., N.Y., 62-72.
143. Krefting, Laurits W. 1951. What is the future of the Isle Royale moose herd? 16th N. Amer. Wildl. Conf. Trans., 461-470.
144. ———. 1963. Beaver of Isle Royale. Naturalist 14(2):2-11.
145. ———. 1969. The rise and fall of the coyote on Isle Royale. Naturalist 20(4):24-31.
146. ———. 1974. The ecology of the Isle Royale moose, with special reference to the habitat. Univ. Minnesota Agr. Exp. Sta., Forestry Ser. 15, 75pp.
147. ———, H. L. Hansen, and M. P. Meyer. 1970. Vegetation type map of Isle Royale National Park. USDI, Bur. Sport Fisheries and Wildl.
148. ———, Forrest B. Lee, Philip C. Shelton, and Karl T. Gilbert. 1966. Birds of Isle Royale in Lake Superior. USDI, Fish and Wildl. Service, Spec. Sci. Rept. Wildl., 94, 56pp.
149. Kruuk, Hans. 1972. Surplus killing by carnivores. J. Zool. Soc. London 166:233-244.
150. ———. 1972. The spotted hyena. Univ. Chicago Press, 335pp.
151. Kuyt, E. 1972. Food habits and ecology of wolves on barren-ground caribou range in the Northwest Territories. Canadian Wildl. Service, Rept. Ser. 21, 36pp.
152. Lagler, Karl F., and Charles R. Goldman. 1959. Fishes and sport fishing in Isle Royale National Park. Isle Royale Nat. Hist. Assoc., 46pp.
153. Larpenteur, Charles. 1933. Forty years a fur trader on the upper Missouri. R. R. Donnelley and Sons, Chicago, 388pp.
154. Lawrence, William H., L. Dale Fay, and Samuel A. Graham. 1956. A report on the beaver die-off in Michigan. J. Wildl. Mgt. 20(2):184-187.
155. Lay, Daniel W. 1957. Browse quality and the effects of prescribed burning in southern pine forests. J. For. 55(5):342-347.
156. LeResche, Robert E., and James L. Davis. 1973. Importance of non-browse foods to moose on the Kenai Peninsula, Alaska. J. Wildl. Mgt. 37(3):279-287.
157. ———, and Robert A. Rausch. 1974. Accuracy and precision of aerial moose censusing. J. Wildl. Mgt. 38(2):175-182.
158. Linn, Robert M. 1957. The spruce-fir, maple-birch transition in Isle Royale National Park, Lake Superior. Duke Univ., Ph.D. Thesis (unpub.), 101pp.
159. ———. 1962. Forests and trees of Isle Royale National Park. Isle Royale Nat. Hist. Assoc., 34pp.
160. Longstreth, T. Morris. 1924. The Lake Superior country. The Century Co., N.Y., 360pp.
161. Markgren, Gunnar. 1975. Winter studies on orphaned moose calves in Sweden. Viltrevy 9(4):193-219.
162. Marsh, George P. 1907. The earth as modified by human action. Charles Scribner's Sons, N.Y., 629pp.
163. Martin, Alexander C., Herbert S. Zim, and Arnold L. Nelson. 1951.

American wildlife and plants. McGraw-Hill Book Company, N.Y., 500pp.
164. Matray, Paul. 1976. The broad-winged hawk. The Conservationist 31(2):20-22.
165. Maximilian, Prince of Wied-Neuwied. 1843. Travels in the interior of North America, 1832-1834. Part II. Ackerman and Co., London.
166. McDermott, John F., ed. 1951. Up the Missouri with Audubon, the journal of Edward Harris. Univ. Oklahoma Press, 222pp.
167. McMillan, John F. 1953. Some feeding habits of moose in Yellowstone Park. Ecol. 34(1):102-110.
168. ———. 1954. Summer home range and population size of moose in Yellowstone National Park. Univ. Wichita Bul. 29(2):1-16.
169. ———. 1954. Some observations on moose in Yellowstone Park. Amer. Midl. Nat. 52(2):392-399.
170. Meagher, Margaret M. 1973. The bison of Yellowstone National Park. USDI, Nat. Park Service, Sci. Mon. Series, 1, 161.
171. Mech, L. David. 1962. The ecology of the timber wolf (*Canis lupus* Linnaeus) in Isle Royale National Park. Purdue Univ., Agr. Exp. Sta., Ph.D. Thesis (unpub.), 284pp.
172. ———. 1966. The wolves of Isle Royale. USDI, Nat. Park Service, Fauna Ser. 7, 210pp.
173. ———. 1966. Hunting behavior of timber wolves in Minnesota. J. Mamm. 47(2):347-348.
174. ———. 1970. The wolf: the ecology and behavior of an endangered species. Nat. Hist. Press, Garden City, N.Y., 384pp.
175. ———. 1973. Wolf numbers in the Superior National Forest of Minnesota. USDA, For. Service, N. Cent. For. Exp. Sta., Res. Pap. NC-97, 10pp.
176. ———. 1977. Population trend and winter deer consumption in a Minnesota wolf pack. *In* 1975 Predator Symposium Proc., Robert L. Phillips and Charles Jonkel, eds., Montana For. and Cons. Exp. Sta. 55-83.
177. ———. 1977. A recovery plan for the eastern timber wolf. Nat. Parks and Cons. Mag. 51(1):17-21.
178. ———. 1977. Productivity, mortality, and population trends of wolves in northeastern Minnesota. J. Mamm. 58(4):559-574.
179. ———, and L. D. Frenzel, Jr. 1971. An analysis of the age, sex, and condition of deer killed by wolves in northeastern Minnesota. *In* Ecological studies of the timber wolf in northeastern Minnesota, Mech and Frenzel, eds., USDA, For. Service, N. Cent. Forest Exp. Sta., Res. Pap. NC-52, 35-51.
180. ———, L. D. Frenzel, Jr., and Patrick D. Karns. 1971. The effect of snow conditions on the vulnerability of white-tailed deer to wolf predation. USDA, For. Service, N. Cent. For. Exp. Sta., Res. Pap. NC-52, 51-62.
181. ———, L. D. Frenzel, Jr., Robert R. Ream, and John W. Winship. 1971. Movements, behavior, and ecology of timber wolves in northeastern Minnesota. *In* Ecological studies of the timber wolf in northeastern Minnesota, Mech and Frenzel, eds., USDA, For. Service, N. Cent. Forest Exp. Sta., Res. Pap. NC-52, 1-35.
182. ———, and Patrick D. Karns. 1978. Role of the wolf in a deer decline in the Superior National Forest. USDA, For. Service, N. Cent. Forest Exp. Sta., Res. Pap. NC-148, 23pp.

183. Medjo, Dennis C., and L. David Mech. 1976. Reproductive activity in nine- and ten-month-old wolves. J. Mamm. 57(2):406–408.
184. Merriam, C. Hart. 1882. The vertebrates of the Adirondack region. Linnaean Soc. New York, Trans., I, 233pp.
185. Meslow, E. Charles, and Lloyd B. Keith. 1968. Demographic parameters of a snowshoe hare population. J. Wildl. Mgt. 32(4):812-834.
186. Mitchell, Brian. 1967. Growth layers in dental cement for determining the age of red deer (*Cervus elephus* L.). J. Anim. Ecol. 36(2):279-293.
187. Mitchell, H. B. 1970. Rapid aerial sexing of antlerless moose in British Columbia. J. Wildl. Mgt. 34(3):645-646.
188. Moll, Don, and B. Kaye Moll. 1976. Moose triplets on Isle Royale. Illinois Acad. Sci. Trans. 69(2):151-152.
189. Montague, Fredrick H. 1975. The ecology and recreational value of the red fox in Indiana. Purdue Univ. Agr. Exp. Sta., Ph.D. Thesis (unpub.), 268pp.
190. Morgan, Dale L. 1953. Jedediah Smith and the opening of the west. Bobbs-Merrill Co., N.Y., 458pp.
191. Morgan, Lewis H. 1868. The American beaver and his works. J. B. Lippincott Co., Philadelphia, 330pp.
192. Murie, Adolph. 1934. The moose of Isle Royale. Univ. Michigan Mus. Zool., Misc. Pub. 25, 44pp.
193. ———. 1940. Ecology of the coyote in the Yellowstone. USDI, Nat. Park Service, Fauna Ser. 4, 206pp.
194. ———. 1944. The wolves of Mount McKinley. USDI, Nat. Park Service, Fauna Series 5, 238pp.
195. Murie, Olaus J. 1951. The elk of North America. Stackpole Co., Harrisburg, Pa., 376pp.
196. Neff, Edwin D. 1951. Can we save the gray wolf? Nat. Hist. 60(9):392-396, 432.
197. Oldemeyer, J. L. 1974. Nutritive value of moose forage. Naturaliste Canadien 101:217-226.
198. Olson, Sigurd F. 1938. A study in predatory relationship with particular reference to the wolf. Sci. Monthly 46:323-336.
199. ———. 1938. Organization and range of the pack. Ecol. 19(1):168-170.
200. ———. 1957. The singing wilderness. Alfred A. Knopf, N.Y., 245pp.
201. Passmore, R. C., R. L. Peterson, and A. T. Cringen. 1955. A study of mandibular tooth-wear as an index to age of moose. *In* Randolph L. Peterson, North American Moose. Univ. Toronto Press, 223-246.
202. Peek, James M. 1962. Studies of moose in the Gravelly and Snowcrest Mountains, Montana. J. Wildl. Mgt. 26(4):360-365.
203. ———. 1970. Wilderness moose. Naturalist 21(4):38-41.
204. ———, Robert E. LeResche, and David R. Stevens. 1974. Dynamics of moose aggregations in Alaska, Minnesota, and Montana, J. Mamm. 55(1):126-137.
205. ———, David L. Urich, and Richard J. Mackie. 1976. Moose habitat selection and relationships to forest management in northeastern Minnesota. Wildl. Mon. 48, 65pp.
206. Peters, Roger P., and L. David Mech. 1975. Scent-marking in wolves. Amer. Sci. 63(6):628-637.
207. Peterson, Randolph L. 1947. A record of a timber wolf attacking a man. J. Mamm. 28(3):294-295.
208. ———. 1953. Studies of the food habits and the habitat of moose in Ontario. Royal Ontario Mus. Zool. and Paleontol. Contrib. 36, 49pp.

209. ———. 1955. North American Moose. Univ. Toronto Press, 280pp.
210. ———, and Vincent Crichton. 1949. The fur resources of Chapleau District, Ontario. Canadian J. Res. D 27:68-84.
211. Peterson, Rolf O. 1974. Wolf Ecology and prey relationships on Isle Royale. Purdue Univ., Agr. Exp. Sta., Ph.D. Thesis (unpub.), 368pp.
212. ———. 1975. Wolf response to increased moose vulnerability on Isle Royale. 11th N. Amer. Moose Conf. and Workshop Proc. 344–368.
213. ———. 1976. The role of wolf predation in a moose population decline. *In* First Conference on Scientific Research in the National Parks, Robert M. Linn, ed., 1979. USDI, Nat. Park Service Proc. Ser. 5, vol. I, 329–333.
214. ———. 1977. Wolf ecology and prey relationships on Isle Royale. USDI, Nat. Park Service, Sci. Mon. Ser. 7, 210 pp.
215. ———, and Durward L. Allen. 1974. Snow conditions as a parameter in moose-wolf relationships. Naturaliste Canadien 101:481-492.
216. Pfeiffer, John E. 1969. The emergence of man. Harper and Row, N.Y., 550pp.
217. Phillips, R. L., W. E. Berg, and D. B. Siniff. 1973. Moose movement patterns and range use in northwestern Minnesota. J. Wildl. Mgt. 37(3):266-278.
218. Pimlott, Douglas H. 1953. Newfoundland moose. 18th N. Amer. Wildl. Conf. Trans., 563-581.
219. ———. 1959. Reproduction and productivity of Newfoundland moose. J. Wildl. Mgt. 23(4):381-401.
220. ———. 1967. Wolf predation and ungulate populations. Amer. Zool. 7:267-278.
221. ———. 1975. The ecology of the wolf in North America. *In* The Wild Canids, Michael W. Fox, ed., Van Nostrand Reinhold Co., N.Y., 280-285.
222. ———, J. A. Shannon, and G. B. Kolenosky. 1969. The ecology of the timber wolf in Algonquin Provincial Park. Ontario Dept. Lands and For., Res. Rept. (Wildl.) 87, 92pp.
223. Porter, Clyde, and Mae Reed Porter [LeRoy R. Hafen, ed.] 1950. Ruxton of the Rockies. Univ. Oklahoma Press, 325pp.
224. Potzger, J. E. 1954. Post-Algonquin and post-Nipissing forest history of Isle Royale, Michigan. Butler Univ. Bot. Studies 11:199-208.
225. Pruitt, W. O., Jr. 1960. Animals in the snow. Sci. Amer. 202(1):60-68.
226. Pulliainen, Erkki. 1965. Studies on the wolf (*Canis Lupus* L.) in Finland. Ann. Zool. Fennici 2:215-259.
227. ———. 1967. A contribution to the study of the social behavior of the wolf. Amer. Zool. 7:221-222.
228. Rabb, George B., Jerome H. Woolpy, and Benson E. Ginsburg. 1967. Social relationships in a group of captive wolves. Amer. Zool. 7:305-311.
229. Rakestraw, Lawrence. 1965. Historic mining on Isle Royale. Isle Royale Nat. Hist. Assoc., 20pp.
230. ———. 1968. Commercial fishing on Isle Royale. Isle Royale Nat. Hist. Assoc., 24pp.
231. Rausch, Robert A. 1967. Some aspects of the population ecology of wolves, Alaska. Amer. Zool. 7(2):253-265.
232. Rausch, Robert L. 1952. Hydatid disease in boreal regions. Arctic 5(3):157-174.
233. ———, and Francis S. L. Williamson. 1959. Studies on the helminth

fauna of Alaska, XXXIV. The parasites of wolves, *Canis lupus,* L. J. Parasitol. 45(4):395-403.
234. Reiger, John F. 1975. American sportsmen and the origins of conservation. Winchester Press, N.Y., 316pp.
235. Rothman, Russell J. 1977. Scent marking in lone wolves and newly formed packs. Univ. Minnesota, M.S. Thesis (unpub.), 69pp.
236. Ruhl, H. D. 1934. Isle Royale moose situation [a report to Conservation Commission]. Michigan Dept. Cons. Proc. 14:69-70.
237. Russell, Osborne. 1955. Journal of a trapper. Aubrey L. Haines, ed. Univ. Oklahoma Press, 191pp.
238. Rutter, Russell J., and Douglas H. Pimlott. 1968. The world of the wolf. J. B. Lippincott Co., N.Y., 202pp.
239. Ruxton, George Frederick. 1849. Life in the Far West. Harper and Bros., N.Y., 235pp.
240. Schaller, George B. 1967. The deer and the tiger. Univ. Chicago Press, 370pp.
241. ———. 1972. The Serengeti lion. Univ. Chicago Press, 480pp.
242. Schenkel, Rudolf. 1947. Ausdruchs-Studien an Wölfen. [Translation by Agnes Klasson, by courtesy of Douglas Pimlott, University of Toronto.] Behaviour 1(2):81-129.
243. ———. 1967. Submission: Its features and function in the wolf and dog. Amer. Zool. 7(2):319-329.
244. Schoolcraft, Henry R. 1851. Personal memoirs of a residence of thirty years with the Indian tribes on the American frontiers. Lippincott, Granbo, and Co., Philadelphia, 703pp.
245. Schramm, Donald L. 1968. A field study of beaver behavior in East Barnard, Vermont. Dartmouth Col., M.S. Thesis (unpub.), 83pp.
246. Scott, John P. 1967. The evolution of social behavior in dogs and wolves. Amer. Zool. 7:373-381.
247. ———, and John L. Fuller. 1965. Genetics and the social behavior of the dog. Univ. Chicago Press, 468pp.
248. Scott, William P. 1925. Reminiscences of Isle Royale. Michigan Hist. Mag. 9:398-412.
249. Seal, Ulysses S., L. David Mech, and Victor Van Ballenberghe. 1975. Blood analyses of wolf pups and their ecological and metabolic interpretation. J. Mamm. 56(1):64-75.
250. Seton, Ernest Thompson. 1909. Life histories of northern animals. Chas. Scribner's Sons, N.Y., 2 vol.
251. Shelton, Philip C. 1966. Ecological studies of beavers, wolves, and moose in Isle Royale National Park. Purdue Univ. Agr. Exp. Sta., Ph.D. Thesis (unpub.), 308pp.
252. Shiras, George, III. 1921. The wild life of Lake Superior, past and present. Nat. Geog. 40(2):113-204.
253. Simkin, D. W. 1965. Reproduction and productivity of moose in northwestern Ontario. J. Wildl. Mgt. 29(4):740-750.
254. Skinner, Milton P. 1927. The predatory and fur-bearing animals of the Yellowstone National Park. Roosevelt Wild Life Bul. 4:163-281.
255. Skunke, Folke. 1949. Algen, studier, jakt och vard. A. Norstedt and Söners, Stockholm. 400 pp. [Translation by D. E. Sergeant, courtesy Wildl. Div., Ontario Dept. Lands and For.]
256. Smits, Lee J. 1952. True voice of the wilderness. Michigan Out-of-Doors 2(12):2-7.
257. ———. 1955. Man's best friend, the wolf. Paper read before the Prismatic

Club (unpub.), Detroit, Michigan, 5 March 1955. 28pp.
258. Snyder, John D., and Robert A. Janke. 1976. Impact of moose browsing on boreal-type forests of Isle Royale National Park. Amer. Midl. Nat. 95(1):79–92.
259. Stanwell-Fletcher, John F. 1942. Three years in the wolves' wilderness. Nat. Hist. 49(3):136–147.
260. Stebler, A. M. 1939. The tracking technique in the study of the larger predatory mammals. 4th N. Amer. Wildl. Conf. Trans., 203–208.
261. ———. 1944. The status of the wolf in Michigan. J. Mamm. 25(1):37–43.
262. Stelfox, John G. 1962. Liver, lungs and larvae. Land, For., Wildl. 5(4):5–12.
263. Stenlund, Milton H. 1953. Report of Minnesota beaver die-off, 1951–1952. J. Wildl. Mgt. 17(3)376–377.
264. ———. 1955. A field study of the timber wolf (*Canis lupus*) on the Superior National Forest, Minnesota. Minnesota Dept. Cons., Tech. Bul. 4, 55pp.
265. Stephenson, Robert O., and Robert T. Ahgook. 1975. The Eskimo hunter's view of wolf ecology and behavior. *In* The Wild Canids, Michael W. Fox, ed., Van Nostrand Reinhold Co., N.Y., 286–291.
266. Stokes, Allen W. 1954. Population studies of the ring-necked pheasants on Pelee Island, Ontario. Ontario Dept. Lands and For., Wildl. Ser., 4, 154pp.
267. Surber, Thaddeus. 1932. The mammals of Minnesota. Minnesota Dept. Cons., Div. Game and Fish, 84pp.
268. Talbot, Lee M., and Martha H. Talbot. 1963. The wildebeest in western Masailand, East Africa. Wildl. Mon. 12, 88pp.
269. Tansley, A. G. 1935. The use and abuse of vegetational concepts and terms. Ecol. 16(3):284–307.
270. Tembrock, Günter. 1957. Das verhalten des rotfuchses. Handbuch der Zool., Berlin 10(15):1–20.
271. ———. 1957. Zur ethologie des rotfuchses (*Vulpes vulpes* L.) unter besonderer berucksichtigung der fortpflanzung. Zool. Garten 23:289–532.
272. Tew, Ronald K. 1970. Seasonal variation in the nutrient content of aspen foliage. J. Wildl. Mgt. 34:(2):475–478.
273. Theberge, John B. 1969. Observations of wolves at a rendezvous site in Algonquin Park. Canadian Field-Nat. 83(2):122–128.
274. ———. 1975. Wolves and wilderness. J. M. Dent and Sons, Canada, 159pp.
275. ———, and J. Bruce Falls. 1967. Howling as a means of communication in timber wolves. Amer. Zool. 7(2):331–338.
276. Thompson, Daniel Q. 1952. Travel, range, and food habits of timber wolves in Wisconsin. J. Mamm. 33(4):429–442.
277. Thompson, W. K. 1949. Observations of moose courting behavior. J. Wildl. Mgt. 13(3):313.
278. Thoreau, Henry David. 1950. The Maine woods. Bramhall House, N.Y., 340pp.
279. ———. 1951. Walden. W. W. Norton and Co., 354pp.
280. Thwaites, Reuben G. 1902. The French regime in Wisconsin-I, 1634–1727. Wisconsin Hist. Col. 16, 72–76.
281. Trefethen, James B. 1975. An American crusade for wildlife. Winchester Press and Boone and Crockett Club, 409pp.

282. Van Ballenberghe, Victor. 1974. Wolf management in Minnesota: an endangered species case history. 39th N. Am. Wildl. and Nat. Res. Conf. Trans., 313-320.
283. ———, Albert W. Erickson, and David Byman. 1975. Ecology of the timber wolf in northeastern Minnesota. Wildl. Mon., 43, 43pp.
284. ———, and L. David Mech. 1975. Weights, growth, and survival of timber wolf pups in Minnesota. J. Mamm. 56(1):44-63.
285. van Lawick-Goodall, Hugo and Jane. 1971. Innocent killers. Houghton Mifflin Co., Boston, 222pp.
286. Verme, Louis J. 1970. Some characteristics of captive Michigan moose. J. Mamm. 51(2):403-405.
287. Walls, Gordon L. 1963. The vertebrate eye and its adaptive radiation. Hafner Publ. Co., N.Y., 785pp.
288. Weise, Thomas F., William L. Robinson, Richard A. Hook, and L. David Mech. 1975. An experimental translocation of the eastern timber wolf. USDI, Fish and Wildl. Service, Reg. 3, Audubon Cons. Rept. 5, 28pp.
289. Wolfe, Michael L. 1969. Age determination in moose from cemental layers of molar teeth. J. Wildl. Mgt. 33(2):428-431.
290. ———. 1977. Mortality patterns in the Isle Royale moose population. Amer. Midl. Nat. 97(2):267-279.
291. ———, and Durward L. Allen. 1973. Continued studies of the status, socialization, and relationships of Isle Royale wolves. J. Wildl. Mgt. 54(3):611-635.
292. Woolpy, Jerome H. 1967. Socially controlled systems of mating among wolves and other gregarious mammals and their implications for the genetics of natural populations. Univ. Chicago, Ph.D. Dissertation (unpub.), iv + 32pp.
293. ———. 1968. The social organization of wolves. Nat. Hist. 77(5):46-55.
294. ———. 1968. Socialization of wolves. Sci. and Psychoanal. XII, 82-94.
295. Wright, George M., Joseph S. Dixon, and Ben H. Thompson. 1933. A preliminary survey of faunal relations in national parks. USDI, Nat. Park Serv., Fauna Ser. 1, 157pp.
296. Young, Stanley P. 1946. The wolf in North American history. Caxton Printers, Caldwell, Idaho, 149pp.
297. ———, and Edward A. Goldman. 1944. The wolves of North America. Amer. Wildl. Inst., xx+636pp.
298. Zimen, Erik. 1971. Wölfe und Königspudel. R. Piper and Co., München, 257pp.
299. ———. 1975. Social Dynamics of the wolf pack. *In* The Wild Canids, Michael W. Fox, ed., Van Nostrand Reinhold Co., N.Y., 336-362.
300. ———. 1976. On the regulation of pack size in wolves. Z. Tierpsychol. 40:300-341.

INDEX

Index

Ann Arbor Paperbacks

Waddell, *The Desert Fathers*
Erasmus, *The Praise of Folly*
Donne, *Devotions*
Malthus, *Population: The First Essay*
Berdyaev, *The Origin of Russian Communism*
Einhard, *The Life of Charlemagne*
Edwards, *The Nature of True Virtue*
Gilson, *Héloïse and Abélard*
Aristotle, *Metaphysics*
Kant, *Education*
Boulding, *The Image*
Duckett, *The Gateway to the Middle Ages* (3 vols.): *Italy; France and Britain; Monasticism*
Bowditch and Ramsland, *Voices of the Industrial Revolution*
Luxemburg, *The Russian Revolution* and *Leninism or Marxism?*
Rexroth, *Poems from the Greek Anthology*
Zoshchenko, *Scenes from the Bathhouse*
Thrupp, *The Merchant Class of Medieval London*
Procopius, *Secret History*
Fine, *Laissez Faire and the General-Welfare State*
Adcock, *Roman Political Ideas and Practice*
Swanson, *The Birth of the Gods*
Xenophon, *The March Up Country*
Trotsky, *The New Course*
Buchanan and Tullock, *The Calculus of Consent*
Hobson, *Imperialism*
Pobedonostsev, *Reflections of a Russian Statesman*
Kinietz, *The Indians of the Western Great Lakes 1615–1760*
Bromage, *Writing for Business*
Lurie, *Mountain Wolf Woman, Sister of Crashing Thunder*
Leonard, *Baroque Times in Old Mexico*
Meier, *Negro Thought in America, 1880–1915*
Burke, *The Philosophy of Edmund Burke*
Michelet, *Joan of Arc*
Conze, *Buddhist Thought in India*
Arberry, *Aspects of Islamic Civilization*
Chesnutt, *The Wife of His Youth and Other Stories*
Gross, *Sound and Form in Modern Poetry*
Zola, *The Masterpiece*
Chesnutt, *The Marrow of Tradition*
Aristophanes, *Four Comedies*
Aristophanes, *Three Comedies*
Chesnutt, *The Conjure Woman*
Duckett, *Carolingian Portraits*
Rapoport and Chammah, *Prisoner's Dilemma*
Aristotle, *Poetics*
Boulding, *Beyond Economics*
Peattie, *The View from the Barrio*
Duckett, *Death and Life in the Tenth Century*
Langford, *Galileo, Science and the Church*
McNaughton, *The Taoist Vision*
Anderson, *Matthew Arnold and the Classical Tradition*
Milio, *9226 Kercheval*
Weisheipl, *The Development of Physical Theory in the Middle Ages*
Breton, *Manifestoes of Surrealism*
Gershman, *The Surrealist Revolution in France*
Burt, *Mammals of the Great Lakes Region*
Lester, *Theravada Buddhism in Southeast Asia*
Scholz, *Carolingian Chronicles*
Marković, *From Affluence to Praxis*
Wik, *Henry Ford and Grass-roots America*
Sahlins and Service, *Evolution and Culture*
Wickham, *Early Medieval Italy*
Waddell, *The Wandering Scholars*
Rosenberg, *Bolshevik Visions* (2 parts in 2 vols.)
Mannoni, *Prospero and Caliban*
Aron, *Democracy and Totalitarianism*
Shy, *A People Numerous and Armed*
Taylor, *Roman Voting Assemblies*
Goodfield, *An Imagined World*
Hesiod, *The Works and Days; Theogony; The Shield of Herakles*
Raverat, *Period Piece*
Lamming, *In the Castle of My Skin*
Fisher, *The Conjure-Man Dies*
Strayer, *The Albigensian Crusades*
Lamming, *The Pleasures of Exile*
Lamming, *Natives of My Person*
Glaspell, Lifted Masks *and Other Works*
Wolff, *Aesthetics and the Sociology of Art*
Grand, *The Heavenly Twins*
Cornford, *The Origin of Attic Comedy*
Allen, *Wolves of Minong*
Brathwaite, *Roots*
Fisher, *The Walls of Jericho*
Lamming, *The Emigrants*